STATPHYS-KOLKATA VIII

Journal of Physics: Conference Series Volume 638

Kolkata, India
1 – 5 November 2014

ISBN: 978-1-5108-1196-6
ISSN: 1742-6588

STATPHYS-KOLKATA VIII

Journal of Physics: Conference Series Volume 638

Kolkata, India
1 – 5 November 2014

IOP
Institute of Physics

Printed from e-media with permission by:

Curran Associates, Inc.
57 Morehouse Lane
Red Hook, NY 12571

Some format issues inherent in the e-media version may also appear in this print version.

Printed by Curran Associates, Inc. (2015)

Institute of Physics
Dirac House, Temple Back
Bristol BS1 6BE UK

Phone: 44 1 17 929 7481
Fax: 44 1 17 920 0979

techtracking@iop.org

Additional copies of this publication are available from:

Curran Associates, Inc.
57 Morehouse Lane
Red Hook, NY 12571 USA
Phone: 845-758-0400
Fax: 845-758-2634
Email: curran@proceedings.com
Web: www.proceedings.com

TABLE OF CONTENTS

STATISTICAL INFERENCE, SPIN GLASS AND DATA-DRIVEN MODELING

Preface

Anirban Chakraborti
School of Computational and Integrative Sciences, Jawaharlal Nehru University, New Delhi-110067, India
E-mail: anirban@jnu.ac.in

Sakuntala Chatterjee
S. N. Bose National Centre for Basic Sciences, Block-JD, Sector-III, Salt Lake, Kolkata-700098, India
E-mail: sakuntala.chatterjee@bose.res.in

Punyabrata Pradhan
S. N. Bose National Centre for Basic Sciences, Block-JD, Sector-III, Salt Lake, Kolkata-700098, India
E-mail: punyabrata.pradhan@bose.res.in

Abstract. A series of international conferences on Statistical Physics, going by the name Statphys-Kolkata, have been organized in Kolkata (erstwhile Calcutta) at regular intervals of two-three years, the first one being held in 1992-93. The eighth of this series, Statphys-Kolkata VIII (http://newweb.bose.res.in/Conferences/STATPHYSKOLKATAVIII/) was held during December 1-5, 2014, at the Satyendra Nath Bose National Centre for Basic Sciences, Kolkata. This volume contains selected papers[1] by the speakers and participants of the Conference.

1. Introduction

Statistical physics, which deals with understanding macroscopic properties of many-particle systems, encompasses a vast field of research in all fundamental branches of science. It is widely taught as a core subject of undergraduate level courses in physics, since it has become central for the understanding of the other specialized branches of physics. Interestingly, over the last few decades, standard techniques and tools of statistical physics have been applied to topics such as opinion dynamics, voter models, distribution of wealth in a population, minority games, stock-market analysis, socio-economic networks, etc., in the domain of social sciences, far away from the conventional domains of physics. As a result, various intimate connections across the disciplines have emerged.

[1] See peer review statement: All papers published in this volume of Journal of Physics: Conference Series have been peer reviewed through processes administered by the proceedings Editors. Reviews were conducted by expert referees to the professional and scientific standards expected of a proceedings journal published by IOP Publishing.

Journal of Physics: Conference Series **638** (2015) 011001 doi:10.1088/1742-6596/638/1/011001

The purpose of this conference was to bring together leading researchers, young scientists and students who have been working in different disciplines of science and have been primarily using tatistical physics techniques for their research. The choice of the subjects and problems that were covered in the conference reflected precisely this multidisciplinary statistical physics approach to understanding collective phenomena in general, where complete theoretical understanding is still lacking. This conference was held to encourage collaboration among researchers belonging to different communities, binding together theoretical, numerical and experimental expertise.

In Statphys-Kolkata VIII, the themes:

- Fluctuation Relations away from Equilibrium,
- Structure and Pattern Formation in Non-equilibrium Systems,
- Interaction and Transport in Biological Systems,
- Co-operative Dynamics in Socio-economical Systems, and
- Statistical Inference, Spin Glass and Data Driven Modeling,

were given special emphasis, while various other issues of Statistical Physics were also addressed. We were glad to note that the conference did attract a large number of young participants, and the talks and poster presentations generated a lot of discussion, interaction and collaboration. The articles appearing in this proceedings are based on the invited talks and a few selected poster presentations.

The group photo taken at the conference is shown in Figure 1.

Figure 1. Group photo at the conference.

One plenary session was dedicated to felicitate the 60th birthday of Prof. Hidetoshi Nishimori, Tokyo Institute of Technology, Japan, a distinguished statistical physicist and our beloved colleague. He is reputed for his contributions in the areas of spin glasses, information processing, quantum annealing, etc. with many books, research articles and publications. Figure 2 shows a photo of the felicitation ceremony.

Figure 2. The felicitation of Prof. Hidetoshi Nishimori (Tokyo Institute of Technology) on his 60th birthday, by Prof. Bikas K. Chakrabarti (SINP) and the research scholars of SNBNCBS.

It came as a rude shock as we learnt that Dr. Jun-ichi Inoue, had suddenly passed away on May 17, 2015. It was in his office in July, 2013, that the idea of organizing the Statphys-Kolkata VIII meeting was conceived. He was one of the convenors of the meeting, as well as one of the editors of this proceedings volume. We dedicate this volume in his memory!

Figure 3 shows the ever cheerful and dedicated scientist, Dr Jun-ichi Inoue, at the inaugural session of the Statphys-Kolkata VIII conference. We reproduce below some of the remarks as tribute to Dr. Inoue, made by his supervisor, colleagues and collaborators:

> *"Jun-ichi did his thesis work under my supervision. He was one of the most energetic and talented (or rather, he was THE most energetic and talented) student I ever supervised. He proposed and solved problems after problems week by week, and I had no other choice than to try my best to follow his explosion of ideas and computations. His enthusiasm and devotion to physics should have continued to the last day of his life. I am so proud to call him my former student and colleague. He will never be extinguished from our minds."*
> – **Prof. Hidetoshi Nishimori**, Tokyo Institute of Technology, Tokyo, Japan

> *"Jun-ichi Inoue first contacted us in 1999-2000 after obtaining his Ph. D. degree in 1998 from the Tokyo Institute of Technology, where his thesis supervisor was Prof. Hidetoshi Nishimori. He had expressed his interest in collaborative research on quantum phase transitions and other current topics in statistical mechanics.*
> *He had first visited us during March 2003, in connection with an international conference on Unconventional Applications of Statistical Physics, held in Saha Institute*

Figure 3. Late Dr. Jun-ichi Inoue, convening the inaugural session of the conference at the S. N. Bose auditorium.

of Nuclear Physics, Kolkata. He was then already a faculty member of the Hokkaido University and invited us to visit him in his University. Since then, most of us in our group had visited him several times in Hokkaido (usually with full financial support from his own research fund). I believe, in the last ten years, he visited us in Kolkata more than fourteen times. Recently he had a frequency of visiting us almost twice a year!

Jun-ichi was prolific in calculations and Monte Carlo computations. In general, he was interested in memory models, quantum glasses and econophysics. He had also personal interest in J. C. Bose's electro-physiological model of plant responses and visited several times the J. C. Bose museum (courtesy Prof. Indrani Bose) of the Bose Institute, Kolkata. I had the good fortune to have collaborated with him in several fields and publish seven papers and one book (Quantum Ising Phases and Transitions in Transverse Ising Models, S. Suzuki, J.-I. Inoue & B. K. Chakrabarti, Springer, Heidelberg (2013)). There is even a contract with the Cambridge University Press for a book on Quantum Spin Glasses & Annealing, with Jun-ichi as one of the co-authors!

Jun-ichi showed remarkable passion for collaborative researches with us in India, the history of which he had deep and profound interest in. He travelled through various parts of India, with keen interest in the people and their culture. Although he was not a mystique, he once told me of his "supernatural" feelings while sitting on the ghats of Ganges in Benaras.

We are devastated, such an energetic and lively colleague of ours died so young; I understand, he was just about forty-eight years in age! Over the years, he became like my younger brother, ready to discuss on any issue at hand.

Spirit of Jun-ichi will be with us for ever."

– **Bikas K. Chakrabarti**, Saha Institute of Nuclear Physics, Kolkata, India

Journal of Physics: Conference Series **638** (2015) 011001 doi:10.1088/1742-6596/638/1/011001

"Jun-ichi remained my friend since 2002. From scientific point of view, what I appreciated most in him was his perseverance and precision that behooves in an earnest bearer of Japanese scientific culture. Jun-Ichi kept himself engaged in the difficult task of refining and perfecting quantum algorithms for practical applications, and was making significant progresses in this field. He was a true gentleman - humble, kind, and friendly. I consider his untimely demise is a deep personal loss for me."
– **Dr. Arnab Das**, Indian Association for the Cultivation of Science, Kolkata, India

Many other colleagues shared their condolences and grief – too many to be reproduced here. Truly, we will miss him!

We would like to thank the Journal of Physics Conference Series, IOP, for publishing the proceedings of the conference, and the referees for their prompt and active support. We would like to specially thank Prof. Yoshiyuki Kabashima, Tokyo Institute of Technology, Japan, for arranging promptly the funds for the publications costs at a very late stage.

The proceedings of the past Statphys-Kolkata conferences have appeared in Journal of Physics Conference Series, vol: 297 (2011); Physica A, vol: 384 (2007); Physica A, vol: 346 (2005); Physica A, vol: 318 (2003); Physica A, vol: 270 (1999); Physica A, vol: 224 (1996); and Physica A, vol: 186 (1992).

We would like to take this opportunity to thank all the members of the Organizing committee and the Advisory committee. We also acknowledge:

- S. N. Bose National Centre for Basic Sciences, Kolkata
- Japan Society for the Promotion of Science, Japan
- Institute of Mathematical Sciences, Chennai
- Tata Institute of Fundamental Research, Mumbai
- Indian Association for the Cultivation of Science, Kolkata
- Hokkaido University, Japan

for their financial support.

Finally, we are grateful to the statistical physics community in India and abroad for their continuous participation and invaluable support in keeping the series of meetings alive!

Peer review statement

All papers published in this volume of *Journal of Physics: Conference Series* have been peer reviewed through processes administered by the proceedings Editors. Reviews were conducted by expert referees to the professional and scientific standards expected of a proceedings journal published by IOP Publishing.

Journal of Physics: Conference Series **638** (2015) 012001 doi:10.1088/1742-6596/638/1/012001

Nonequilibrium Response and Frenesy

Urna Basu and Christian Maes
Institute for Theoretical Physics, KU Leuven
E-mail: `urna.basu@fys.kuleuven.be`

Abstract. We present examples of how time-symmetric kinetic factors contribute to the response either in nonlinear order around equilibrium or in linear order around nonequilibrium. The phenomenology we associate to that so called frenetic contribution are negative differential conductivity, changes in the Einstein relation between friction and noise, and population inversion.

1. Introduction

Understanding how a system responds to an external stimulus is both interesting and important. Questions include how to predict it and to identify on what it depends. We could for example suspect from general intuition that the response of an observable would be related to its variance; the system is more susceptible to external stimuli if it shows already larger fluctuations in its unperturbed condition. An early example of that connection was the Sutherland–Einstein relation which gives the mobility proportional to the diffusion constant for particles suspended in an equilibrium fluid [1, 2]. The equilibrium fluctuation—dissipation theorem generalizes that idea and the (Green–)Kubo formulæ relate the linear response coefficients of an equilibrium system to correlation functions in the unperturbed system [3, 4]. More precisely, they tell us that to linear order around equilibrium the response is given by a correlation of the observable with the entropy flux generated by the perturbation. Let us start by a newer view on that result.

When dynamics enters into the statistical description of physical systems it appears useful to consider dynamical ensembles [5]. On the formal side those are very reminiscent of what we more generally call path-space integration. One considers a system on a coarse-grained level with states x and we denote the possible trajectories over time-interval $[0, t]$ by $\omega = (x_\tau, 0 \leq \tau \leq t)$. The states x can be values of macroscopic variables or they could denote the position and momenta of some particles etc. The state evolves in time and we can first consider an equilibrium process where the path probabilities are time-reversal symmetric: the probability of each trajectory satisfies

$$\text{Prob}[\omega] = \text{Prob}[\theta\omega]$$

where $\theta\omega$ denotes the time reversed trajectory obtained by inverting the the time-sequence of states and, if the state involves velocities, also flipping the sign of all velocities. The above relation gets often described as microscopic reversibility or detailed balance.

We now start the system in equilibrium at time zero but then by a perturbation the system is pushed out and its statistics follows another dynamical ensemble with probabilities

$$\text{Prob}^h[\omega] = e^{-A(\omega)}\,\text{Prob}[\omega] \tag{1}$$

Journal of Physics: Conference Series **638** (2015) 012001 doi:10.1088/1742-6596/638/1/012001

where the superscript h stands for the amplitude of the (unspecified) perturbation and $A = A^h$ is called the action of order h. We want to know the value of some path-observable $O(\omega)$ during $[0, t]$. We can compute its expectation value as follows:

$$\begin{aligned}
\langle O(\omega)\rangle^h &= \sum_\omega O(\omega)\, e^{-A(\omega)}\, \text{Prob}[\omega] \quad \text{and similarly} \\
\langle O(\theta\omega)\rangle^h &= \sum_\omega O(\theta\omega) e^{-A(\omega)}\, \text{Prob}[\omega] \\
&= \sum_\omega O(\omega) e^{-A(\theta\omega)}\, \text{Prob}[\omega]
\end{aligned}$$

(The sums over the trajectories ω are a bit formal but that does not matter here.) By subtracting the last line from the first line we thus obtain

$$\langle O(\omega)\rangle^h - \langle O(\theta\omega)\rangle^h = \sum_\omega O(\omega) \left[e^{-A(\omega)} - e^{-A(\theta\omega)}\right] \text{Prob}[\omega] \tag{2}$$

Such formulæ are useful for response theory. For example, we can take for observable O a particle current (where $O\theta = -O$) or even simpler we can choose $O(\omega) = \delta(x_t - x)$, the indicator to be in state x at the final time t. In the last case we find

$$\text{Prob}^h[x_t = x] - \text{Prob}^h[x_0 = x] = \sum_{\omega: x_t = x} \left[e^{-A(\omega)} - e^{-A(\theta\omega)}\right] \text{Prob}[\omega] \tag{3}$$

Using that in the reference equilibrium, $\text{Prob}[x_t = x] = \text{Prob}[x_0 = x]$ and since at time zero $\text{Prob}^h[x_0 = x] = \text{Prob}[x_0 = x]$, the perturbation is not yet visible, formula (3) gives the change by the perturbation as

$$\text{Prob}^h[x_t = x] - \text{Prob}[x_t = x] = \sum_{\omega: x_t = x} \left[e^{-A(\omega)} - e^{-A(\theta\omega)}\right] \text{Prob}[\omega] \tag{4}$$

Response theory was first systematically developed some 60 years ago for dynamical ensembles that have a time-reversal symmetry and when the perturbation can be taken very small. That is the context of linear response theory around equilibrium. While that is not the usual textbook presentation, the above can be applied immediately. We should expand the exponential to first order in h (linear order perturbation theory). Then (4) simplifies to

$$\text{Prob}^h[x_t = x] - \text{Prob}[x_t = x] = \sum_{\omega: x_t = x} [A(\theta\omega) - A(\omega)]\, \text{Prob}[\omega] \tag{5}$$

and we see that the linear response formula around equilibrium is completely expressed by the time-reversal antisymmetric part $A(\theta\omega) - A(\omega)$ of the action that describes the perturbation. That is interesting because in many situations that antisymmetric part has a clear physical meaning. That measure of time-reversal breaking is indeed related to dissipation and entropy fluxes, at least if the environment of the system remains in local thermal equilibrium. That makes the relation between linear response around equilibrium and entropy fluxes, as summarized in the fluctuation–dissipation theorem one of the manifestations of what we can call the miracle of equilibrium: there is basically a unique entropy, having many faces in terms of different thermodynamic potentials, that governs heat exchange with the environment via Clausius theorem, appears as a time-increasing function in Boltzmann's H-theorem, characterizes macroscopic fluctuations in the system as pioneered by Planck and Einstein, specifies the statistical forces as clarified by Onsager and, as explained above, also decides the linear response as developed by e.g. Kubo.

We can hope for more life (but maybe less luck) if the reference system is out of equilibrium or when considering nonlinear response around equilibrium. Energy–entropy considerations only are most likely not enough to characterize nonequilibrium systems. In fact it would be very surprising if the details of the dynamics would not matter much more for stationary nonequilibrium conditions. Kinetic factors encoded in friction or reactivities will show up in the stationary state weights and no guarantees exist about a purely thermodynamic meaning of fluctuation or variational functionals. This immediately implies that nonequilibrium theory cannot just be nonequilibrium thermodynamics and in general one has to go beyond the concept of entropy and work to construct nonequilibrium statistical mechanics.

Indeed, (5) will be false and the physics of the time-symmetric part in the action A comes to the foreground. That corresponds to a different physics; when a perturbation is applied around a nonequilibrium condition, kinetic factors, apart from entropy, start to play a key role [6, 7]. Similarly for higher order response around equilibrium, these kinetic factors appear explicitly [8]. By 'kinetic factors' we generically refer to the non-thermodynamic aspects like exact details of couplings of a system with a reservoir or time–symmetric Arrhenius prefactors in transition rates — quantities which are irrelevant in the study of static equilibrium linear response. *A priori*, this seems a formidable scenario — there could be many possible ways to perturb a system which are different kinetically but equivalent thermodynamically. Is there a generic method to describe response theory beyond the equilibrium fluctuation-dissipation relation? Luckily, it turns out that the answer is yes. The notion of 'frenesy' or dynamical activity, which is a time-symmetric quantity, allows us to provide a unified framework to study nonequilibrium response [6].

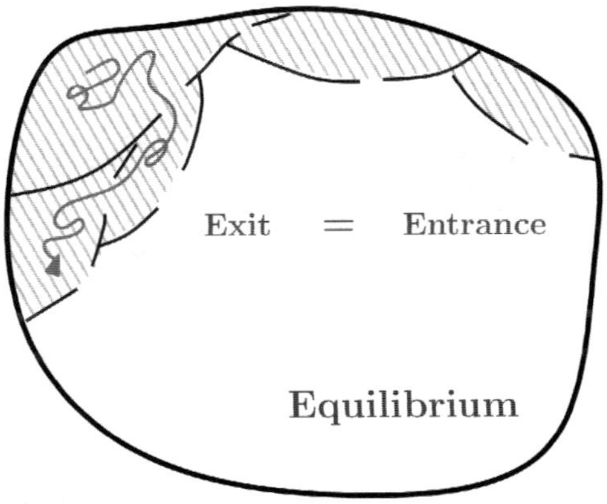

Figure 1. Schematic representation of the phase space of a macroscopic system. In equilibrium, which corresponds to the largest region, only volume (entropy) is relevant while surface effects (reactivity and escape rate) are substantial when the system is in a smaller region corresponding to some nonequilibrium stationary state. Reactivities are symmetric, suggested here by having exits equal to entrances between any two regions.

Frenetic contributions have already been described in studies of nonequilibrium system in various contexts. It is the key quantity leading to negative differential mobility for particle transport in disordered systems or constrained geometries [9, 10]. The dynamical activity is also an important player in studies of kinetically constrained models [11] or glass transitions [12]. Experimental studies of micro-swimmers also pick up the signature of this time–symmetric quantity [13]. On a more theoretical level we see the frenetic contribution as a complement to entropic considerations. The latter ultimately refer to estimates on phase space volumes for a physical coarse-graining of the energy surface corresponding to the appropriate closed and isolated universe containing our system. Volumes in phase space are especially important if the differences are large, with the equilibrium condition taking in fact almost all volume in phase space (see Fig. 1). When well away from that largest volume, it does not seem unreasonable to think that also "surfaces" of phase space regions get important in the sense that the exit and entrance rates of the regions start much more there to influence response and fluctuations. That time-symmetric traffic between different conditions (phase space regions) is what we have in mind with dynamical activity and the frenetic contribution manifests itself via correlation functions with such a traffic or dynamical activity.

The aim of this article is not to provide a comprehensive review of these studies, rather to illustrate how frenetic aspects lead to some interesting phenomena with the help of some simple examples. Relevant references are provided for more detailed studies, wherever necessary. In the next section we give a brief account of the theoretical background of response away from equilibrium; the relevant response formulæ, derived from a path-ensemble formalism are quoted. The precise meaning and definition of dynamical activity is also illustrated in the context of Markov jump processes. In Sec. 3 we move on to a few chosen examples starting with the toy model of a biased random walker. How negative differential mobilities can arise in presence of 'trapping' in the system is discussed next. Section 3.3 focuses on higher order response around equilibrium which picks up kinetic details in the coupling with the reservoir. The last example in Sec. 3.4 explores the effect of population inversion on energy transport in a simple setting of a multilane stochastic particle system. The population inversion is again caused by a suitable kinetic aspect. We conclude with some general remarks about nonequilibrium response theory.

2. Response formulae

In this section we give a brief review of the relevant response formulæ. That is meant to establish the notation and to explain the concepts rather than to prove them. We refrain from giving the derivations but the references will point to more reviews.

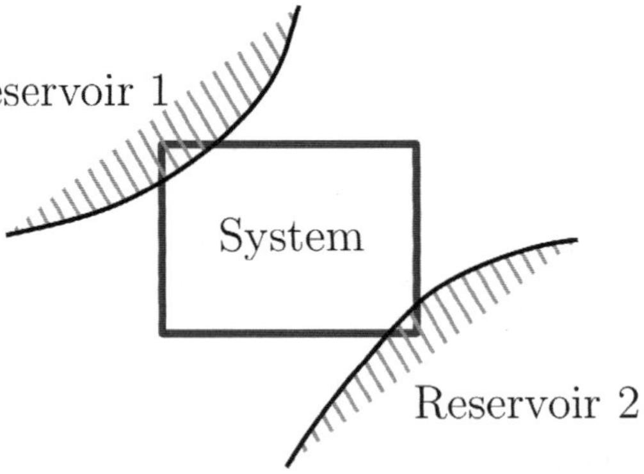

Figure 2. An open system in contact with spatially separated equilibrium reservoirs satisfies local detailed balance.

(1) makes sense also outside equilibrium, i.e., for a reference dynamical ensemble that is not

time-reversal symmetric. In a way (1) defines what are the possible perturbations that we can consider via the path-space formalism. For the physics we move now to the context of open systems in contact with multiple spatially well-separated equilibrium reservoirs (see Fig. 2). For systematics it is then useful to decompose the action into its time-antisymmetric and time-symmetric components,

$$S_h := A\theta - A, \quad D_h := \frac{1}{2}(A + A\theta), \quad A = D_h - \frac{1}{2}S_h$$

They come with a physical interpretation. That is most clear and familiar for

$$S_h(\omega) = \log\frac{\text{Prob}^h[\omega]}{\text{Prob}^h[\theta\omega]} - \log\frac{\text{Prob}[\omega]}{\text{Prob}[\theta\omega]}$$

giving a measure of the excess in time-reversal breaking by applying the perturbation. When the system is in weak contact with spatially separated equilibrium reservoirs, that time-antisymmetric part can be identified with the entropy flux (per k_B) into that environment along the path ω. It is in other words a thermodynamic quantity related to heat and dissipation of work, and the reason is summarized as the condition of local detailed balance; see [5, 14, 15, 16, 17, 18]. Also the dynamical activity $D_h(\omega)$ is identified for each trajectory ω; it is time-symmetric and its role and meaning is part of the questions in today's attempts towards constructing nonequilibrium statistical mechanics. Also this paper will contribute to illustrating its nature. For being more specific we illustrate below, using the example of a Markov jump process, how these quantities can be computed for a given dynamical ensemble [6, 7, 10].

2.1. Linear response around nonequilibrium

Formula (1) implies that $\langle O(\omega)\rangle^h = \langle e^{-A(\omega)}\, O(\omega)\rangle^0$. That immediately gives the linear differential response to the perturbation,

$$\frac{\mathrm{d}}{\mathrm{d}h}\langle O(\omega)\rangle^h = \frac{1}{2}\left\langle O(\omega); S_h'(\omega)\right\rangle^h - \left\langle O(\omega); D_h'(\omega)\right\rangle^h \tag{6}$$

$\langle f; g\rangle$ denotes the covariance between the observables f, g and S_h', D_h' are first derivatives with respect to the drive h. The averages $\langle\cdot\rangle^h$ are over trajectories including possibly the initial conditions and depending on the driving field h. We will often drop the explicit dependence on h in the notation. Thus, the first term in (6) signifies the covariance or the connected correlation of the observable O with the linear excess of entropy generated due to the perturbation and the second term arises from the correlation with the change in dynamical activity. Note also from (6) that whenever O only depends on the initial state x_0, then the left-hand side is zero and so must the right-hand side

$$\frac{1}{2}\left\langle O(x_0); S_h'(\omega)\right\rangle^h = \left\langle O(x_0); D_h'(\omega)\right\rangle^h \tag{7}$$

which relates the entropy flux with the dynamical activity. Also observe that we can take as a special case $O(\omega) = S_0'(\omega)$ in (6), to find

$$\frac{\mathrm{d}}{\mathrm{d}h}\langle S_0'(\omega)\rangle^h = \frac{1}{2}\left\langle S_0'(\omega); S_0'(\omega)\right\rangle^h - \left\langle S_0'(\omega); D_h'(\omega)\right\rangle^h \tag{8}$$

where the first term is always positive. In fact, the observable $S_0'(\omega)$ is time-antisymmetric and refers to an excess of entropy flux carried by the different currents in the system. The formula (8) is therefore an extended Green-Kubo formula (in the Helfand form) for the differential transport coefficients. We remark also that from (6)-(8) we can obtain the Harada-Sasa equality connecting energy dissipation with the violation of the usual fluctuation–response relation, [16, 19]. Finally we feel that it makes sense to make a connection here with the notion of effective temperature. Effective temperature from the response formula (6) can be defined by dividing that formula by $\langle O(\omega); S'_h(\omega)\rangle^h$. In the case of equilibrium that would give the (real thermodynamic) temperature; for nonequilibrium, there is a correction involving the frenetic contribution.

Let us look again at the special case of perturbing around equilibrium $h = 0$. In that case, if the observable O is antisymmetric under time-reversal, e.g. being a time-integrated energy or particle current $O(\omega) = J(\omega)$, then $\langle J(\omega); D'_0(\omega)\rangle^0 = 0$ because the dynamical activity $D_h(\omega)$ in (6) is itself time-symmetric and the equilibrium process is time-reversal invariant. Thence,

$$\frac{\mathrm{d}}{\mathrm{d}h}\langle J(\omega)\rangle^h\Big|_{h=0} = \frac{1}{2}\left\langle J(\omega); S'_0(\omega)\right\rangle^0 \tag{9}$$

That equilibrium result is the well–known Green–Kubo relation for equilibrium conductivities. We will see it more explicitly in later examples. Another special case is when the observable is a state function $O(\omega) = O(x_t)$. Then, in equilibrium, (6) reduces to the Kubo formula,

$$\frac{\mathrm{d}}{\mathrm{d}h}\langle O(x_t)\rangle^h\Big|_{h=0} = \left\langle O(x_t); S'_0(\omega)\right\rangle^0 = \left\langle O(x_t)S'_0(\omega)\right\rangle^0 \tag{10}$$

The first equality follows from

$$\langle O(x_t)D'_0(\omega)\rangle^0 = \langle O(x_0)D'_0(\omega)\rangle^0 = \frac{1}{2}\langle O(x_0)S'_0(\omega)\rangle^0 = -\frac{1}{2}\langle O(x_t)S'_0(\omega)\rangle^0$$

where we have used (7) for the middle equality. All the other equalities follow by time-reversal invariance of the equilibrium process $\langle\cdot\rangle^0$.

For the response in (9)–(10) around equilibrium it suffices to know how the perturbation affects the entropy flux (or, excess dissipation in the environment). Away from equilibrium, however, the frenetic contribution $\langle O(\omega); D'_h(\omega)\rangle^h$ in (6) involving the dynamical activity $D(\omega)$ plays a major role. As we will see later, this $D(\omega)$ explicitly involves 'kinetic factors' like escape rates, symmetric prefactors or diffusion constants, which implies that the nonequilibrium response formula carries more detailed information about the system than the equilibrium one.

2.2. Higher order response

Using dynamical ensembles allows one also to easily investigate higher order responses. In particular, to second order the general nonequilibrium response formula takes the form

$$\begin{aligned}\frac{\mathrm{d}^2}{\mathrm{d}h^2}\langle O(\omega)\rangle^h &= -\langle D''_h(\omega)O(\omega)\rangle^h + \langle (D'_h(\omega))^2 O(\omega)\rangle^h \\ &\quad + \frac{1}{4}\langle (S'_h(\omega))^2 O(\omega)\rangle^h - \langle S'_h(\omega)D'_h(\omega)O(\omega)\rangle^h \end{aligned}\tag{11}$$

Here we have assumed that $S''_h = 0$, i.e., the entropy flux S_h is linear in the field h which is true for most physically relevant examples of perturbations. Unsurprisingly, the dynamical activity appears explicitly in this nonequilibrium second order response formula. What is interesting is that the frenetic effects become important even when one looks at response around equilibrium from second order onwards. In particular for a time-antisymmetric observable $J(\omega)$, and the perturbation around equilibrium $h = 0$ one obtains the second order correction to the Green-Kubo formula [8],

$$\frac{\mathrm{d}^2}{\mathrm{d}h^2}\langle J(\omega)\rangle^h\Big|_{h=0} = -\langle S_0'(\omega)D_0'(\omega)J(\omega)\rangle^0 \tag{12}$$

The extension of the Kubo formula for a state observable $O(x_t)$ is the second order response around equilibrium,

$$\frac{\mathrm{d}^2}{\mathrm{d}h^2}\langle O(x_t)\rangle^h\Big|_{h=0} = -2\langle S_0'(\omega)D_0'(\omega)O(x_t)\rangle^0 \tag{13}$$

Indeed the above expressions depend explicitly on the excess in the dynamical activity D_0' generated by the perturbation while the first order, as given by (10), does not. The direct implication is that perturbations which are thermodynamically equivalent — generating the same entropy fluxes but differing in kinetic details, would have different second order responses.

Before we discuss the frenetic aspects of response in more details with the help of several physically motivated examples let us make more explicit the quantities $S_h(\omega)$ and $D_h(\omega)$ for systems modelled by Markov jump processes. We consider a Markov jump process specified by transition rates $k(x, y)$ for jumps $x \to y$ between states x, y. These are parametrized by the time-antisymmetric and symmetric $s(x, y)$ and $\psi(x, y)$ respectively

$$\begin{aligned} k(x,y) &= \psi(x,y)\, e^{s(x,y)/2}, \\ \psi(x,y) &= \psi(y,x) \geq 0,\ s(x,y) = -s(y,x) \end{aligned} \tag{14}$$

all possibly depending on the field or potential with amplitude h. A trajectory $\omega := (x_\tau, 0 \leq \tau \leq t)$ over time interval $[0, t]$ is characterized by discrete jumps at times τ_i and by exponentially distributed waiting times $\tau_{i+1} - \tau_i$. The entropy flux and dynamical activity associated with this trajectory (up to some additive constants which depends on the reference process) are

$$\begin{aligned} S_h(\omega) &= \sum_i s(x_{\tau_i}, x_{\tau_{i+1}}) \\ D_h(\omega) &= \int_0^t \mathrm{d}\tau\, \xi(x_\tau) - \sum_i \log\psi(x_{\tau_i}, x_{\tau_{i+1}}) \end{aligned} \tag{15}$$

where $\xi(x) = \sum_y k(x, y)$ is the escape rate at state x. The dynamical activity is time-symmetric; reversing the time over the trajectory in $[0, t]$ does not change it. It consists of two contributions, the reactivities ψ and the escape rates ξ. These are the *kinetic factors* that become particularly significant away from equilibrium. One of the interesting questions is to discover how ψ is influenced by the nonequilibrium condition and how it changes with h. The response will teach us about some of that dependence. On the other hand, S_h is time-antisymmetric and corresponds to the *thermodynamic* entropy flux over $[0, t]$ when the condition of local detailed balance is satisfied. Then $s(x, y)$ denotes the entropy flux to the environment (per k_B) in the transition $x \to y$. The excess entropy and dynamical activity produced by the perturbation is calculated from the above Eq. (15) and must be used in the relevant response formulae (6), (11) or (13). Note that under (global) detailed balance $s(x, y) = \beta[U(x) - U(y)]$ for some potential U, and then the stationary equilibrium distribution $\sim \exp[-\beta U]$ is completely independent of the prefactors $\psi(x, y)$.

3. Examples

In this section we discuss several examples where kinetic factors play an important role in the response behaviour.

3.1. Toy model: 1D Random walk

Our first example is a toy model of a biased one-dimensional random walk on a lattice which, despite being extremely simple, serves as a paradigmatic example to illustrate the notions of kinetic factors and frenetic contributions appearing in nonequilibrium response.

Let us consider a one-dimensional nearest neighbour continuous time random walk specified by rates p and q of jumping to the right, respectively left neighbour (see Fig. 3). In the parametrization (14),

$$\psi(x, x \pm 1) = \sqrt{pq}, \quad s(x, x \pm 1) = \pm \log \frac{p}{q}, \quad x \in \mathbb{Z}$$

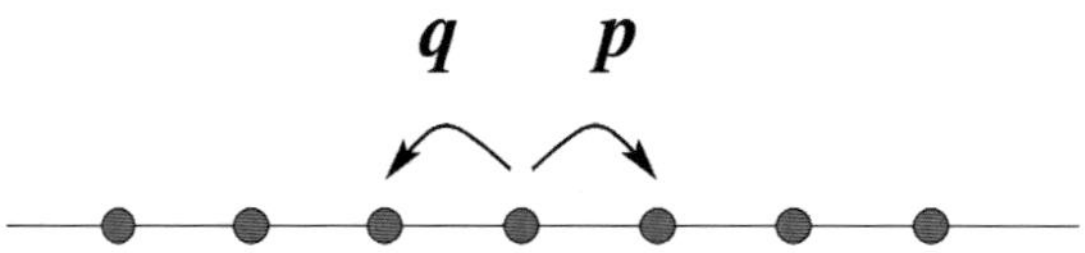

Figure 3. The usual parametrization for a continuous time biased random walk.

Equilibrium dynamics corresponds to $p = q$. Away from equilibrium, the local detailed balance condition characterizes the ratio of rates,

$$\log \frac{p}{q} = \beta E$$

for an environment at temperature β^{-1} and with external field E to the right. The symmetric prefactor or the reactivity ψ however is not specified yet by that condition. In general, one can imagine that $\psi = \psi_\beta(E)$ is also affected by the field E; the exact functional form would depend on the specific way the system is coupled to the environment and the driving agency. The walker is just an effective or reduced description of a possibly much more complicated process.

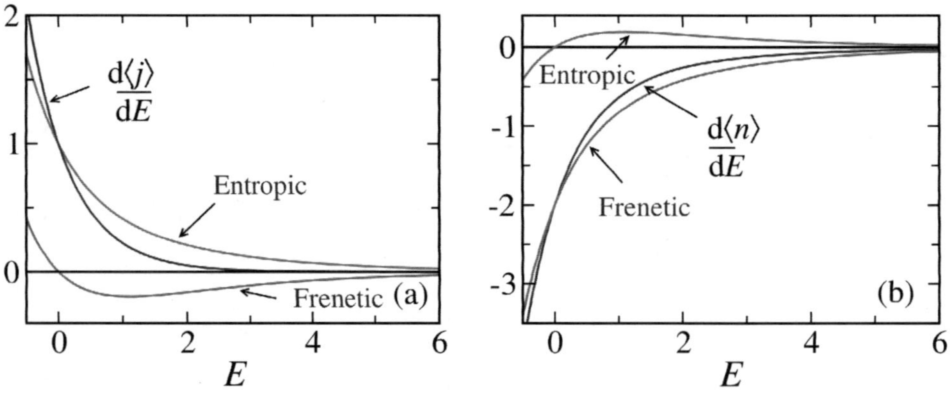

Figure 4. Linear response for 1-dimensional biased random walk. (a) The differential conductivity $\frac{\mathrm{d}\langle j \rangle}{\mathrm{d}E}$ along with its entropic and frenetic components are plotted as a function of the bias E. (b) The same for the time-symmetric traffic $\langle n \rangle$. Here $\psi_\beta(E) = \exp(-\beta E)$.

We look now at the differential response when the drive $E \to E + \mathrm{d}E$. It is needless to say that for this simple random walk model all the physical properties can be calculated exactly, and

one does not require linear response theory to compute the effect of the perturbation. Again, this biased random walker acts as an effective model for studying negative differential response in particle transport.
The entropy flux and dynamical activity for a given trajectory ω are given by (15),

$$\begin{aligned} S(\omega) &= (N_+ - N_-)\log\frac{p}{q} = \beta E J \\ D(\omega) &= (p+q)t - \frac{1}{2}(N_+ + N_-)\log pq \\ &= 2\psi_\beta(E)\cosh\frac{\beta E}{2}t - N\log\psi_\beta(E) \end{aligned}$$

where N_+ and N_- denote the number of jumps in $[0,t]$ to the right and the left respectively. $J = N_+ - N_-$ is the time-integrated current along the field and $N = N_+ + N_-$ is the 'traffic' *i.e.* the total number of jumps during the interval $[0,t]$. The primary observable of interest is the current J; the differential conductivity is then predicted using (6) and writing $h = E$,

$$\begin{aligned} \frac{\mathrm{d}}{\mathrm{d}E}\langle J\rangle &= \frac{1}{2}\big\langle J; S'(\omega)\big\rangle - \big\langle J; D'(\omega)\big\rangle \\ &= \frac{\beta}{2}\langle J;J\rangle + \frac{\psi'_\beta(E)}{\psi_\beta(E)}\langle N;J\rangle \qquad (16) \\ &= \left[\beta\psi_\beta(E)\cosh\frac{\beta E}{2} + 2\psi'_\beta(E)\sinh\frac{\beta E}{2}\right]t \qquad (17) \end{aligned}$$

The first term in (16), the variance of the current J, is the positive entropic contribution. The correlation $\langle N;J\rangle$ arises from the reactivity and note that it vanishes in equilibrium ($E = 0$). In other words the change in the prefactor $\psi_\beta(E)$ as function of E remains unnoticed in linear response around equilibrium. The interesting scenario is when the second term contributes sufficiently negatively yielding a negative differential response. From (17) we see what is needed — $\psi'_\beta(E)$ should be 'sufficiently' negative. Fig. 4(a) shows the differential conductivity for such a choice of $\psi_\beta(E) = \exp(-\beta E)$. For the sake of convenience we have used the current per unit time $j = J/t$ as the observable; the entropic and frenetic contributions are also shown separately.
Similarly, one can also look for the response in the traffic $N = N_+ + N_-$,

$$\begin{aligned} \frac{\mathrm{d}}{\mathrm{d}E}\langle N\rangle &= \frac{\beta}{2}\langle N;J\rangle + \frac{\psi'_\beta(E)}{\psi_\beta(E)}\langle N;N\rangle \\ &= \left[\beta\psi_\beta(E)\sinh\frac{\beta E}{2} + 2\psi'_\beta(E)\cosh\frac{\beta E}{2}\right]t \qquad (18) \end{aligned}$$

Note that there the variance of N appears in the frenetic term. That was seen already in the context of linear response for the boundary driven zero range process [20]. In fact, since N is time-symmetric the role of entropic and frenetic contribution gets reversed; even for $E = 0$ we get only the frenetic part. That is reminiscent of the situation for the momentum current in shear experiments. The viscosity is then in fact frenetic and not entropic. That similarity has been discussed in e.g. [21].
Fig. 4(b) shows the response and the entropic and frenetic components separately for $\langle n\rangle = \langle N\rangle/t$, the average number of undirected jumps per unit time. We see that the differential response for $\langle n\rangle$ is negative for this specific choice of $\psi(E)$.
The second order response for the current J around equilibrium also immediately follows from Eq. (12),

$$\frac{\mathrm{d}^2}{\mathrm{d}E^2}\langle J(\omega)\rangle\bigg|_{E=0} = -\langle S'_0(\omega)D'_0(\omega)J(\omega)\rangle^0$$

$$= -\beta \left[2\psi_\beta'(0)t\langle J^2\rangle^0 - \frac{\psi_\beta'(0)}{\psi_\beta(0)}\langle J^2 N\rangle^0 \right]$$
$$= 2\beta\psi_\beta'(0)t$$

Clearly that second order response now depends explicitly on how the reactivity ψ changes with the driving E. In particular, as already mentioned in the previous section, perturbations which are equivalent thermodynamically *i.e.*, produce the same entropy fluxes but differ in their effect upon the kinetic prefactor would generate separate second order responses.

3.2. Trapping effect: negative differential mobility

Mobility of a particle quantifies how its velocity changes when subjected to an external mechanical force. Requirements for thermodynamic stability ensure that the mobility is strictly positive around equilibrium *i.e* the velocity or current increases when the applied force is larger. This is a consequence of the fact that the equilibrium mobility, as given by the Sutherland-Einstein relation, is nothing but the variance of the current in the unperturbed state. This statement is no longer true when the system is driven far away from equilibrium, where the frenetic contribution becomes relevant, [22]. There one speaks about the differential mobility which characterizes how the current changes with the drive. One of the consequences of the modified Kubo formula is that negative differential mobilities are allowed to exist, *i.e.*, the current need not be a monotonically increasing function of the drive.

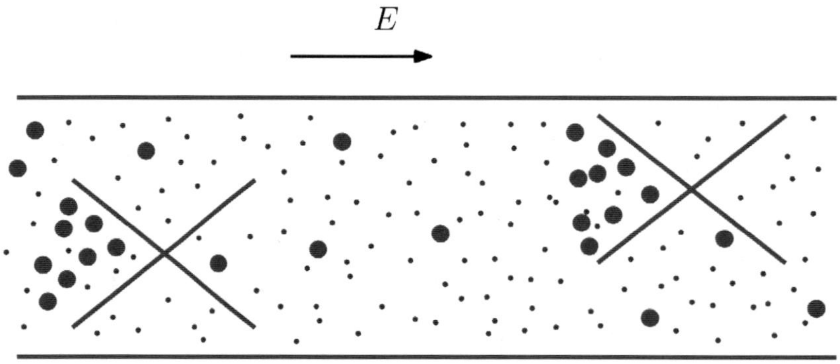

Figure 5. Driven particle motion in constrained geometry giving rise to trapping. Colloid particles (dark green discs) tend to reside inside the cages created by the obstacles when pushed hard. The blue dots represent the equilibrium medium.

Several studies of negative differential mobility in nonequilibrium particle systems can be found in the literature. Examples include particles diffusing in restrictive or confined geometries [10, 23, 24], driven lattice Lorentz gas [25, 26], driven particle in an environment of mobile obstacles [9, 27] and exclusion processes with constrained dynamics [21].
To understand the origin of such negative differential mobility from a general viewpoint let us look at Eq. (6) where $O = J$, the time-integrated current, in what we imagine as a driven particle system. The excess entropy is essentially proportional to the current itself implying the positivity of the entropic term. A large negative contribution from the frenetic term, however, could result in an overall negative differential response of the current.

It was shown in [10, 23] that this indeed might happen if there is a possibility of trapping in the system. Intuitively that is quite easy to understand — if the particles get trapped then the dynamical activity is affected and if it so happens that the trapping or caging increases when the perturbing field becomes stronger then negative differential mobility can occur. A simple example is the driven motion of colloids in a narrow channel in presence of directed 'cages' or obstacles (Fig. 5); the particles get more and more trapped when the driving force is increased resulting in a decrease of the current. It is important, however, to note that 'trapping' need not always mean a real trap or cage in the physical space, it might as well refer to effective trapping in some region of the phase space (cf. Fig. 1).

The biased random walk discussed in the previous section can be seen as a paradigmatic example of this phenomenon — if the driving affects the reactivities in such a way that the escape rate is a decreasing function of the drive, then the current shows a non-monotonic behaviour. This was discussed in detail in [9, 10], where it was shown that indeed the frenetic component is responsible for the negative differential mobility. Here it suffices to mention that several models of interacting particle systems can effectively be mapped to this simple picture of biased random walk.

3.3. Coupling with reservoir

Our next example illustrates how the second order response picks up kinetic details of the coupling to a particle reservoir.

We consider a tracer particle of mass M with position x_t coupled to an equilibrium bath, in the sense that it obeys the Newton equation

$$M\ddot{x}_t = -\nabla_x U(x_t,\eta_t) =: F(x_t,\eta_t) \tag{19}$$

where the η denotes a family of N degrees of freedom making the equilibrium reservoir. η would typically contain position r_i and velocity $\dot{r}_i$ of the reservoir particles. The force $F(x_t,\eta_t)$ acting on the particle is "noisy" as the η_t gets sampled from the equilibrium process which in back-reaction depends on the history $(x_s,\ s\leq t)$ of the particle.

The bath degrees of freedom evolve under a dynamics where each position $r_i, i=1,\ldots,N$ is coupled to the probe via the same $U(x,\eta)$ and possibly with other interactions while on the whole satisfying detailed balance at fixed temperature β^{-1}. For greater simplicity we take a coupling (in one-dimensional notation)

$$U(x,\eta) = \frac{\lambda}{2}\sum_i (x-u_i(\eta))^2, \quad F(x,\eta) = \lambda\sum_i (u_i(\eta)-x) \tag{20}$$

so that for some function $u_i(\eta)$

$$\begin{aligned} -\frac{\partial}{\partial r_j}U(x,\eta) &= -\lambda\sum_i u_i(\eta)\frac{\partial}{\partial r_j}u_i(\eta) + x\frac{\partial}{\partial r_j}V(\eta) \\ V(\eta) &:= \lambda\sum_i u_i(\eta) \end{aligned} \tag{21}$$

That choice implies that for the $\eta-$dynamics, changing the position $x\rightarrow x+\mathrm{d}x$ means simply to add the potential V with small amplitude $\mathrm{d}x$ with respect to its original dynamics. The response below will be written in terms of that perturbing potential V. Note that $0<\lambda\sim 1/N$ to mimic a mean field coupling in which the particle couples equally with all the bath degrees of freedom.

Integrating out the equilibrium bath from (19) should give rise to an effective and generalized Langevin dynamics for the (tracer) particle,

$$M\,\ddot{x}_t = -\lambda N x + \langle V(\eta_t)\rangle^{(x_s,s\leq t)} + \xi_t \tag{22}$$

Journal of Physics: Conference Series **638** (2015) 012001 doi:10.1088/1742-6596/638/1/012001

with the noise $\xi_t := V(\eta_t) - \langle V(\eta_t)\rangle^{(x_s,s\leq t)}$ by construction dependent on the whole history $(x_s,\, s \leq t)$. The expectation $\langle\cdot\rangle^{(x_s,s\leq t)}$ is in the $\eta-$process conditioned on a fixed particle trajectory up to time t. In other words, we treat the particle trajectory as a fixed protocol in the dynamics for η. The regime where our perturbation approach makes sense is when there occurs a time-scale separation between the characteristic times τ_x of the particle and τ_η of the sea. To evaluate the expectation in (22) perturbatively we take as reference the stationary $\eta-$dynamics with fixed $x_s \equiv x_t, s \leq t$, the particle position at time t. In other words we have a time-dependent perturbation amplitude $h_s = x_s - x_t$. We apply the response formalism of [8] to second order,

$$\begin{aligned}\langle V(\eta_t)\rangle^{(x_s,s\leq t)} &= \langle V\rangle_{x_t} \\ &+ \beta\int_{-\infty}^{t}\frac{\mathrm{d}}{\mathrm{d}s}\langle V(\eta_s)\,V(\eta_t)\rangle_{x_t}\cdot(x_s-x_t)\,\mathrm{d}s \\ &+ \frac{\beta^2}{2}\int_{-\infty}^{t}\mathrm{d}s\int_{-\infty}^{t}\mathrm{d}s'\,\langle V(\eta_s)\,L_{x_t}V(\eta_{s'})\,V(\eta_t)\rangle_{x_t}\,\dot{x}_s\,(x_{s'}-x_t)\end{aligned} \tag{23}$$

where L_{x_t} is the backward generator of the η-equilibrium process at fixed particle position x_t. The first line contains the time-local term which is the statistical force in the infinite time-scale separation limit between bath and particle. The expectation $\langle\cdot\rangle_x$ is in the stationary equilibrium for the reservoir at fixed particle position x. That statistical force will be given by the gradient of the equilibrium free energy constrained to x. The second line arises from the standard Kubo-term in linear response. It gives rise, via partial integration, to the usual friction which is proportional to the noise covariance (what gives the second fluctuation–dissipation relation, at least in the absence of the last line). The last line indeed gives the correction coming from second order (nonlinear) response, containing frenetic aspects. The point is that L_x generates the time-evolution of the reservoir (at fixed particle position x) and thus contains the microscopic interaction between the reservoir degrees of freedom.

Equation (23) must be substituted into (22) for obtaining the Langevin-type effective dynamics of the particle. The reservoir kinetics sits already in the time-correlation functions that make the usual friction and noise term, but the last line in (23) adds significantly to that. The details of the dynamics will enter explicitly even in the Markov approximation. To the best of our knowledge that additional frenetic aspect has not been studied before, but we hope it will be experimentally visible. A similar study can be done when starting with nonequlibrium reservoirs. We then need to apply formula (6) for obtaining the friction; see [28, 29].

We conclude this discussion with the remark that if the bath is linear, say consisting of harmonic oscillators like in Section 1.6 of Zwanzig's book, [30], we can do an exact calculation with independent bath particles. Then the linear (Kubo) order for response suffices and there are in fact no corrections of second order. Therefore, saying it differently, forgetting about second order (kinetic) features in deriving the Langevin equation is like forgetting nonharmonic effects in the equilibrium bath.

3.4. Population inversion

Our last example centres around another typical nonequilibrium phenomenon, namely population inversion. Population inversion is about being able to select a certain (output) stationary statistical distribution over the various available states of the system. In equilibrium, temperature and energy decide and we are stuck with the Maxwell-Boltzmann statistics. Driving conditions add flexibility. It is not however the driving itself or only the driving that decides the (new) stationary distribution — the reactivities or time-symmetric factors in the transition rates are (also) essential. A very simple example is to take a ring with L sites and a biased walker with rates $k(x, x+1) = a_x e^{E/2}, k(x+1, x) = a_x e^{-E/2}$. For $a_x \equiv 1$ the stationary distribution is uniform for no matter what E. But for large E we can basically concentrate the stationary distribution on whatever site x we want by a suitable choice of a_x. We believe that such a mechanism of kinetic selection is widespread in bio-chemical systems for better efficiency in getting the right output product. Since that mechanism is time-symmetric and works (only) out-of-equilibrium we call it again frenetic. Similarly, for the origin of laser working one needd to produce a population inversion. It is well–known that at low temperatures, a suitable choice of symmetric prefactors can select any desired (even high) energy level of a multilevel system [31], so as to establish an inversion of the population with respect to the usual Boltzmann statistics. Next we investigate how this population inversion affects the energy transport in a spatially extended system. Such energy transport can now show unusual features which ultimately go back to frenetic aspects playing a role in the energy reservoirs. In fact we will drive one energy bath out-of-equilibrium and enforce in it a population inversion.

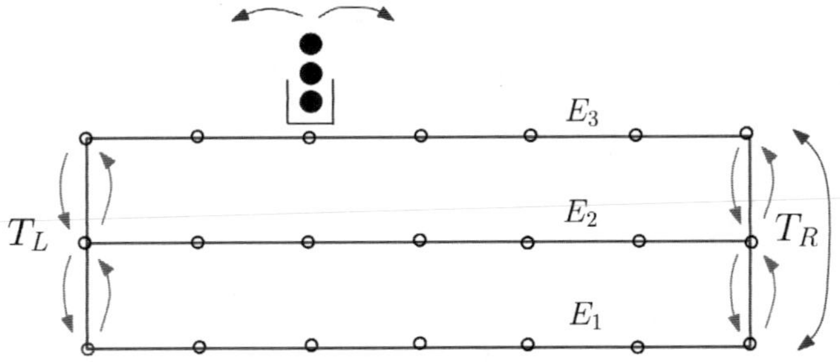

Figure 6. Schematic representation of the multilane particle transport model coupled to two thermal reservoirs (with $K = 3$).

The model is defined on a two–dimensional lattice; the horizontal direction is interpreted as space and the vertical one as energy. The different energy levels are spatially invariant and we refer to them as lanes. Following this convention a $L \times K$ lattice corresponds to a spatially extended one–dimensional system with L sites, each site having K possible energy levels having energies $E_1 < E_2 \cdots < E_K$. Each site on this lattice can accommodate any number of particles; the configuration of the system is fully specified by the set $\{n_{i\ell}; i \in [L], \ell \in [K])\}$ where $n_{i\ell}$ is the number of particles at the ℓth level of the ith site. We suppose that energy is conserved in the bulk, *i.e.*, particles cannot change lanes there; it can only hop to its neighboring sites on the same lane with some rate which we take to be unity here. A particle can jump across the energy lane at the left and right boundaries where the system is connected to two thermal reservoirs with temperatures T_L and T_R respectively (see Fig. 6); the system can gain or lose energy when a particle changes the lane. We assume free particle motion at the boundary sites so that the jump rate is proportional to the number of particles n at the source site. (The detailed dynamics is specified later.)

This system can be thought of as a simple discrete model for energy transport in an extended system; the particles represent energy packets which are transported across the system without scattering. The situation is somewhat similar to the phenomenon of stimulated Raman adiabatic passage [32]. When the temperatures $T_L > T_R$ are different, there will be an energy current flowing through the system which is proportional to the temperature difference (Fourier law).

Here we are interested in the effect of population inversion on the thermal current. To this

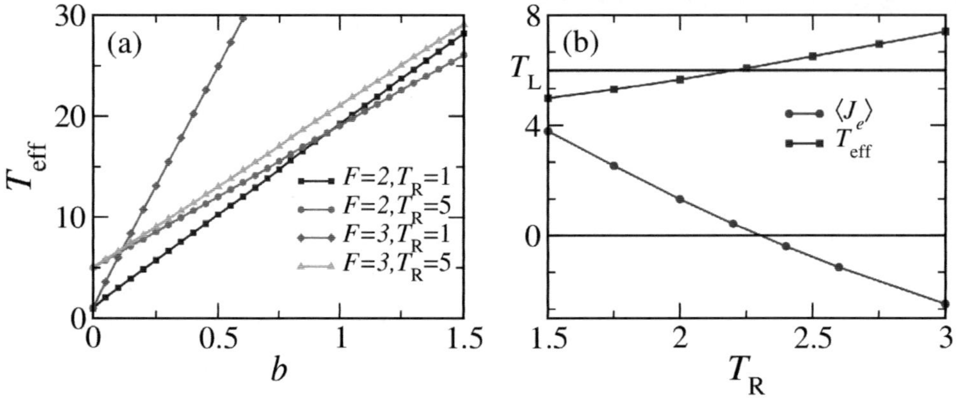

Figure 7. (a) The effective temperature T_{eff} defined in (25) for a three-level system, increasing with equalizing strength b. The case $b = 0$ corresponds to equilibrium where $T_{\text{eff}} = T$. The crossing of two curves at the same F but with different T_R is a signature of the resulting population inversion. (b) Illustrating the meaning of effective temperature: when the right reservoir has an effective temperature that equals the temperature of the left reservoir, the energy current gets close to zero and then reverses sign. Here $b = 0.5, F = 1$ and the left temperature $T_L = 6$. The current $\langle J_e\rangle$ has been amplified by a factor for better visibility.

end we choose the jump rates at, say, the right boundary in a way so that the highest energy level K becomes most populated at low temperature T_R. We also assume free particle motion at the boundaries. The complete dynamics is then characterized by the transition rates $k(i,\ell;j,m)$ of a single particle from position (i,ℓ) to position (j,m) :

$$\begin{array}{llll}
\text{Bulk} & k(i,\ell;i\pm 1,\ell) &=& 1 \quad \forall i\neq 1,L \\
\text{Left Boundary} & k(1,\ell;2,\ell) &=& n_{1\ell} \\
 & k(1,\ell;1,\ell+1) &=& n_{1\ell}\, e^{-\frac{1}{T_L}(E_{\ell+1}-E_\ell)} \quad \forall \ell\neq K \\
 & k(1,\ell;1,\ell-1) &=& n_{1\ell} \quad \forall \ell\neq 1 \\
\text{Right Boundary} & k(L,\ell;L-1,\ell) &=& n_{L\ell} \\
 & k(L,\ell;L,\ell+1) &=& n_{L\ell}\, e^{-\frac{1}{T_R}(E_{\ell+1}-E_\ell)} \quad \forall \ell\neq K-1,K \\
 & k(L,\ell;L,\ell-1) &=& n_{L\ell} \quad \forall \ell\neq K \\
 & k(L,K-1;L,K) &=& n_{L,K-1}\, e^{-\frac{F}{T_R}} e^{-\frac{1}{T_R}(E_K-E_{K-1})} \\
 & k(L,K;L,K-1) &=& n_{L,K}\, e^{-\frac{F}{T_R}} \\
 & k(L,1;L,K) &=& b\, n_{L,1} \\
 & k(L,K;L,1) &=& b\, n_{L,K}
\end{array} \qquad (24)$$

The dynamics conserves the particle density ρ in the system but not the total energy.

The equalizing symmetric prefactor $b > 0$ between the highest and lowest energy level along with the factor e^{-F/T_R} at the right boundary gives rise to the desired population inversion [31]. To illustrate that effect we first concentrate on the right reservoir itself — decoupled from the bulk. It is equivalent to a system with K energy levels where the lowest and highest levels are connected by an equalizer and an additional energy barrier F exists between the K and $K-1^{th}$ level. A good indicator of the resulting population inversion is the effective temperature measured from the stationary state probability of the highest and lowest levels, here defined as

$$T_{\text{eff}} = (E_K - E_1)\left[\log \frac{\rho_1}{\rho_K}\right]^{-1} \tag{25}$$

Fig. 7 shows that effective temperature T_{eff}, increasing with the strength of the equalizer b. For a fixed F, the curve of T_{eff} corresponding to a lower thermodynamic temperature T_R crosses that of a higher temperature from below signifying population inversion. The barrier F facilitates this phenomenon — crossing occurs for smaller b when F is increased.

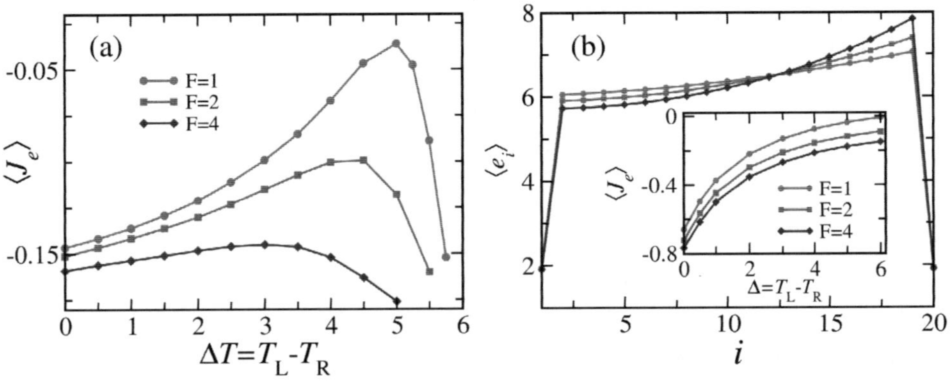

Figure 8. (a) Energy current $\langle J_e \rangle$ as a function of the temperature difference $\Delta T = T_L - T_R$ for different values of F. Here the temperature gradient is increased by decreasing T_R while $T_L = 6$ is kept fixed. (b) The energy profile $\langle e_i \rangle$ for different values of F for fixed $T_L = 6, T_R = 1$. The inset shows $\langle J_e \rangle$ as a function of the temperature difference $\Delta T = T_L - T_R$ when $T_R = 1$ is fixed. Here $b = 1$ and the number of energy levels $K = 3$ and particle density $\rho = 2.0$. Also, we have taken $E_\ell = (l-1)$.

One can expect that when this system serves as a thermal reservoir it would drive a current opposite to the thermal gradient which is present even when $T_L = T_R$.
The quantity of interest is the total energy current through the system,

$$J_e = \sum_{i=1}^{L-1}\sum_{\ell=1}^{K} E_\ell J_\ell^{i,i+1} \tag{26}$$

where $J_\ell^{i,i+1}$ denotes the particle current (net number of jumps) from $i \to i+1$ at the ℓth level. We have adopted the notation where the superscript to the current J indicates the spatial position and the subscript indicates the energy level. Fig. 7(b) illustrates the relevance of the effective temperature for the full system. The current flowing through the system vanishes when $T_{\text{eff}} \simeq T_L$ *i.e* when the effective bias becomes zero.

To calculate the response we need the usual path dependent entropy and dynamical activity,

$$\begin{aligned} S_{T_R}(\omega) &= -\frac{1}{T_R}\sum_{\ell=1}^{K-1} J^R_{\ell,\ell+1}\Delta E_\ell \\ D_{T_R}(\omega) &= \frac{1}{2T_R}\left[\sum_{\ell=1}^{K-1} I_{\ell,\ell+1}\Delta E_\ell + 2F I_{K-1,K}\right] \end{aligned}$$

Journal of Physics: Conference Series **638** (2015) 012001 doi:10.1088/1742-6596/638/1/012001

$$+ \int_0^t \mathrm{d}\tau \left[\sum_{\ell=1}^{K-2} e^{\left[-\frac{\Delta E_\ell}{T_R}\right]} n_{L,l}(\tau) + e^{\left[-\frac{\Delta E_{K-1}+F}{T_R}\right]} n_{L,K-1}(\tau) + e^{\left[-\frac{F}{T_R}\right]} n_{L,K}(\tau)\right]$$

Here $J^R_{\ell,\ell+1}$ and $I_{\ell,\ell+1}$ are the net particle current (directed from ℓ to $\ell+1^{th}$ level) and traffic (total number of jumps) respectively at the right boundary between ℓ^{th} energy level and $\ell+1^{th}$ level; $\Delta E_\ell = E_{\ell+1} - E_\ell$ is the corresponding energy difference between these levels. Note that since we are interested in the thermal response where the thermal gradient is changed by varying the right bath temperature T_R we have kept only those terms which depend on T_R. The thermal conductivity can now be calculated from formula (6) where S' and D' would be the derivatives with respect to $\Delta = T_L - T_R$. Once more we see that the quantity F which is related to the reactivity of a particular transition appears explicitly only in the dynamical activity.

It is easy to see that this dynamics is essentially an inhomogeneous zero range process [33] on a two–dimensional lattice where the hop rate $u_{\alpha\gamma}(n_\alpha)$ from site α to γ depends only on the particle number at the departure site. It is known that the stationary distribution has a product structure when the hop rate has the general form

$$u_{\alpha\gamma}(n_\alpha) = u_\alpha(n_\alpha) W_{\alpha\gamma} \tag{27}$$

where $W_{\alpha\gamma}$ gives the probability that γ is chosen as the target site; $\sum_\gamma W_{\alpha\gamma} = 1$. The dynamics (24) satisfies this condition with $u_\alpha(n) = \xi_\alpha$ when $\alpha = \{i,\ell\}$ refers to a site in the bulk and $u_\alpha(n) = n\xi_\alpha$ for boundary sites; ξ_α being the n–independent part of the escape rate of site α, containing the energy dependent component of the transition rates. For example $\xi_{i,\ell} = 2$ for bulk sites, $\xi_{L,1} = 1 + b + \exp(-\Delta E_1/T_R)$ and so on. With this identification and following [33] one can write the individual site weights,

$$f_\alpha(n) = \prod_{m=1}^{n} \left[\frac{q_\alpha}{u_\alpha(m)}\right] \tag{28}$$

where q_α are the solutions of the linear system of equations $q_\alpha = \sum_\gamma W_{\gamma\alpha} q_\gamma$.
The stationary current and density profile $\rho_{i\ell}$ (satisfying $\frac{1}{LK}\sum_{i,\ell}\rho_{i\ell} = \rho$) can easily be obtained from this exact solution. The average spatial energy profile, defined as

$$\langle e_i \rangle = \sum_{\ell=1}^{K} E_\ell \rho_{i\ell} \tag{29}$$

is plotted in Fig. 8(b).

The average energy current $\langle J_e \rangle$ as a function of the temperature gradient Δ for different values of F is shown in Fig. 8(a). While the thermal gradient drives a current from the left boundary to the right one, because of the population inversion, there is an opposing current due to the density gradient at the highest level. Now, the population inversion becomes more efficient as the right temperature T_R goes down, hence if the thermal gradient is increased by decreasing T_R the opposing drive becomes stronger and hence the current goes down. On the other hand we also expect from the above explanation that there is no negative response when the left temperature T_L increases. That is confirmed and shown in the inset of Fig. 8(b).

4. Conclusion

Non-thermodynamic features appear naturally under nonequilibrium conditions. We have given examples how time-symmetric kinetic aspects contribute to interesting phenomenology. The new unifying quantity here is the dynamical activity which via its excess due to the perturbation correlates from second order onwards with the (excess) entropy flux and produces the frenetic contribution to response. Frenetic aspects arise then for example from reactivities that are modified by external fields or by the nonequilibrium condition in general.

Acknowledgments

We very much thank the organizers for a great conference and stimulating environment.

References

[1] Einstein A 1905 *Ann. der Physik* **17** 549
[2] Sutherland W 1905 *Phil. Mag. S.* **9** 781
[3] Green M S 1954 *J. Chem. Phys.* **22** 398
[4] Kubo R 1957 *J. Phys. Soc. Jpn.* **12** 570
[5] Bergman P and Lebowitz J 1955 *Phys. Rev.* **99** 578
[6] Baiesi M, Maes C and Wynants B 2009 *Phys. Rev. Lett.* **103** 010602
[7] Baiesi M and Maes C 2013 *New J. Phys.* **15** 013004
[8] Basu U, Krüger M, Lazarescu A and Maes C 2015 *Phys. Chem. Chem. Phys.* **17** 6653
[9] Bénichou O, Illien P, Oshanin G, Sarracino A and Voituriez R 2014 *Phys. Rev. Lett.* **113** 268002
[10] Baerts P, Basu U, Maes C and Safaverdi S 2013 *Phys. Rev. E* **88** 052109
[11] Garrahan J P, Jack R L, Lecomte V, Pitard E, van Duijvendijk K and van Wijland F 2009 *J. Phys. A.: Math. Gen.* **42** 075007; Jack R, Garrahan J P, Chandler D, 2006 *J. Chem. Phys.* **125**, 184509
[12] Speck T, Royall C P and Williams S R 2014 *arXiv:1409.0751*
[13] Volpe G, Buttinoni I, Vogt D, Kümmerer H J and Bechinger C 2011 *Soft Matter* **7** 8810
[14] Katz S, Lebowitz J L and Spohn H 1984 *J. Stat. Phys.* **34** 497
[15] Maes C and Netočný K 2003 *J. Stat. Phys.* **110** 269
[16] Harada T and Sasa S-I 2005 *Phys. Rev. Lett.* **95** 130602
[17] Derrida B 2007 *J. Stat. Mech.* **2007** P07023
[18] Tasaki H 2007 *arXiv:0706.1032v1*
[19] Lippiello E, Baiesi M and Sarracino A 2014 *Phys. Rev. Lett.* **112** 140602
[20] Maes C and Salazar A 2014 *Chaos, Solitons & Fractals* **64** 78
[21] Sellitto M 2008 *Phys. Rev. Lett.* **101** 048301
[22] Baiesi M, Maes C and Wynants B 2011 *Proc. R. Soc. A.* vol 467 p 2792
[23] Zia R, Præstgaard E L and Mouritsen O 2002 *Am. J. Phys.* **70** 384
[24] Safaverdi S 2013 *Transport and diffusion in nonequilibrium Langevin systems* (Ph.D. Thesis: Katholieke Universiteit Leuven, ISBN 978-90-8649-687-7)
[25] Barma M and Dhar D 1983 *J. Phys. C* **16** 1451
[26] Leitmann S and Franosch T 2013 *Phys. Rev. Lett.* **111** 190603
[27] Basu U and Maes C 2014 *J. Phys. A: Math. Theor.* **47** 255003
[28] Maes C 2014 *J. Stat. Phys.* **154** 705
[29] Maes C and Steffenoni S 2015 *Phys. Rev. E* **91** 022128
[30] Zwanzig R 1992 *J. Phys. Chem.* **96** 3926
[31] Maes C, Netočný K and de Galway W O 2014 *J. Phys. A: Math. Theor.* **47** 035002
[32] Gaubatz U, Rudecki P, Schiemann S and Bergmann K 1990 *J. Chem. Phys.* **92** 5363
[33] Evans M R and Hanney T 2005 *J. Phys. A: Math. Gen.* **38** R195

Journal of Physics: Conference Series **638** (2015) 012002 doi:10.1088/1742-6596/638/1/012002

Nonequilibrium thermodynamics and a fluctuation theorem for individual reaction steps in a chemical reaction network

Krishnendu Pal, Biswajit Das, Kinshuk Banerjee and Gautam Gangopadhyay

S.N. Bose National Centre For Basic Sciences
Block-JD, Sector-III, Salt Lake, Kolkata-700098, India

E-mail: gautam@bose.res.in

Abstract. We have introduced an approach to nonequilibrium thermodynamics of an open chemical reaction network in terms of the propensities of the individual elementary reactions and the corresponding reverse reactions. The method is a microscopic formulation of the dissipation function in terms of the relative entropy or Kullback-Leibler distance which is based on the analogy of phase space trajectory with the path of elementary reactions in a network of chemical process. We have introduced here a fluctuation theorem valid for each opposite pair of elementary reactions which is useful in determining the contribution of each sub-reaction on the nonequilibrium thermodynamics of overall reaction. The methodology is applied to an oligomeric enzyme kinetics at a chemiostatic condition that leads the reaction to a nonequilibrium steady state for which we have estimated how each step of the reaction is energy driven or entropy driven to contribute to the overall reaction.

1. Introduction

Approach of chemical kinetics of bulk reaction systems needs serious modification [1, 2, 3] to apply for small a system containing a few number of molecules to accommodate the significant fluctuation of the species [1, 2, 3, 4, 5, 6, 7, 8, 9]. Stochastic approach becomes the theoretical tool for the description of the inherent statistical nature of the fluctuating dynamics of reaction system [1, 2, 10, 11, 12], as well as chemical master equation [4, 5, 6]. In single molecule kinetics [13, 14], the concentration terms are replaced by the corresponding probabilities and number of molecules of a reacting species is a fluctuating quantity [1, 2, 11, 12, 15, 16, 17].

In terms of master equation the stochastic thermodynamic description of such systems are given starting from a definition of the system entropy in terms of state probabilities [10, 18, 19, 20, 21]. It has been established that closed systems tend to equilibrium whereas open systems can reach a nonequilibrium steady state(NESS) characterized by positive total entropy production rate (EPR) [10, 12]. In chemical reaction systems, NESS can be attained in presence of chemiostatic condition where concentrations of some species are kept constant throughout the reaction [10]. Such situation can arise in a living cell where, *e.g.*, an enzyme catalysis occurs with substrate and product concentrations inside the cell maintained at a constant value through some cellular mechanism that continuously supplies the substrate to and withdraws the product from the system [22]. Recently, the master equation approach has been used to

Journal of Physics: Conference Series **638** (2015) 012002 doi:10.1088/1742-6596/638/1/012002

formulate the nonequilibrium thermodynamics describing generally the first and second laws in terms of various entropy production rates [23, 24, 25] and energetics of reactions occurring far from equilibrium [10, 11, 12, 22].

Here our aim is to determine the role of individual elementary reactions in governing the nonequilibrium thermodynamics of open chemical reaction. We have formmulated a theory of nonequilibrium thermodynamics of a chemical reaction using Kullback-Leibler(KL) distance in terms of propensity [4, 5] of each elementary reaction. We are interested to know the work done and various energy functions of the nonequilibrium chemical system, like heat dissipation and work done as well as state functions like internal energy, free energy and system entropy. We have studied how the kinetics and energetics are coupled in this nonequilibrium system, specially at NESS and how the fluctuation relations are connected with it. For this purpose the total epr is formulated using the propensities of an elementary reaction and its reverse reaction assuming that microscopic reversibility holds, similar to the relative entropy or Kullback-Leibler(KL) distance formalism of the dissipation function in terms of forward and (time-)reversed phase space trajectories [26, 27, 28, 29]. We have applied the theory on a single oligomeric enzyme catalysis reaction reaching NESS [11, 12, 22, 15, 30] and studied using real parameters of a single Escherichia coli β- galactosidase enzyme under chemiostatic condition.

The paper is organized as follows. In Sec. II, we have given the scheme of an open nonequilibrium reaction network described by a chemical master equation. Expressions of different entropy production rates corresponding to the irreversible processes are given from this master equation. In Sec. III, a method of calculation of total entropy production rate is devised in terms of the individual reactions based on propensity distance. The theory is applied for the kinetics of a single oligomeric enzyme catalysis in Sec. IV. We have discussed on propensity approach to total EPR of individual sub-reaction in Section V. A Fluctuation Theorem is introduced in Section VI for the total EPR of elementary reactions. A thorough analysis of the contributions of elementary reactions are performed for the single molecule kinetics of a Escherichia coli β-galactosidase enzyme in Sec. VII. The paper is concluded Sec. VIII.

2. Master equation for nonequilibrium reaction

We consider a nonequilibrium steady state reaction network

$$\sum_{\mathrm{i}=1}^{\mathrm{a}} \nu_{<\mu}^{\mathrm{i}} \mathrm{A_i} + \sum_{\mathrm{j}=1}^{\mathrm{b}} \nu_{<\mu}^{\mathrm{j}} \mathrm{Z_j} \underset{\mathrm{k}_{-\mu}}{\overset{\mathrm{k}_\mu}{\rightleftharpoons}} \sum_{\mathrm{i}=1}^{\mathrm{a}} \nu_{>\mu}^{\mathrm{i}} \mathrm{A_i} + \sum_{\mathrm{j}=1}^{\mathrm{b}} \nu_{>\mu}^{\mathrm{j}} \mathrm{Z_j}\,. \tag{1}$$

Here Z_j are the intermediate species and A_i are species whose concentrations are kept constant by connecting the reaction system to reservoirs, known as chemiostatic condition. The system is assumed to be at thermal equilibrium with the reservoir kept at some temperature, T. μ is the elementary reaction index with $\mu = \pm1, \pm2, \ldots, \pm k$. So every transition in the population state space due to the reaction μ has a counterpart in the reverse transition by the reaction $-\mu$. The stoichiometric coefficient, ν_μ^j of the species Z_j involved in the μ-th reaction is defined as

$$\nu_\mu^j = \nu_{>\mu}^j - \nu_{<\mu}^j, \quad (j = 1, \ldots, b), \tag{2}$$

where $\nu_{<\mu}^j$ and $\nu_{>\mu}^j$ are the corresponding coefficient of the j-th species for reactant and product, respectively for the μ reaction. The number of molecules of different intermediate species are collectively denoted by the vector $\mathbf{Z} = \{Z_j\}, (j = 1, \ldots, b)$ and the corresponding vector of the stoichiometric coefficients associated with the μ-th reaction is denoted by $\boldsymbol{\nu_\mu}$. The chemical master equation governing the stochastic reaction event is given by [10, 19, 20, 21]

$$\frac{dP(\mathbf{Z},t)}{dt} = \sum_{\mu=\pm1}^{\pm k} [w_\mu(\mathbf{Z} - \boldsymbol{\nu_\mu}|\mathbf{Z})P(\mathbf{Z} - \boldsymbol{\nu_\mu}, t) - w_{-\mu}(\mathbf{Z}|\mathbf{Z} - \boldsymbol{\nu_\mu})P(\mathbf{Z}, t)]. \tag{3}$$

Here $P(\mathbf{Z},t)$ is the probability of having population vector $\mathbf{Z}$ of the intermediate species at time t and $w_\mu(\mathbf{Z}-\boldsymbol{\nu}_\mu|\mathbf{Z})$ is the transition probability for the jump $\mathbf{Z}-\boldsymbol{\nu}_\mu \to \mathbf{Z}$ in the population state space via reaction μ. The rate of entropy production from the above master equation (Eq.3), from system's Shannon entropy [10, 16, 18, 19, 20, 21]

$$S_{sys}(t) = -\sum_{\mathbf{Z}} P(\mathbf{Z},t)\ln P(\mathbf{Z},t), \tag{4}$$

and using Eq.(3), we get the system entropy production rate (EPR) as

$$\dot{S}_{sys}(t) = \frac{1}{2}\sum_{\mathbf{Z},\mu}[w_\mu(\mathbf{Z}-\boldsymbol{\nu}_\mu|\mathbf{Z})P(\mathbf{Z}-\boldsymbol{\nu}_\mu,t) - w_{-\mu}(\mathbf{Z}|\mathbf{Z}-\boldsymbol{\nu}_\mu)P(\mathbf{Z},t)] \times \ln\frac{P(\mathbf{Z}-\boldsymbol{\nu}_\mu,t)}{P(\mathbf{Z},t)}. \tag{5}$$

Then the system EPR can be splitted as $\dot{S}_{sys}(t) = \dot{S}_{tot}(t) - \dot{S}_m(t)$, where the first term in the r.h.s. gives the total EPR and the second term denotes the medium EPR due to the entropy flux into the surroundings given by

$$\dot{S}_{tot}(t) = \frac{1}{2}\sum_{\mathbf{Z},\mu}[w_\mu(\mathbf{Z}-\boldsymbol{\nu}_\mu|\mathbf{Z})P(\mathbf{Z}-\boldsymbol{\nu}_\mu,t) - w_{-\mu}(\mathbf{Z}|\mathbf{Z}-\boldsymbol{\nu}_\mu)P(\mathbf{Z},t)] \times \ln\frac{w_\mu(\mathbf{Z}-\boldsymbol{\nu}_\mu|\mathbf{Z})P(\mathbf{Z}-\boldsymbol{\nu}_\mu,t)}{w_{-\mu}(\mathbf{Z}|\mathbf{Z}-\boldsymbol{\nu}_\mu)P(\mathbf{Z},t)} \tag{6}$$

and

$$\dot{S}_{m}(t) = \frac{1}{2}\sum_{\mathbf{Z},\mu}[w_\mu(\mathbf{Z}-\boldsymbol{\nu}_\mu|\mathbf{Z})P(\mathbf{Z}-\boldsymbol{\nu}_\mu,t) - w_{-\mu}(\mathbf{Z}|\mathbf{Z}-\boldsymbol{\nu}_\mu)P(\mathbf{Z},t)] \times \ln\frac{w_\mu(\mathbf{Z}-\boldsymbol{\nu}_\mu|\mathbf{Z})}{w_{-\mu}(\mathbf{Z}|\mathbf{Z}-\boldsymbol{\nu}_\mu)}. \tag{7}$$

It is evident from Eq.(6) that $\dot{S}_{tot}(t) \geq 0$. The medium EPR is identified as the temperature-scaled heat dissipation rate that can be positive or negative. Recently, Esposito *et al.* [24, 25] showed that the total EPR, $\dot{S}_{tot}(t)$ can be separated into adiabatic and nonadiabatic parts, $\dot{S}_a(t)$ and $\dot{S}_{na}(t)$, respectively which for this reaction network case can be given by

$$\dot{S}_{a}(t) = \frac{1}{2}\sum_{\mathbf{Z},\mu}[w_\mu(\mathbf{Z}-\boldsymbol{\nu}_\mu|\mathbf{Z})P(\mathbf{Z}-\boldsymbol{\nu}_\mu,t) - w_{-\mu}(\mathbf{Z}|\mathbf{Z}-\boldsymbol{\nu}_\mu)P(\mathbf{Z},t)] \times \ln\frac{w_\mu(\mathbf{Z}-\boldsymbol{\nu}_\mu|\mathbf{Z})P^{st}(\mathbf{Z}-\boldsymbol{\nu}_\mu)}{w_{-\mu}(\mathbf{Z}|\mathbf{Z}-\boldsymbol{\nu}_\mu)P^{st}(\mathbf{Z})} \tag{8}$$

and

$$\dot{S}_{na}(t) = \frac{1}{2}\sum_{\mathbf{Z},\mu}[w_\mu(\mathbf{Z}-\boldsymbol{\nu}_\mu|\mathbf{Z})P(\mathbf{Z}-\boldsymbol{\nu}_\mu,t) - w_{-\mu}(\mathbf{Z}|\mathbf{Z}-\boldsymbol{\nu}_\mu)P(\mathbf{Z},t)] \times \ln\frac{P(\mathbf{Z}-\boldsymbol{\nu}_\mu,t)P^{st}(\mathbf{Z})}{P^{st}(\mathbf{Z}-\boldsymbol{\nu}_\mu)P(\mathbf{Z},t)}, \tag{9}$$

where $P^{st}(\mathbf{Z})$ is the steady state probability of having the population vector $\mathbf{Z}$. Unlike the system and medium contributions, the adiabatic and nonadiabatic parts characterize the rates

Journal of Physics: Conference Series **638** (2015) 012002 doi:10.1088/1742-6596/638/1/012002

of internal energy(U) change, free energy(F) change and first law can be given in terms of rates of heat dissipation and irreversible work done [23]. It is also shown that the nonadiabatic EPR is negative of the temperature-scaled free energy dissipation rate, $\dot{S}_{na}(t) = -\dot{F}(t)/T$ and the adiabatic EPR, $\dot{S}_a(t)$ is the rate of irreversible work done(temperature-scaled) on the system. The rate of internal energy change behaves as $\dot{U}(t)/T = \dot{S}_a(t) - \dot{S}_m(t)$ like first law.

3. Propensity distance for individual reactions

We consider a transition along a stochastic trajectory in the population state space of the chemical reaction network as $(\mathbf{Z}, t) \to (\mathbf{Z} + \boldsymbol{\nu}_{\boldsymbol{\mu}}, t + \Delta t)$. This can be described as a transition from the state $(\mathbf{Z})$ to $(\mathbf{Z} + \boldsymbol{\nu}_{\boldsymbol{\mu}})$ at time t and then considering a time interval Δt such that no further transition takes place in this interval. The system entropy production [31, 32, 33] for the transition due to the elementary reaction μ over the time interval Δt can be written as

$$\Delta s_{sys}^{(\mu)} = -\ln P(\mathbf{Z} + \boldsymbol{\nu}_{\boldsymbol{\mu}}, t + \Delta t) - (-\ln P(\mathbf{Z}, t)) = \ln \frac{P(\mathbf{Z}, t)}{P(\mathbf{Z} + \boldsymbol{\nu}_{\boldsymbol{\mu}}, t + \Delta t)}. \tag{10}$$

The medium entropy production or entropy flow associated with this transition along the trajectory is defined as [31, 32, 33, 25]

$$\Delta s_m^{(\mu)} = \ln \frac{w_\mu(\mathbf{Z}|\mathbf{Z} + \boldsymbol{\nu}_{\boldsymbol{\mu}})}{w_{-\mu}(\mathbf{Z} + \boldsymbol{\nu}_{\boldsymbol{\mu}}|\mathbf{Z})}. \tag{11}$$

Thus we write the total entropy production as

$$\Delta s_{tot}^{(\mu)} = \Delta s_{sys}^{(\mu)} + \Delta s_m^{(\mu)} = \ln \frac{P_\mu(\mathbf{Z}, t|\mathbf{Z} + \boldsymbol{\nu}_{\boldsymbol{\mu}})}{P_{-\mu}(\mathbf{Z} + \boldsymbol{\nu}_{\boldsymbol{\mu}}, t + \Delta t|\mathbf{Z})}. \tag{12}$$

Here we introduce the propensity, $P_\mu(\mathbf{Z}, t|\mathbf{Z} + \boldsymbol{\nu}_{\boldsymbol{\mu}}) = w_\mu(\mathbf{Z}|\mathbf{Z} + \boldsymbol{\nu}_{\boldsymbol{\mu}})P(\mathbf{Z}, \mathbf{t})$ which is the reaction probability density discussed by Gillespie [4, 5, 6]. Then $P_\mu(\mathbf{Z}, t|\mathbf{Z} + \boldsymbol{\nu}_{\boldsymbol{\mu}})dt$ is the probability that the system at time t will go from state $\mathbf{Z}$ to state $\mathbf{Z} + \boldsymbol{\nu}_{\boldsymbol{\mu}}$ due to the occurrence of an elementary reaction μ in the next infinitesimal time interval dt.

Now all the above expressions of entropy production are for a specific transition as indicated by the lower case 's'. Thus we need to average over all the transitions due to the μ-th reaction considering all possible values of the population vector $\mathbf{Z}$ to get the overall entropy production due to this elementary reaction. This averaging is done with the probability, $P_\mu(\mathbf{Z}, t|\mathbf{Z} + \boldsymbol{\nu}_{\boldsymbol{\mu}})\Delta t$. Here Δt is considered to be small enough so that over this interval P_μ remains constant. Then

$$\Delta S_{tot}^{(\mu)} = \langle \Delta s_{tot}^{(\mu)} \rangle = \sum_{\mathbf{Z}} (P_\mu(\mathbf{Z}, t|\mathbf{Z} + \boldsymbol{\nu}_{\boldsymbol{\mu}})\Delta t) \ln \frac{P_\mu(\mathbf{Z}, t|\mathbf{Z} + \boldsymbol{\nu}_{\boldsymbol{\mu}})}{P_{-\mu}(\mathbf{Z} + \boldsymbol{\nu}_{\boldsymbol{\mu}}, t + \Delta t|\mathbf{Z})}. \tag{13}$$

Thus the expression of the total entropy production rate becomes

$$\lim_{\Delta t \to 0} \frac{\Delta S_{tot}^{(\mu)}}{\Delta t} = \dot{S}_{tot}^{(\mu)} = \sum_{\mathbf{Z}} P_\mu(\mathbf{Z}, t|\mathbf{Z} + \boldsymbol{\nu}_{\boldsymbol{\mu}}) \ln \frac{P_\mu(\mathbf{Z}, t|\mathbf{Z} + \boldsymbol{\nu}_{\boldsymbol{\mu}})}{P_{-\mu}(\mathbf{Z} + \boldsymbol{\nu}_{\boldsymbol{\mu}}, t|\mathbf{Z})}. \tag{14}$$

The structure of Eq.(14) is similar to the relative entropy or the Kullback-Leibler(KL) distance of two probability distributions [26, 27, 28, 34, 35]. Hence a KL distance type relation given in terms of the forward and reverse propensities, P_μ and $P_{-\mu}$ gives rise to the total entropy production rate due to the μ-th reaction in the forward direction which is similar to the description of the total entropy production given by the KL distance of forward and backward trajectories in phase space [26, 28].

Journal of Physics: Conference Series **638** (2015) 012002 doi:10.1088/1742-6596/638/1/012002

4. Single oligomeric enzyme catalysis

To apply this formulation we consider an example of a prototype chemical reaction network as a single oligomeric enzyme kinetics where the reaction of an active site is written as [15, 17]

$$\mathrm{E} + \mathrm{S} \underset{\mathrm{k}_{-1}}{\overset{\mathrm{k}_1'}{\rightleftharpoons}} \mathrm{ES} \underset{\mathrm{k}_2'}{\overset{\mathrm{k}_{-2}}{\rightleftharpoons}} \mathrm{E} + \mathrm{P}. \tag{15}$$

Under chemiostatic condition [10, 12, 22] with concentrations of substrate, S and product, P maintained constant, the above reaction scheme can be reduced to

$$\mathrm{E} \underset{\mathrm{K}_2}{\overset{\mathrm{K}_1}{\rightleftharpoons}} \mathrm{ES}, \tag{16}$$

where $K_1 = k_1 + k_2$ and $K_2 = k_{-1} + k_{-2}$. The pseudo first order rate constants k_1 and k_2 are given by $k_1 = k_1'[S]$ and $k_2 = k_2'[P]$. We consider the enzyme to be consisting of n_T number of identical non-interacting subunits where each subunit has one active site. During the progress of the reaction, the active sites which have already formed the ES-complex are referred to as occupied sites and those lying vacant at that moment are called the vacant sites. The reaction system can thus be viewed in terms of the number of occupied sites present at a particular time. Due to the random occurrence of the four elementary reactions, the number of occupied sites of a single oligomeric enzyme becomes a fluctuating quantity. The state of the system is denoted by the random variable, n where n is the number of occupied sites present at time t. After a small time τ, the system can move to a new state, $(n + \nu_\mu)$ through any one of the four elementary reactions. Here ν_μ is the stoichiometric coefficient of the μ-th reaction with rate constant k_μ. Then the stochastic master equation can be written as [15, 16, 17]

$$\frac{dP(n,t)}{dt} = \sum_{\mu=\pm 1}^{\pm 2} [w_\mu(n - \nu_\mu|n)P(n - \nu_\mu, t) - w_{-\mu}(n|n - \nu_\mu)P(n,t)]. \tag{17}$$

Here $P(n,t)$ is the probability of having n number of occupied states at time t with $\nu_\mu = 1$ for $\mu > 0$ and $\nu_\mu = -1$ for $\mu < 0$. The transition probabilities are defined as follows,

$$w_\mu(n - \nu_\mu|n) = k_\mu(n_T - (n - \nu_\mu)),\ \mu > 0$$

and

$$w_\mu(n - \nu_\mu|n) = k_\mu(n - \nu_\mu),\ \mu < 0. \tag{18}$$

Then Eq.(17) can be rewritten as

$$\frac{dP(n,t)}{dt} = K_1(n_T - n + 1)P(n-1,t) + K_2(n+1)P(n+1,t)$$

$$-K_1(n_T - n)P(n,t) - K_2 n P(n,t). \tag{19}$$

Solving the master equation with the standard approach of generating function method [36, 37] we get the probability distribution function and average number of occupied site [15, 16] assuming that initially all the active sites are vacant.

5. Propensity approach to Total EPR of elementary sub-reaction

Here we shall introduce the approach to calculate the total EPR of individual sub-reactions in terms of propensities. We shall describe the mater equation in terms of propensity equation of each sub-reaction. By this procedure one can obtain the property of the irreversible chemical flow system using the microscopic reversibility in the chemical reaction steps.

Now we give a description of the total entropy production rate, where the elementary reactions $\mu = \pm 1$ and $\mu = \pm 2$ form the reaction-'counter'reaction pairs on the basis of ES-complex formation and dissociation. The transition between two states of the system, say state m and state l, are governed by Eq.(17). As $\nu_\mu = \pm 1$, so $l = m \pm 1$. In terms of state variable, n we can write $m, l = (n-1), n$ and $(n+1)$, for example, for a forward transition, $l > m$ and for a backward transition, $l < m$. Now a specific reaction is not responsible for all the possible transitions between the system states. For example, the system can go from the state $(n-1)$ to state n by the elementary reaction $\mu = 1$ or $\mu = 2$ but not by $\mu = -1$ or $\mu = -2$. More clearly, for $\mu > 0$, 'l' must be greater than 'm' with $m = (n-1), n$. Similarly for $\mu < 0$, 'l' is less than 'm' with $m = n, (n+1)$. If one considers transition from state m to state l at time t via the reaction μ with the propensity $P_\mu(m, t|l)$ (having the dimension of rate), then the reverse transition occurs through the reaction channel $-\mu$ with the corresponding propensity being $P_{-\mu}(l, t|m)$.

The master equation Eq.(17)can be cast into the form of propensity equation as

$$P_\mu(m, t|l) = w_\mu(m|l) P(m, t) \tag{20}$$

where we call $P_\mu(m, t|l)dt$ is the probability at time time t that the system will go from state m to state l in the time interval dt through the μ-th reaction channel. Then one can define the total entropy production rate at time t due to the reaction μ as

$$\begin{aligned}\dot{S}_{tot}^{(\mu)}(t) &= \sum_{m,l} P_\mu(m, t|l) \ln\left(\frac{P_\mu(m, t|l)}{P_{-\mu}(l, t|m)}\right) \delta_{l,m+1},\ \mu > 0 \\ &= \sum_{m,l} P_\mu(m, t|l) \ln\left(\frac{P_\mu(m, t|l)}{P_{-\mu}(l, t|m)}\right) \delta_{l,m-1},\ \mu < 0.\end{aligned} \tag{21}$$

We now come to specific cases. For the reaction $\mu = 1$, using Eq.(18) along with its solution in Eq.(21), we get

$$\begin{aligned}\dot{S}_{tot}^{(1)}(t) &= \sum_m P_1(m, t|m+1) \ln\left(\frac{P_1(m, t|m+1)}{P_{-1}(m+1, t|m)}\right) \\ &= \frac{1}{2}\left[\sum_{n-1} P_1(n-1, t|n) \ln\left(\frac{P_1(n-1, t|n)}{P_{-1}(n, t|n-1)}\right) + \sum_n P_1(n, t|n+1) \ln\left(\frac{P_1(n, t|n+1)}{P_{-1}(n+1, t|n)}\right)\right] \\ &= (k_1 n_T Y) \ln\left(\frac{k_1 n_T Y}{k_{-1} n_T X}\right).\end{aligned} \tag{22}$$

The entropy production rate for other reactions can be calculated similarly. They are given in a compact form as

$$\begin{aligned}\dot{S}_{tot}^{(\mu)}(t) &= (k_\mu n_T Y) \ln\left(\frac{k_\mu n_T Y}{k_{-\mu} n_T X}\right), \mu > 0 \\ &= (k_\mu n_T X) \ln\left(\frac{k_\mu n_T X}{k_{-\mu} n_T Y}\right), \mu < 0,\end{aligned} \tag{23}$$

Journal of Physics: Conference Series **638** (2015) 012002 doi:10.1088/1742-6596/638/1/012002

and we can obviously write the overall total EPR as

$$\dot{S}_{tot}(t) = \sum_{\mu=\pm 1}^{\pm 2} \dot{S}_{tot}^{(\mu)}(t). \tag{24}$$

One can easily verify that this is the same expression one will get using the master equation description starting from the definition of Shannon entropy, as given in Eq.(6). However, Eq.(6) implies that contribution of reaction μ towards $\dot{S}_{tot}(t)$ is the same as that of reaction $-\mu$. Now according to Eq.(21), the individual reaction contributions to the total EPR are distinct from each other following the propensity distance formalism. Using this approach, it is also straightforward to calculate the reaction contributions to the various other entropy production rates. This approach has subtle physical connection with the microscopic description of the second law of thermodynamics and the macroscopic law of irreversibility of arrow of time [26, 27, 28].

Now from Eq.(21), we can easily calculate the medium EPR, $\dot{S}_m^{(\mu)}(t)$ and system EPR, $\dot{S}_{sys}^{(\mu)}(t)$ for individual reaction and they can be expressed as

$$\begin{aligned}\dot{S}_m^{(\mu)}(t) &= \sum_{m,l} P_\mu(m,t|l) ln\left(\frac{w_\mu(m,t|l)}{w_{-\mu}(l,t|m)}\right)\delta_{l,m+1}, \mu > 0 \\ &= \sum_{m,l} P_\mu(m,t|l) ln\left(\frac{w_\mu(m,t|l)}{w_{-\mu}(l,t|m)}\right)\delta_{l,m-1}, \mu < 0.\end{aligned} \tag{25}$$

and

$$\begin{aligned}\dot{S}_{sys}^{(\mu)}(t) &= \sum_{m,l} P_\mu(m,t|l) ln\left(\frac{P(m,t)}{P(l,t)}\right)\delta_{l,m+1}, \mu > 0 \\ &= \sum_{m,l} P_\mu(m,t|l) ln\left(\frac{P(m,t)}{P(l,t)}\right)\delta_{l,m-1}, \mu < 0.\end{aligned} \tag{26}$$

The medium EPR, $\dot{S}_m^{(\mu)}(t)$ is related with the heat dissipation rate, h_d as $h_d = T\dot{S}_m^{(\mu)}(t)$, where T is the equilibrium temperature of the surrounding. Now following Esposito *et al.* [24, 25], the total EPR, $\dot{S}_{tot}^{(\mu)}(t)$ can be separated into adiabatic and nonadiabatic EPRs, $\dot{S}_a^{(\mu)}(t)$ and $\dot{S}_{na}^{(\mu)}(t)$, respectively and they can be expressed as

$$\begin{aligned}\dot{S}_a^{(\mu)}(t) &= \sum_{m,l} P_\mu(m|l,t) ln\left(\frac{P_\mu^{(NESS)}(m|l)}{P_{-\mu}^{(NESS)}(l|m)}\right)\delta_{l,m+1}, \mu > 0, \\ &= \sum_{m,l} P_\mu(m|l,t) ln\left(\frac{P_\mu^{(NESS)}(m|l)}{P_{-\mu}^{(NESS)}(l|m)}\right)\delta_{l,m-1}, \mu < 0\end{aligned} \tag{27}$$

and

$$\begin{aligned}\dot{S}_{na}^{(\mu)}(t) &= \sum_{m,l} P_\mu(m|l,t) ln\left(\frac{P(m)P(l)^{(NESS)}}{P(l)P(m)^{(NESS)}}\right)\delta_{l,m+1}, \mu > 0, \\ &= \sum_{m,l} P_\mu(m|l,t) ln\left(\frac{P(m)P(l)^{(NESS)}}{P(l)P(m)^{(NESS)}}\right)\delta_{l,m-1}, \mu < 0.\end{aligned} \tag{28}$$

6. Fluctuation Theorem for the total EPR of elementary reactions

Here we focus entirely on the entropy production rates at NESS of elementary reactions. As the system EPR and the nonadiabatic EPR are zero at NESS both being state functions, so we have $\dot{S}_{tot}^{NESS} = \dot{S}_m^{NESS} = \dot{S}_a^{NESS}$. We estimate here the individual reaction contributions to the total EPR and the fluctuation relations in the total entropy production and its possible implications.

The total entropy production during a time interval from t to $t+\Delta t$ when the system remains at NESS is simply given by

$$\Delta S_{tot}^{(NESS)} = \int_t^{t+\Delta t} \dot{S}_{tot}^{(NESS)} dt = \sum_\mu \dot{S}_{tot}^{\mu(NESS)} \Delta t = \sum_\mu \Delta S_{tot}^{\mu(NESS)}. \tag{29}$$

$\Delta S_{tot}^{\mu(NESS)}$ is the total entropy production due to the elementary reaction μ at NESS defined as

$$\begin{aligned} \Delta S_{tot}^{\mu(NESS)} &= \sum_{m,l} (P_\mu^{(NESS)}(m|l)\Delta t) \ln\left(\frac{P_\mu^{(NESS)}(m|l)}{P_{-\mu}^{(NESS)}(l|m)}\right) \delta_{l,m+1},\ \mu > 0 \\ &= \sum_{m,l} (P_\mu^{(NESS)}(m|l)\Delta t) \ln\left(\frac{P_\mu^{(NESS)}(m|l)}{P_{-\mu}^{(NESS)}(l|m)}\right) \delta_{l,m-1},\ \mu < 0. \end{aligned} \tag{30}$$

Now for a particular transition, say from state m to state l ($l = m \pm 1$) via the elementary reaction μ at NESS, the entropy production per step (written with lowercase s) can be written following Eq.(12) as

$$\Delta s_{tot}^{\mu(NESS)}(\Delta t) = \ln\left(\frac{P_\mu^{(NESS)}(m|l)\Delta t}{P_{-\mu}^{(NESS)}(l|m)\Delta t}\right), \tag{31}$$

where $\Delta S_{tot}^{\mu(NESS)} = \langle \Delta s_{tot}^{\mu(NESS)} \rangle$, the average being taken with respect to the time-integrated reaction propensity, $P_\mu^{(NESS)}(m|l)\Delta t$. One can check that the average over all possible transitions for the reaction channel μ in terms of the reaction probability, $P_\mu^{(NESS)}\Delta t$ gives back $\langle \Delta s_{tot}^{\mu(NESS)} \rangle$. The fluctuation theorem for a particular transition $m \to l$ via the reaction channel μ with the entropy production, $\Delta s_{tot}^{\mu(NESS)}(\Delta t)$ for some arbitrary time, Δt at NESS follows

$$\frac{P_\mu^{(NESS)}(m|l)}{P_{-\mu}^{(NESS)}(l|m)} = e^{\Delta s_{tot}^{\mu(NESS)}(\Delta t)}. \tag{32}$$

As an example, we consider a transition from state m to state $(m+1)$ via the elementary reaction $\mu = 1$. From Eq.(18), at NESS we get

$$\Delta s_{tot}^{1(NESS)} = \ln\left(\frac{k_1 K_2}{k_{-1} K_1}\right). \tag{33}$$

The entropy production per step due to reaction $\mu = 1$ is thus the same for all the reaction steps at NESS and this is also true for all the other elementary reactions. The above expression can be splitted as

$$\Delta s_{tot}^{1(NESS)} = \ln\left(\frac{k_1(n_T - m)}{k_{-1}(m+1)}\right) + \ln\left(\frac{K_2(m+1)}{K_1(n_T - m)}\right). \tag{34}$$

The first term in the right hand side of Eq.(34) contains the logarithm of the ratio of the propensities of the reactions $\mu = 1$ and $\mu = -1$ for a transition between states m and $(m+1)$. The second term gives the logarithm of the ratio of the propensities for the overall dissociation

$(K_2 = k_{-1} + k_{-2})$ and formation $(K_1 = k_1 + k_2)$ reactions and it is easy to see that this term is common in all the $\Delta s_{tot}^{\mu(NESS)}$ expressions except for a sign inversion. Hence the per step entropy production due to an elementary reaction μ at NESS can be separated into local and global contributions.

The deeper significance of the above fluctuation theorem can be shown to be reflected in the relation between the net velocity and total EPR of elementary reactions at NESS. It will be very important to have a relationship between the experimentally measurable $v_{net}^{(NESS)}$ and the entropy production rates of the elementary reactions.

By setting $k_2' \approx 0$, the enzyme kinetics in Eq.(15) tends to the traditional Michaelis-Menten scheme. In this limit, $k_2 \approx 0$ with finite values of [P] and $K_1 \approx k_1$. The net velocity of the Michaelis-Menten enzyme catalysis at NESS can be defined as $v_{net}^{(NESS)} = k_{-2}\langle n\rangle^{NESS}$. Using the expression of $\langle n\rangle$ at NESS, we have

$$v_{net}^{(NESS)} = n_T k_{-2}\frac{K_1}{K_1 + K_2} = \frac{n_T k_{-2}[S]}{K_M + [S]}. \tag{35}$$

This is the celebrated Michaelis-Menten equation of enzyme catalysis which is shown here for the single oligomeric enzyme case with n_T number of active sites with $K_M = \frac{k_{-1}+k_{-2}}{k_1'} = \frac{K_2}{k_1'}$ is the Michaelis-Menten constant. Now at NESS, from Eq.(22) we get

$$\dot{S}_{tot}^{1(NESS)} = (k_1 n_T Y^{NESS})\ln\left(\frac{k_1 n_T Y^{NESS}}{k_{-1} n_T X^{NESS}}\right)$$

$$= \frac{n_T K_2 \ln f[S]}{K_M + [S]}, \quad f = \frac{K_2}{k_{-1}}, \; k_2' \approx 0. \tag{36}$$

Similar treatment for the other reactions gives

$$\dot{S}_{tot}^{-1(NESS)} = -\frac{n_T k_{-1}\ln f[S]}{K_M + [S]}, \tag{37}$$

$$\dot{S}_{tot}^{2(NESS)} = \frac{n_T k_2' K_M \ln f'[P]}{K_M + [S]}, \tag{38}$$

and

$$\dot{S}_{tot}^{-2(NESS)} = -\frac{n_T k_{-2}\ln f'[S]}{K_M + [S]}, \tag{39}$$

where $f' = \frac{k_2' K_2 [P]}{k_1' k_{-2}[S]}$. From the above equations it is clear that the total EPR for the reactions $\mu = \pm 1$ show hyperbolic dependence on substrate concentration at NESS whereas for the reactions $\mu = \pm 2$, the dependence is more complicated as the factor f' is a function of [S] and [P]. The Eqs.(36)-(39) can be designated as the 'Michaelis-Menten analogue' for the total entropy production rates due to the elementary reactions.

Again from Eq.(36) and Eq.(37), we have

$$\dot{S}_{tot}^{1(NESS)} + \dot{S}_{tot}^{-1(NESS)} = v_{net}^{(NESS)}\ln f \tag{40}$$

and from Eq.(38) and Eq.(39)

$$\dot{S}_{tot}^{2(NESS)} + \dot{S}_{tot}^{-2(NESS)} \approx \dot{S}_{tot}^{-2(NESS)} = -v_{net}^{(NESS)}\ln f'. \tag{41}$$

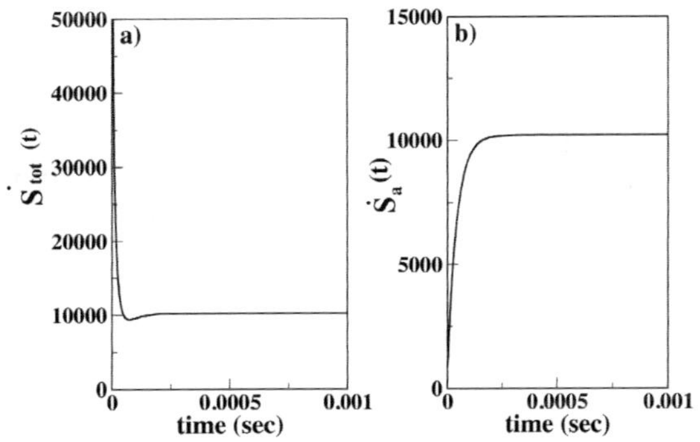

Figure 1. Total EPR, $\dot{S}_{tot}$ and adiabatic EPR, $\dot{S}_a$ are plotted as a function of time at substrate concentration $[S] = 100\mu M$. The rate constants are: $k_1 = 5 \times 10^7 M^{-1}s^{-1}$, $k_{-1} = 18,300s^{-1}$, $k_{-2} = 730s^{-1}$, $k_2 = 1 \times 10^{-5}M^{-1}s^{-1}$.

As a consequence, we obtain

$$\dot{S}_{tot}^{(NESS)} = v_{net}^{(NESS)} \ln\gamma, \tag{42}$$

where $\gamma = \frac{k_1 k_{-2}}{k_{-1} k_2}$. Thus the total entropy production rate often signifies as an arrow of time which is directly measured from the deviation from equilibrium where for $\gamma = 1$, $\dot{S}_{tot}^{(NESS)} = 0$ corresponds to equilibrium.

The method is based on the analogy of a trajectory in phase space of a Hamiltonian system with the path of elementary reactions in a network of chemical process which in general can be an open system. The analogy is very qualitative in nature as it serves here as a motivation for study. It is also interesting to understand more quantitive meaning of this analogy which may provide some new methodology as well as some insight into problems of nonequilibrium thermodynamics of chemical reaction where kinetics and time dependent energetics of the process are highly coupled. Here we have applied the nonequilibrium thermodynamic relations including the fluctuation theorem for a model of oligomeric enzyme kinetics as it is an useful class of reactions and equally applicable to other reactions. Another important point is that the condition of microscopic reversibility is the fundamental requirement for the theory to hold as it needs to consider the time-reversed process. So any rate constant of the elementary reactions can be small but can not be taken exactly zero which may appear as limitation in some cases.

7. Result and Discussion

Here we have studied the individual elementary reaction contributions to various thermodynamic quantities at chemiostatic condition for a single Escherichia coli β- galactosidase enzyme at NESS as well as their time variation. The enzyme is tetrameric with four independent catalytic sites [40, 41, 42] and follows the Michaelis-Menten kinetics [13, 14]. To calculate the thermodynamic quantities for individual reactions of this system, we have taken the rate constants, $k_1 = 5 \times 10^7 M^{-1}s^{-1}$, $k_{-1} = 18,300s^{-1}$ and $k_{-2} = 730s^{-1}$, from the single molecule experiment [13]. We set the vanishly small rate constant $k_2 = 1 \times 10^{-5}M^{-1}s^{-1}$ as the calculation of the nonequilibrium thermodynamic quantities as the formalism does not permit any rate constant to be exactly zero. Here k_1 and k_2 are pseudo first order rate constants depending linearly on the substrate and product concentrations, respectively.

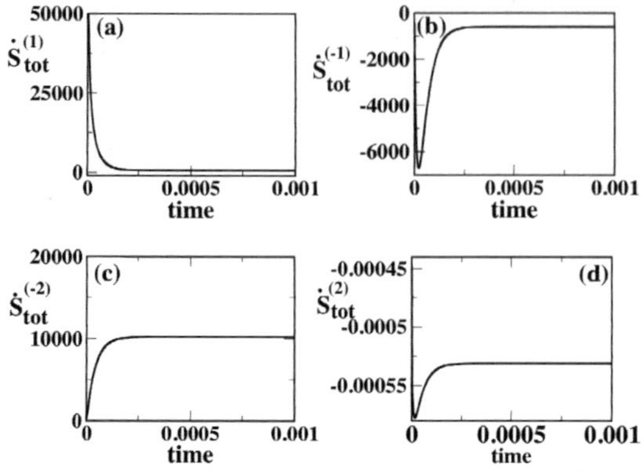

Figure 2. Total EPR, $\dot{S}_{tot}$ and adiabatic EPR, $\dot{S}_a$ are plotted as a function of time at substrate concentration $[S] = 100\mu M$. The rate constants are: $k_1 = 5 \times 10^7 M^{-1}s^{-1}$, $k_{-1} = 18,300 s^{-1}$, $k_{-2} = 730 s^{-1}$, $k_2 = 1 \times 10^{-5} M^{-1}s^{-1}$.

In Fig.1(a) and (b), we have plotted the total EPR, $\dot{S}_{tot}(t)$ and the adiabatic EPR, $\dot{S}_a(t)$, respectively, as a function of time at substrate concentration $[S] = 100\mu M$. From Fig.1(a), it is observed that $\dot{S}_{tot}(t)$ shows a sharp decay with time, passes through a minimum and finally attains a constant non-zero value as the system reaches NESS. From Fig.1(b), we see that $\dot{S}_a(t)$ increases with time and also saturates at NESS characterizing the irreversible work that has to be done on the system to sustain the NESS by continuous injection of substrate and ejection of product maintaining the chemiostatic condition. It follows from Eq.(8), that if the system obeys detailed balance, $\dot{S}_a(t)$ always remains zero. When detailed balance is broken, $\dot{S}_a(t)$ remains positive for all time; it becomes zero only when some species is absent, *e.g.*, in our case at $t = 0$ when no ES-complex is formed.

Next we have plotted the time evolution of total EPR due to the elementary reactions, $\mu = \pm 1, \pm 2$ in Fig.2(a)-(d). The plots reveal four main features: (i) $\dot{S}^{(1)}_{tot}(t)$, $\dot{S}^{(-2)}_{tot}(t)$ remain positive over the time course whereas $\dot{S}^{(-1)}_{tot}(t)$, $\dot{S}^{(2)}_{tot}(t)$ remain negative throughout. (ii) Initially, the major contribution to $\dot{S}_{tot}(t)$(see Fig.1) comes from $\dot{S}^{(1)}_{tot}(t)$. This is because contributions of elementary reactions $\mu = -1, -2$ to total EPR starts from zero. $\dot{S}^{(2)}_{tot}(t)$ is throughout negligible compared to others and plays no part in governing the thermodynamics. (iii) At NESS, $\dot{S}^{(1)}_{tot}$ and $\dot{S}^{(-1)}_{tot}$ almost cancel each out each other and hence the total EPR is governed by $\dot{S}^{(-2)}_{tot}$. (iv) At intermediate times, $\dot{S}^{(1)}_{tot}(t)$ and $\dot{S}^{(-2)}_{tot}(t)$ continuously decreases and increases, respectively, whereas $\dot{S}^{(-1)}_{tot}(t)$ passes through a minimum.

To investigate the individual reaction contributions in total EPR at NESS, we have calculated $\dot{S}^{\mu(NESS)}_{tot}$ as a function of substrate concentration, $[S]$ from Eqs.(36)-(39). We have also determined the quantities by numerically solving the master equation (Eq.19). The theoretical and numerical results tally very well as depicted in Fig.3(a)-(d). In Fig.3(a), (b) and (c), it is observed that the variation of $\dot{S}^{1(NESS)}_{tot}$, $\dot{S}^{-1(NESS)}_{tot}$ and $\dot{S}^{-2(NESS)}_{tot}$ as a function of substrate concentration are hyperbolic in nature with $\dot{S}^{-1(NESS)}_{tot}$ being negative throughout, its magnitude almost same as that of $\dot{S}^{1(NESS)}_{tot}$. $\dot{S}^{2(NESS)}_{tot}$ also remains negative throughout and at higher substrate concentration, it approaches zero as shown in Fig.3(d). However, its value is negligible. Therefore, as $\dot{S}^{1(NESS)}_{tot}$ and $\dot{S}^{-1(NESS)}_{tot}$ almost cancel out each other, $\dot{S}^{(NESS)}_{tot}$ is governed by

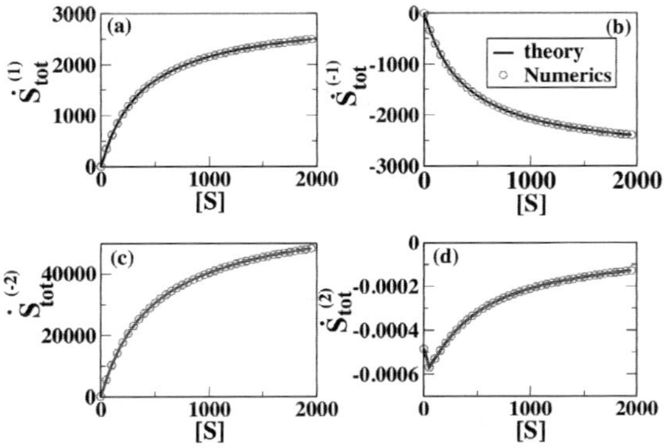

Figure 3. The individual reaction contributions in total entropy production rate, $S_{tot}^{(\mu)}$ are plotted as a function of substrate concentration, $[S]$ at NESS. The rate constants are: $k_1 = 5 \times 10^7 M^{-1}s^{-1}$, $k_{-1} = 18,300s^{-1}$, $k_{-2} = 730s^{-1}$, $k_2 = 1 \times 10^{-5}M^{-1}s^{-1}$.

the entropy production rate of reaction $\mu = -2$ only.

So far we have discussed about the time-dependence of the total EPR as well as its variation with substrate concentration at NESS, analyzing in detail the role of each elementary reaction. Now we determine various entropy productions by integrating the corresponding rates over the time interval starting from $t = 0$ to a fixed final time when the system resides in NESS. We have plotted the individual reaction contributions in these entropy productions as well as the free enrgy and internal energy changes as a function of $[S]$ at NESS shown in Fig.4(a)-(d). In the figures we have not shown the contribution of reaction $\mu = 2$ due to its negligible value. Keeping in mind the special role of adiabatic entropy production, ΔS_a as a measure of the (temperature-scaled) irreversible work done on the system that drives the system to NESS and then sustains it.

To investigate how the irreversible work is utilized by the reactions, we have plotted $\Delta S_{tot}^{(\mu)}$ as it is equal to $\Delta S_a^{(\mu)}$ at NESS which is depicted in Fig.4(a). In the figure it is observed that $\Delta S_{tot}^{(1)}$ has positive contribution whereas $\Delta S_{tot}^{(-1)}$ has negative contribution towards ΔS_{tot} and they are close in magnitude. Thus the irreversible work done on the system due to reaction $\mu = 1$ is almost compensated by the work performed by reaction $\mu = -1$. It becomes evident from Fig.4(a) that the irreversible work done on the system is mostly consumed by the reaction $\mu = -2$, more so at higher substrate concentration. Now first law suggests that some amount of the irreversible work done on the system changes the internal energy of the system and rest of it dissipates as heat to the surroundings. To find out the contribution of individual reactions in the change of free energy and internal energy of the system as a function of $[S]$ in Fig.4(b) and (c), respectively.

The difference between $\Delta U^{(\mu)}/T$ and $\Delta S_{sys}^{(\mu)}$ gives the temperature-scaled free energy change, $\Delta F^{(\mu)}/T$(which is also the nonadiabatic entropy production, $\Delta S_{na}^{(\mu)}$) due to each elementary reaction as shown in Fig.4(b). The magnitude of $\Delta F^{(\mu)}/T$ is small compared to the other quantities due to the quantitative similarity of $\Delta U^{(\mu)}/T$ and $\Delta S_{sys}^{(\mu)}$. In Fig.4(b), we see that the contribution of reactions $\mu = -1, -2$ are positive as well as small and the major, negative contribution in the total free energy change comes from the reaction $\mu = 1$. This indicates that the irreversible work consumed by the reaction $\mu = -2$ almost entirely dissipates as heat and therefore, the contribution of this reaction to internal energy change, $\Delta U^{(-2)}/T$ is little as

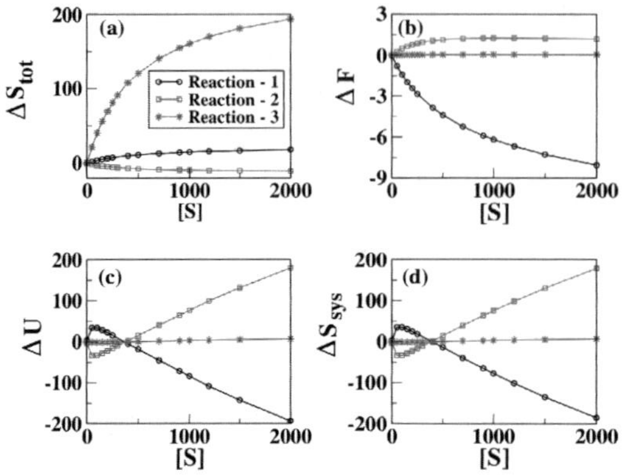

Figure 4. The individual reaction contributions in the system and total entropy productions along with the internal energy change are plotted as a function of $[S]$. The rate constants are: $k_1 = 5 \times 10^7 M^{-1}s^{-1}$, $k_{-1} = 18,300s^{-1}$, $k_{-2} = 730s^{-1}$, $k_2 = 1 \times 10^{-5}M^{-1}s^{-1}$

shown in Fig.4(c). The variation of the corresponding contributions of reactions $\mu = \pm 1$ towards the internal energy change, $\Delta U^{(\pm 1)}/T$ show the opposite trend as a function of $[S]$ compared to the variation of $\Delta S_tot^{(\pm 1)}$(see Fig.4(c)). This is expected as heat absorption by the system for an elementary reaction leads to an increase in internal energy due to that reaction whereas heat released by the system results in fall of its internal energy. Same is true for the reaction contributions in the system entropy change, $\Delta S_{sys}^{(\mu)}$ plotted in Fig.4(d).

8. Conclusion

We have formulated the total entropy production rate due to a complex chemical reaction in the spirit of Gillespie's [4, 5] propensity-concept for each elementary reaction. The formalism is based on the relative entropy or Kullback-Leibler distance of probability distributions for a chemical reaction associated with an elementary reaction and its microscopic reversible 'counter' reaction similar to the approach of forward and time-reversed trajectories in phase space. A general chemical reaction network is discussed in presence of chemiostatic condition that leads the open reaction system to a nonequilibrium steady state. To understand how the overall work done on the reaction system is used up in changing the internal energy of the reactants and heat dissipation, we have thoroughly analyzed path functions like heat dissipation and work done as well as state functions like internal energy, free energy and system entropy for each elementary steps. As a real case study, we have taken a single molecule kinetics of the tetrameric Escherichia coli β- galactosidase enzyme under chemiostatic condition. The variation of total entropy production rate for individual reactions are analyzed with time as well as with substrate concentration at steady state.

To understand the role of each elementary sub-reactions in governing the overall kinetics and thermodynamics, we have shown that it is the fluctuation relations which maintains an intrinsic interdependence of each step. As an implication of fluctuation theorem arises for each sub-reaction at nonequilibrium steady state, we have established relations for total entropy production rate due to each elementary reaction analogous to the Michaelis-Menten equation of enzymatic reaction rate. Total entropy production due to each reaction step is determined and the ratio of the total entropy production rate due to a reaction and its reverse reaction is shown to be constrained by this quantity at steady state. One can also estimate how each step of the

Journal of Physics: Conference Series **638** (2015) 012002 doi:10.1088/1742-6596/638/1/012002

reaction is energy driven or entropy driven to contribute to the overall reaction. Therefore, one can control the overall reaction more effectively by the choice of the proper parameters of the individual steps.

References

[1] McQuarrie D A 1963 *J. Chem. Phys.* **38** 433
[2] McQuarrie D A, Jachimowski C J and Russell M E 1964 *J. Chem. Phys.* **40** 2914
[3] McQuarrie D A 1967 *J. Appl. Prob.* **4** 413
[4] Gillespie D T 1976 *J. Comput. Phys.* **22** 403
[5] Gillespie D T 1977 *J. Phys. Chem.* **81** 2340
[6] Gillespie D T 1992 *Physica A* **188** 404
[7] Delbruck M 1940 *J. Chem. Phys.* **8** 120
[8] Ishida K 1960 *Bull. Chem. Soc. Japan* **33** 1030
[9] Bartholomay A 1958 *Bull. Math. Biophys.* **20**175
[10] Gaspard P 2004 *J. Chem. Phys.* **120** 8898
[11] Qian H and Elson E L 2002 *Biophys. Chem.* **101** 565
[12] Qian H 2007 *Annu. Rev. Phys. Chem.* **58** 113
[13] Lu H P, Xun L and Xie X S 1998 *Science* **282** 1877
[14] English B P, Min W, van Oijen A M, Lee K T, Luo G, Sun H, Cherayil B J, Kou S C and Xie X S 2006 *Nat. Chem. Biol.* **2** 87
[15] Das B, and Gangopadhyay G 2010 *J. Chem. Phys.* **132** 135102
[16] Das B, Banerjee K, and Gangopadhyay G 2013 *J. Math. Chem.* **51** 588
[17] Banerjee K, Das B, and Gangopadhyay G 2012 *J. Chem. Phys.* **136** 154502
[18] Jiu-li L, Van den Broeck C and Nicolis G 1984 *Z. Phys. B* **56** 165
[19] Nicolis G and Prigogine I 1977 *Self-Organization in nonequilibrium Systems* (New York: Willey).
[20] Nicolis G and Prigogine I 1971 *Proc. Natl. Acad. Sci. U.S.A.* **68** 2102
[21] Nicolis G 1972 *J. Stat. Phys.* **6** 195
[22] Min W, Jiang L, Yu J, Kou S C, Qian H and Xie X S 2005 *Nano Lett.* **5** 2373
[23] Ge H and Qian H 2010 *Phys.Rev.E* **81** 051133
[24] Esposito M and Van den Broeck C 2010 *Phys. Rev. E* **82** 011143
[25] Esposito M and Van den Broeck C 2010 *Phys. Rev. Lett.* **104** 090601
[26] Kawai R, Parrondo J M R and Van den Broeck C 2007 *Phys. Rev. Lett.* **98** 080602
[27] Feng E H and Crooks G E 2008 *Phys. Rev. Lett.* **101** 090602
[28] Parrondo J M R, Van den Broeck C and Kawai R 2009 *New. J. Phys.* **11** 073008
[29] Xiao T J, Hou Z and Xin H 2008 *J.Chem. Phys.* **129** 114506
[30] Das B, Banerjee K and Gangopadhyay G 2013 *J. Chem. Phys.* **139** 244104
[31] Seifert U 2005 *Phys. Rev. Lett.* **95** 040602
[32] Schmiedl T and Seifert U 2007 *J. Chem. Phys.* **126** 044101
[33] Schmiedl T, Speck T and Seifert U 2007 *J. Stat. Phys.* **128** 77
[34] Kullback S and Leibler R A 1951 *Ann. Math. Stat.* **22** 79
[35] Crooks G E and Sivak D A 2011 *J. Stat. Mech.* P06003
[36] Gardiner C W 1985 *Handbook of stochastic methods for physics, chemistry and the natural sciences* 2nd edn. (New York: Springer).
[37] Vankampen N G 2007 *Stochastic processes in physics and chemistry* 3rd edn (Amsterdam: Elsevier).
[38] Steinfeld J I, Francisco J S and Hase W L 1999 *Chemical Kinetics and Dynamics* 2nd edn (Upper Saddle River: Prentice-Hall).
[39] Rae M and Berberan-Santos M N 2004 *J. Chem. Educ.* **81** 436
[40] Jacobson R H, Zhang X J, DuBose R F and Matthews B W 1994 *Nature* **369** 761
[41] Richard J P, Huber R E, Heo C, Amyes T L and Lin S 1996 *Biochemistry* **35** 1238
[42] Marchesi S L, Steers Jr E and Shifrin S 1969 *Biochim. Biophys. Acta* **181** 20

Journal of Physics: Conference Series **638** (2015) 012003 doi:10.1088/1742-6596/638/1/012003

Mathematical understanding of detailed balance condition violation and its application to Langevin dynamics

Masayuki Ohzeki

Department of Systems Science, Graduate School of Informatics, Kyoto University, 36-1 Yoshida Hon-machi, Sakyo-ku, Kyoto 606-8501, Japan

E-mail: mohzeki@i.kyoto-u.ac.jp

Akihisa Ichiki

Green Mobility Collaborative Research Center, Nagoya University, Fro-cho, Nagoya, 464-8603, Japan

Abstract. We develop an efficient sampling method by simulating Langevin dynamics with an artificial force rather than a natural force by using the gradient of the potential energy. The standard technique for sampling following the predetermined distribution such as the Gibbs-Boltzmann one is performed under the detailed balance condition. In the present study, we propose a modified Langevin dynamics violating the detailed balance condition on the transition-probability formulation. We confirm that the numerical implementation of the proposed method actually demonstrates two major beneficial improvements: acceleration of the relaxation to the predetermined distribution and reduction of the correlation time between two different realizations in the steady state.

1. Introduction

To investigate the statistical properties of a many-body interacting system, often sampling following the predetermined distribution is employed. Because of strong correlation between degrees of freedom, it is usually intractable to compute the expectation, covariance, etc. following the distribution by direct manipulation. Instead, we often use a sampling method. One of the well-known techniques is the Markov-chain Monte Carlo (MCMC) method, which imitates the stochastic dynamics of the discrete-time master equation. This is a promising method used to evaluate the desired quantities, but it sometimes takes a long time to convert the initially prepared system into a suitable state for computation because the method imitates the relaxation of the stochastic dynamics to the predetermined distribution as a steady state. Innovative studies to enhance its sampling efficacy and accelerate the relaxation to the predetermined distribution have been extensively reported [1, 2, 3, 4, 5, 6, 7, 8].

Basically, we construct a transition matrix to relax the initially prepared system to the steady state, which corresponds to the predetermined distribution in the MCMC method. A series of instantaneous realizations is generated by simulating the multiplication of the transition matrix. After sufficient iteration steps, the obtained series of realizations follows the predetermined distribution. This is assured by the balance condition (BC) on the transition matrix. In general, it is difficult to solve the BC to compute the transition matrix. The standard approach for finding the solution is by use of the detailed balance condition (DBC). Because the DBC always satisfies the BC, we may instead solve the DBC for construction of the transition matrix. However, the solution under the DBC is not necessarily an optimal choice of the transition matrix. Recent studies reveal the superior performance, which is confirmed by means of the faster convergence to the predetermined distribution, by using nontrivial solution neglecting the DBC [9, 10, 11]. As detailed also in the present paper, the present authors provided its mathematical validation through an analysis on the eigenvalues of the transition matrix in a previous study [12].

The method for sampling is not limited to the MCMC method. The simplest technique is to implement Langevin dynamics, which is available for the stochastic dynamics of continuous variables in continuous-time evolution. In this alternative, does the analogous procedure for achieving faster convergence to the predetermined distribution exist? The time evolution of the probabilistic distribution for the degrees of freedom driven by Langevin dynamics is written by using the Fokker-Planck equation. This is the continuous equation for the instantaneous distribution function. In the steady state, the divergence of the flow of the distribution must vanish. This is analogous to the BC in the context of the master equation. The condition that the steady flow itself vanishes corresponds to the DBC. We often demand that the force, which drives the system to simulate the natural dynamics, be the gradient of the potential energy. This natural force holds the analogous condition to the BC and can be confirmed to satisfy the DBC in a different formulation. However, it is not necessarily the case if we desire to reach the predetermined distribution as fast as possible. We have room to design a force to speed up the convergence to the desired distribution. In the present study, we propose several forces to violate the condition analogous to the DBC while satisfying the analogous condition to the BC in the steady state.

The rest of the paper consists of the following sections. We review several ways to construct the transition matrix neglecting the DBC in the MCMC method in the next section. After that, we introduce the Langevin equation and the corresponding Fokker-Planck equation in the third section. In this section, we find several nontrivial solutions violating the analogous condition to the DBC. The following section presents several numerical tests of the artificial forces to speed up convergence to the desired distribution. Before closing the present paper, we show the mathematical assurance of the faster convergence by analyzing the eigenvalues of the operators in the Fokker-Planck equation. In the last section, we summarize our study and provided direction for future plans.

2. Sampling method: Monte Carlo simulation

Let us briefly review the method to achieve faster convergence to the desired distribution in an MCMC simulation from its fundamental ingredients [13].

2.1. Master equation

Markov-chain Monte Carlo simulations imitate the discrete-time evolution of the master equation given by

$$P(\mathbf{y}, t+1) = \sum_{\mathbf{x}} P(\mathbf{y}|\mathbf{x}) P(\mathbf{x}, t), \tag{1}$$

where t is a discrete time and $P(\mathbf{x}, t)$ is the instantaneous distribution for the degrees of freedom $\mathbf{x}$. The transition matrix $P(\mathbf{y}|\mathbf{x})$ defines the rule of the stochastic dynamics in the master equation. At least, the conservation of the probability (CP) demands that the transition matrix should satisfy

$$\sum_{\mathbf{y}} P(\mathbf{y}|\mathbf{x}) = 1. \tag{2}$$

In addition, the nondiagonal elements of the transition matrix must be non-negative.

2.2. Balanced condition and its solution

In MCMC simulations, we numerically perform an iterative update of the instantaneous distribution following the master equation. The distribution function in the steady state $P^{(\mathrm{ss})}(\mathbf{x})$ is defined by the balance condition

$$P^{(\mathrm{ss})}(\mathbf{y}) = \sum_{\mathbf{x}} P(\mathbf{y}|\mathbf{x})P^{(\mathrm{ss})}(\mathbf{x}). \tag{3}$$

By use of the transition matrix satisfying the BC, we obtain the desired distribution function, which is set to be $P^{(\mathrm{ss})}(\mathbf{x})$, after sufficient iteration steps. For simplicity, we restrict ourselves to the case aimed at generating the Gibbs-Boltzmann distribution $P^{(\mathrm{ss})}(\mathbf{x}) = \exp(-\beta E(\mathbf{x}) + \beta F)$, where F is the free energy and β is the inverse temperature.

It is not simple to obtain the nontrivial solution under the BC. The most standard way to find the solution is to use the detailed balance condition

$$P(\mathbf{x}|\mathbf{y})P^{(\mathrm{eq})}(\mathbf{y}) = P(\mathbf{y}|\mathbf{x})P^{(\mathrm{eq})}(\mathbf{x}) \tag{4}$$

for $\mathbf{y} \neq \mathbf{x}$. One can confirm that the DBC indeed satisfies the BC by use of CP. However, this is not a unique way to find the nontrivial solution under the BC. Recent studies on DBC violation have led to new way to construct the transition matrix, as introduced below.

2.3. Violation of the detailed balance condition

The construction of the transition matrix can be roughly categorized as 1. optimization following some rule of the transition matrix neglecting the DBC and construction of a duplicate system.

2.3.1. Suwa-Todo method The former type of method was recently proposed by Suwa and Todo [9]. They considered the minimization of the rejection rate, which corresponds to the diagonal elements of the transition matrix. In numerical implementation of the MCMC method, we generate a candidate as the next configuration at each update. Then we accept the candidate following the probability determined by the nondiagonal elements. The average of the rejection rate is thus given by the summation over the diagonal elements of the transition matrix. The diagonal elements are given by the summation of the nondiagonal elements from CP. To eliminate the diagonal elements, the Suwa-Todo method thus allocates the nondiagonal elements while neglecting the DBC but satisfying the BC and CP. Let us rewrite the transition matrix into the probabilistic flow as

$$V(\mathbf{y}|\mathbf{x}) = P(\mathbf{y}|\mathbf{x})P^{(\mathrm{ss})}(\mathbf{x}). \tag{5}$$

Then the BC and CP are given as

$$\sum_{\mathbf{x}} V(\mathbf{y}|\mathbf{x}) = P^{(\mathrm{ss})}(\mathbf{y}), \tag{6}$$

$$\sum_{\mathbf{y}} V(\mathbf{y}|\mathbf{x}) = P^{(\mathrm{ss})}(\mathbf{x}), \tag{7}$$

respectively. The symmetry of the probabilistic flow $V(\mathbf{x}|\mathbf{y}) = V(\mathbf{y}|\mathbf{x})$ reads the DBC. The Suwa-Todo method finds the rejection-free solution violating the symmetry of the probabilistic flow. We do not follow their method in the present paper but we remark on its efficacy from theoretical aspects. To eliminate the diagonal elements of the transition matrix, we may change the nondiagonal elements of the transition matrix. As also shown below in a different formulation, the asymmetry in the transition matrix can accelerate relaxation to the steady state, as shown by the present authors in Reference [12]. We thus conclude that part of the efficacy of the Suwa-Todo method comes from the asymmetry in the transition matrix, which violates the DBC. However, Suwa and Todo simultaneously reduce the rejection rate by decreasing the diagonal part. In the present paper, we restrict ourselves to the case in which the diagonal elements are fixed. In this sense, the efficacy of the Suwa-Todo method is not yet completely understood.

2.3.2. Skewed detailed balance condition The latter type is the skewed detailed balance condition (SDBC) [10]. We prepare a duplicate system by introducing a replica of the original system. For each system, we construct transition matrices $P_1(\mathbf{x}|\mathbf{y})$ and $P_2(\mathbf{x}|\mathbf{y})$. Then we define the following master equation:

$$P_i(\mathbf{y}, t+1) = \sum_{\mathbf{x}} P_i(\mathbf{y}|\mathbf{x}) P_i(\mathbf{x}, t) + Q_i(\mathbf{y}) P_{\bar{i}}(\mathbf{y}) - Q_{\bar{i}}(\mathbf{y}) P_i(\mathbf{y}), \tag{8}$$

where $P_i(\mathbf{x}, t)$ is an instantaneous distribution for each system. The swap transition from i to $\bar{i}$ (the counter system of i) is given by $Q_i(\mathbf{y})$ and $Q_{\bar{i}}(\mathbf{y})$ for the inverse manner. The balanced condition for the duplicate system holds as

$$P^{(\mathrm{ss})}(\mathbf{y}) = \sum_{\mathbf{x}} P_i(\mathbf{y}|\mathbf{x}) P^{(\mathrm{ss})}(\mathbf{x}) + Q_i(\mathbf{y}) P^{(\mathrm{ss})}(\mathbf{y}) - Q_{\bar{i}}(\mathbf{y}) P^{(\mathrm{ss})}(\mathbf{y}). \tag{9}$$

We here impose that the steady state shares the same distribution as $P^{(\mathrm{ss})}(\mathbf{x})$. One of the generic solutions satisfying the BC is known as the SDBC. The transition matrix under the SDBC is given by

$$P_i(\mathbf{y}|\mathbf{x}) P^{(\mathrm{ss})}(\mathbf{x}) = P_{\bar{i}}(\mathbf{x}|\mathbf{y}) P^{(\mathrm{ss})}(\mathbf{y}) \tag{10}$$

for $\mathbf{y} \neq \mathbf{x}$ and the transition between each system should simultaneously satisfy

$$P_1(\mathbf{y}|\mathbf{y}) - P_2(\mathbf{y}|\mathbf{y}) + Q_1(\mathbf{y}) - Q_2(\mathbf{y}) = 0. \tag{11}$$

In other words, the BC for each system in the ordinary form as in Equation (3) is violated. Instead, we demand the BC for the duplicate system, in which the transition between each system is taken into account.

2.4. Ichiki-Ohzeki asymmetric transition matrix

The violation of the DBC can be recast as the asymmetry of the transition matrix. We rescale the transition matrix and the instantaneous distribution function as

$$\begin{aligned} R(\mathbf{x}, t) &= \frac{P(\mathbf{x}, t)}{\sqrt{P^{(\mathrm{ss})}(\mathbf{x})}}, \\ W(\mathbf{y}|\mathbf{x}) &= \frac{1}{\sqrt{P^{(\mathrm{ss})}(\mathbf{y})}} P(\mathbf{y}|\mathbf{x}) \sqrt{P^{(\mathrm{ss})}(\mathbf{x})}. \end{aligned} \tag{12}$$

Then the master equation can be rewritten as

$$R(\mathbf{y}, t+1) = \sum_{\mathbf{x}} W(\mathbf{y}|\mathbf{x}) R(\mathbf{x}, t). \tag{13}$$

Two of the conditions, namely, the BC and CP, that the master equation must satisfy are simply written as

$$\sum_{\mathbf{x}} W(\mathbf{y}|\mathbf{x})\sqrt{P^{(\mathrm{ss})}(\mathbf{x})} = \sqrt{P^{(\mathrm{ss})}(\mathbf{y})}, \tag{14}$$

$$\sum_{\mathbf{y}} \sqrt{P^{(\mathrm{ss})}(\mathbf{y})}W(\mathbf{y}|\mathbf{x}) = \sqrt{P^{(\mathrm{ss})}(\mathbf{x})}, \tag{15}$$

respectively. In this expression, the DBC is interpreted as symmetry of the rescaled transition matrix as $W(\mathbf{y}|\mathbf{x}) = W(\mathbf{x}|\mathbf{y})$. The present authors have proven that the asymmetry of the rescaled transition matrix, violation of the DBC, leads to faster acceleration of relaxation to the steady state then that driven by using the symmetric part of the rescaled transition matrix [12].

We emphasize that the essential aspect of speeding up the relaxation is the asymmetry of the rescaled transition matrix. To introduce the asymmetry, for example in the SDBC, the transition to the duplicate system was considered [10]. In contrast, in the Suwa-Todo method [9], reducing the rejection rate and the nondiagonal elements were changed from those under the DBC as a result.

3. Sampling method: Langevin equation

3.1. Langevin equation and its corresponding Fokker-Planck equation

To generate the stochastic dynamics, instead of the master equation, we may use the overdamped Langevin dynamics defined as

$$d\mathbf{x} = \mathbf{A}(\mathbf{x})dt + \sqrt{2T}d\mathbf{W}, \tag{16}$$

where $\mathbf{x}$ is the vector standing for the N-dimensional degrees of freedom and $d\mathbf{x}$ is the infinitesimal change of $\mathbf{x}$ during dt. In addition, $\mathbf{A}(\mathbf{x})$ is a time-independent force vector, T is the temperature, and $d\mathbf{W}$ is the Wiener process for the N-dimensional degrees of freedom. The force is not necessarily given by $-\mathrm{grad}E(\mathbf{x})$ as in equilibrium. The purpose is hereafter to find an artificial force to realize faster convergence to the desired distribution.

For achieving this goal, let us write the corresponding Fokker-Planck equation to the above Langevin equation:

$$\frac{\partial P(\mathbf{x},t)}{\partial t} = -\mathrm{div}\mathbf{J}(\mathbf{x},t), \tag{17}$$

where $\mathbf{J}(\mathbf{x},t)$ is the probabilistic flow defined as

$$\mathbf{J}(\mathbf{x},t) = (\mathbf{A}(\mathbf{x}) - T\mathrm{grad})\, P(\mathbf{x},t). \tag{18}$$

3.2. Divergence-free condition

When the system is in any steady state, the divergence of the probabilistic flow vanishes. Therefore, we demand the following equality in the steady state:

$$0 = -\mathrm{div}\mathbf{J}^{(\mathrm{ss})}(\mathbf{x}), \tag{19}$$

where

$$\mathbf{J}^{(\mathrm{ss})}(\mathbf{x}) = (\mathbf{A}(\mathbf{x}) - T\mathrm{grad})\, P^{(\mathrm{ss})}(\mathbf{x}). \tag{20}$$

We refer to the above equality as the divergence-free condition. Our problem is to find a solution $\mathbf{J}^{(\mathrm{ss})}(\mathbf{x})$ satisfying the divergence-free condition. The above equality, which holds in any steady state, is analogous to the BC in the context of the master equation. Hereafter the important thing is to find a nontrivial solution under the divergence-free condition. The solution is violating the DBC but under the BC in the same spirit as in the previous studies for acceleration of the MCMC method [9, 10].

3.2.1. Equilibrium solution We can immediately find a trivial solution of the divergence-free condition. When the probabilistic flow in a steady state vanishes, the divergence-free condition is satisfied:

$$\mathbf{J}^{(\mathrm{eq})}(\mathbf{x}) = \mathbf{0}, \tag{21}$$

where the superscript has been changed from "ss" to "eq". This steady state is in particular called the equilibrium state because no current exists. Then the force becomes

$$\mathbf{A}(\mathbf{x}) = -\mathrm{grad}E(\mathbf{x}). \tag{22}$$

This is the trivial solution in terms of our purpose. Indeed, the resulting force does not violate the DBC. This fact can be confirmed through a formulation of the transition probability. To violate the DBC, we add an additional term to the trivial force. Below we show several nontrivial solutions, which can indeed violate the DBC.

3.2.2. Vector-field solutions Let us move on from the case without current. We demand the probabilistic flow in a steady state to take the following form:

$$\mathbf{J}^{(\mathrm{ss})}(\mathbf{x}) = \gamma\mathbf{B}(\mathbf{x})P^{(\mathrm{ss})}(\mathbf{x}), \tag{23}$$

where γ is a control parameter on the degree of violation of the DBC and will be detailed below. Owing to the divergence-free condition, the vector field $\mathbf{B}(\mathbf{x})$ must satisfy

$$0 = \gamma\sum_{k=1}^{N}\left(\frac{\partial[\mathbf{B}]_k}{\partial[\mathbf{x}]_k} - \frac{1}{T}[\mathbf{B}]_k\frac{\partial E}{\partial[\mathbf{x}]_k}\right)P^{(\mathrm{ss})}(\mathbf{x}), \tag{24}$$

where we have used the fact that the distribution function is set to be $P^{(\mathrm{ss})}(\mathbf{x}) \propto \exp(-E(\mathbf{x})/T)$ and the subscript for the bracket denotes the element of the vector. A trivial solution is given by

$$[\mathbf{B}(\mathbf{x})]_k = \exp(E(\mathbf{x})/T - F/T). \tag{25}$$

Then the probabilistic flow becomes a constant current as

$$\mathbf{J}^{(\mathrm{ss})}(\mathbf{x}) = \gamma\mathbf{1}. \tag{26}$$

Here $\mathbf{1}$ denotes a vector with all elements unity. We obtain the resulting force as

$$\mathbf{A}(\mathbf{x}) = -\mathrm{grad}E(\mathbf{x}) + \gamma\exp(E(\mathbf{x})/T - F/T)\mathbf{1}. \tag{27}$$

For practical use of this force, we may rescale $\gamma\exp(-F/T)$ as γ. However, we have to be careful about the instability in using the exponential term, in particular in the low-temperature region.

Let us find an alternative to the above obtained force to remove the instability of the exponential term. We set the vector field as

$$[\mathbf{B}(\mathbf{x})]_k = \left(\frac{\partial E}{\partial[\mathbf{x}]_{k-1}} - \frac{\partial E}{\partial[\mathbf{x}]_{k+1}}\right), \tag{28}$$

where $[\mathbf{x}]_{N+1} = [\mathbf{x}]_1$ and $[\mathbf{x}]_0 = [\mathbf{x}]_N$. Then the divergence-free condition holds. The resulting force is given as

$$[\mathbf{A}(\mathbf{x})]_k = -\frac{\partial E}{\partial[\mathbf{x}]_k} + \gamma\left(\frac{\partial E}{\partial[\mathbf{x}]_{k-1}} - \frac{\partial E}{\partial[\mathbf{x}]_{k+1}}\right). \tag{29}$$

This force drives the system while satisfying the divergence-free condition but neglecting the DBC. In this sense, the resulting force is analogous to the case of using the Suwa-Todo method in the context of the MCMC method [9]. This type of solution is not unique. We may permute the index $j = 1, 2, \ldots, N$ as $j = P(1), P(2), \ldots, P(N)$, where $P(\cdot)$ denotes a permutation of N indices. To reach the best performance, we may pursue an optimal permutation to realize the fastest convergence. This is beyond our scope in the present study. Let us further find any other nontrivial solutions.

Journal of Physics: Conference Series **638** (2015) 012003 doi:10.1088/1742-6596/638/1/012003

3.3. Divergence-free condition for a duplicate system
As in the case of the SDBC [10], let us duplicate the system. Then we may neglect the DBC for each system but hold the BC for the composite system. The analogy to the SDBC leads to the duplicate Fokker-Planck equation

$$\frac{\partial P(\mathbf{x}_1, \mathbf{x}_2, t)}{\partial t} = -\sum_{i=1}^{2} \mathrm{div}_i \mathbf{J}_i(\mathbf{x}_1, \mathbf{x}_2, t), \tag{30}$$

where $\mathbf{x}_i$ is the N-dimensional vector for each system denoted by i and $\mathbf{J}_i(\mathbf{x}_1, \mathbf{x}_2, t)$ is the probabilistic flow defined as

$$\mathbf{J}_i(\mathbf{x}_1, \mathbf{x}_2, t) = \left(\mathbf{A}_i(\mathbf{x}_1, \mathbf{x}_2) - T\mathrm{grad}_i\right) P(\mathbf{x}_1, \mathbf{x}_2, t). \tag{31}$$

The gradient indexed by i is taken for each system. Again we impose the divergence-free condition. The divergence-free condition for the duplicate system can be recast as

$$0 = -\sum_{i=1}^{2} \mathrm{div}_i \mathbf{J}_i^{(\mathrm{ss})}(\mathbf{x}_1, \mathbf{x}_2), \tag{32}$$

where

$$\mathbf{J}_i^{(\mathrm{ss})}(\mathbf{x}_1, \mathbf{x}_2) = \left(\mathbf{A}_i(\mathbf{x}_1, \mathbf{x}_2) - T\mathrm{grad}_i\right) P^{(\mathrm{ss})}(\mathbf{x}_1) P^{(\mathrm{ss})}(\mathbf{x}_2). \tag{33}$$

We impose the steady state of the duplicate system as $P^{(\mathrm{ss})}(\mathbf{x}_1, \mathbf{x}_2) = P^{(\mathrm{ss})}(\mathbf{x}_1)P^{(\mathrm{ss})}(\mathbf{x}_2)$. We can find a nontrivial solution by considering the mixture of the forces on the duplicate system as

$$\mathbf{A}_1(\mathbf{x}_1, \mathbf{x}_2) = -\mathrm{grad}_1 E(\mathbf{x}_1) + \gamma \mathrm{grad}_2 E(\mathbf{x}_2), \tag{34}$$

$$\mathbf{A}_2(\mathbf{x}_1, \mathbf{x}_2) = -\mathrm{grad}_2 E(\mathbf{x}_2) - \gamma \mathrm{grad}_1 E(\mathbf{x}_1). \tag{35}$$

Then the probabilistic flow in the steady state is given as

$$\mathbf{J}_1^{(\mathrm{ss})}(\mathbf{x}_1, \mathbf{x}_2) = \gamma \mathrm{grad}_2 E(\mathbf{x}_2) P^{(\mathrm{ss})}(\mathbf{x}_1) P^{(\mathrm{ss})}(\mathbf{x}_2), \tag{36}$$

$$\mathbf{J}_2^{(\mathrm{ss})}(\mathbf{x}_1, \mathbf{x}_2) = -\gamma \mathrm{grad}_1 E(\mathbf{x}_1) P^{(\mathrm{ss})}(\mathbf{x}_1) P^{(\mathrm{ss})}(\mathbf{x}_2). \tag{37}$$

One can confirm that the divergence of the above probabilistic flows can vanish immediately because the summation becomes zero.

One may recall the exchange Monte Carlo simulation, in which a multiple system with different temperatures is driven simultaneously, while each realization is exchanged several times. The technique exhibits extremely outstanding performance with fast convergence to the equilibrium state. Naturally, we may expect the existence of an analogous Langevin dynamics to the exchange Monte Carlo simulation. We here mention this fascinating problem as a subject of a future study, which will be published elsewhere.

3.4. Path probability and detailed balance condition
As shown in the previous subsection, we find nontrivial solutions satisfying the analogous condition to the BC (19) and (32). We here confirm the violation of the DBC in these solutions. Let us calculate the transition probability during an infinitesimal time interval, $[t, t + dt]$. We obtain

$$L^{\gamma}(\mathbf{x}[t + dt]|\mathbf{x}[t]) \propto \exp\left\{-\frac{1}{4T}(\dot{\mathbf{x}} - \mathbf{A})^2 dt - \frac{1}{2}\mathrm{div}\mathbf{A}dt\right\}, \tag{38}$$

where $\mathbf{x}[t]$ is the location at time t. For the duplicated system, a similar computation yields the transition probability. We here omit the arguments of the quantities on the right-hand side to simplify the expressions. We use the midpoint prescription, $\bar{\mathbf{x}} = (\mathbf{x}[t+dt] + \mathbf{x}[t])/2$; we take the partial derivative with respect to the location. The ratio of the transition probability between the forward and backward processes confirms violation of the DBC owing to the existence of the probabilistic flow as follows:

$$\frac{L^{\gamma}(\mathbf{x}[t+dt]|\mathbf{x}[t])}{L^{\gamma}(\mathbf{x}[t]|\mathbf{x}[t+dt])} = \exp\left(\frac{\gamma}{T}\dot{\mathbf{x}}^{\mathrm{T}}\mathbf{B}dt - \frac{1}{T}\dot{\mathbf{x}}^{\mathrm{T}}\mathrm{grad}E(\bar{\mathbf{x}})dt\right). \tag{39}$$

The first term in the exponential function corresponds to the negative of the time-derivative housekeeping heat $\dot{Q}_{\mathrm{hk}}$, while the second term can be regarded as the negative of the time-derivative excess heat $\dot{Q}_{\mathrm{ex}}$, i.e., [14, 15]

$$\dot{Q}_{\mathrm{hk}} = -\gamma\dot{\mathbf{x}}^{\mathrm{T}}\mathbf{B}, \tag{40}$$

$$\dot{Q}_{\mathrm{ex}} = \dot{\mathbf{x}}^{\mathrm{T}}\mathrm{grad}E(\bar{\mathbf{x}}). \tag{41}$$

The existence of the housekeeping heat signals violation of the DBC and, in this sense, γ, which is the coefficient in the vector field $\mathbf{B}$, controls the degree of violation of the DBC. Also in the case of the duplicate system, we can compute the ratio of the transition probabilities for forward and backward processes as

$$\prod_{i=1,2}\frac{L_i^{\gamma}(\mathbf{x}_i[t+dt]|\mathbf{x}_i[t])}{L_i^{\gamma}(\mathbf{x}_i[t]|\mathbf{x}_i[t+dt])} = \exp\left(\frac{\gamma}{T}(\dot{\mathbf{x}}_1{}^{\mathrm{T}}\mathrm{grad}_2E(\bar{\mathbf{x}}_2) - \dot{\mathbf{x}}_2{}^{\mathrm{T}}\mathrm{grad}_1E(\bar{\mathbf{x}}_1))dt\right)$$
$$\times\prod_{i=1,2}\exp\left(-\frac{1}{T}\dot{\mathbf{x}}^{\mathrm{T}}\mathrm{grad}_iE(\bar{\mathbf{x}}_i)dt\right). \tag{42}$$

Then the time derivatives of the housekeeping heat and excess heat are given as

$$\dot{Q}_{\mathrm{hk}} = \gamma\left(\dot{\mathbf{x}}_2{}^{\mathrm{T}}\mathrm{grad}_1E(\bar{\mathbf{x}}_2) - \dot{\mathbf{x}}_1{}^{\mathrm{T}}\mathrm{grad}_2E(\bar{\mathbf{x}}_1)\right), \tag{43}$$

$$\dot{Q}_{\mathrm{ex}} = \sum_{i=1,2}\dot{\mathbf{x}}_i{}^{\mathrm{T}}\mathrm{grad}_iE(\bar{\mathbf{x}}_i). \tag{44}$$

In the duplicate system, we also confirm the violation of the DBC from the existence of the housekeeping heat. Let us below evaluate the performance of using the artificial force violating the DBC by the above nontrivial solutions.

4. Numerical tests

We implement the above artificial force to introduce the probabilistic flow violating the DBC. To confirm that the artificial force accelerates the relaxation to the steady state, we demonstrate simple numerical tests below. We here restrict ourselves to the case of the additional force to the duplicate system. Let us consider the system with the following double-valley potential energy:

$$U(x) = -\frac{1}{2}x^2 + \frac{1}{4}x^4. \tag{45}$$

The minima of the potential are located at $x = -1$ and 1. The equilibrium point is $x = 0$; that is, the average over independent runs converges to $\langle x[t]\rangle \equiv \sum_{k=1}^{N_{\mathrm{sam}}} x^{(k)}[t]/N_{\mathrm{sam}}$ in $t \to \infty$, where $x[t]$ denotes the realization at the time t. We simulate the independent $N_{\mathrm{sam}} = 1000$ runs. The initial condition is set to be at $x^{(k)}[0] = 1$. To achieve the equilibrium point, the particles must climb up the local maximum and reach the other minimum of the potential energy. Therefore, it takes a relatively long time to reach the equilibrium point. Then we introduce the additional force to accelerate the stochastic dynamics and duplicate the system. The time evolution of the Langevin equation is evaluated by using the ordinary method known as the Heun scheme [16]. We set the infinitesimal time as $dt = 0.0001$. We take an average over independent $N_{\mathrm{sam}} = 1000$ runs, while taking the time average during $\Delta t = 0.1$.

Figure 1 shows a comparison between the cases without and with additional force. We describe the orbit on the plane of the locations of the duplicate system. The averaged locations of the duplicate system are expressed as $\langle x_1 \rangle$ and $\langle x_2 \rangle$. The case without additional force traces the line between $(0, 0)$ and $(1, 1)$ because any difference does not exist in the duplicate system. In contrast, the case with additional force describes a different path from the line. The particles seem to wrap around the swelled area of the potential energy in the plane of (x_1, x_2). For weak γ, the relaxation to the steady state is not necessarily faster. However, an increase of the value of γ leads to faster convergence to the steady state than that for the case without additional force. This is one of the benefits in violating the DBC, as shown in previous studies in terms of the Suwa-Todo method and the SDBC. In addition, we observe the integrated correlation time in the steady state. The integrated correlation time is defined as the collection of the correlation times:

$$\tau_i = \sum_{t=1}^{\infty} \tau_{i,t}^{O} = \frac{\langle O_i O_{i+t} \rangle - \langle O_i \rangle \langle O_{i+t} \rangle}{\langle O^2 \rangle - \langle O \rangle^2}. \tag{46}$$

In the steady state, the dependence of the correlation time on i is lost. In numerical computation, we cannot sum the correlation time $\tau_{i,t}^{O}$ up to an infinite number. Thus the dependence on i remains. We then take an average over i to avoid this dependence. In this computation, we use the location as the physical observable. The results are shown in Figure 2. We confirm the decrease of the integrated correlation time. In addition, the integrated correlation time reaches a lower limit against any further increase of the value of γ. The decrease of the correlation time gives a reduction of the necessary number to efficiently compute the averaged value from the sampling and thus leads to an increase in the efficacy of the sampling. This beneficial point is the same as the one found in the Suwa-Todo method and the SDBC.

In addition, the present authors demonstrated a further application of our method to a classical spin model, the XY model, which has a critical slowing down in the low-temperature region [17]. The proposed method exhibited outstanding performance, removing the critical slowing down of the XY model, which often hampers efficient computation in the steady state.

The remaining problem is the mathematical and theoretical understanding of the faster convergence to the steady state and the reduction of the correlation time owing to violation of the DBC. Fortunately, the former property can be explained by considering the eigenvalues of the operator form of the Fokker-Planck equation and transition-rate matrix in the master equation. In the next section, we demonstrate the property of the operator form of the Fokker-Planck equation in cases without and with additional forces and show the shift of the eigenvalues such that it promotes accelerating the relaxation to the steady state.

5. Mathematical assurance: analysis of eigenvalues

To clarify the origin of the acceleration to the steady state by the additional force, we rewrite the Fokker-Planck equation in operator form [18]. This expression can help us to find a mathematical understanding of the acceleration of the relaxation to the steady state in our technique.

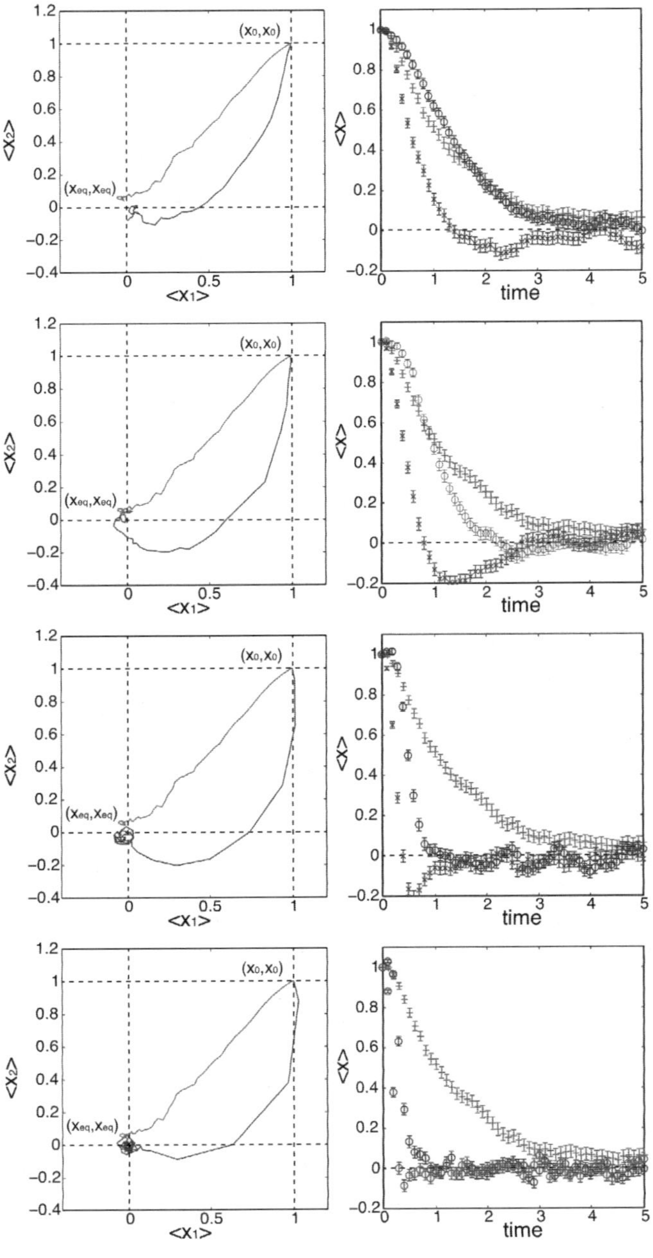

Figure 1. Comparison between the cases with and without additional force. From top to bottom, the values of γ take 1, 2, 5, and 10. In the left panels, the red orbit denotes the case without additional force. The blue ones represent the case with additional force on the plane of x_1 and x_2, which is the location of a particle in the duplicate system. The right panels show the average of the location over independent runs over $N_{\rm sam} = 1000$. Red plots are for the case without additional force and the blue and purple ones stand for each average of the location of the duplicate system with additional force.

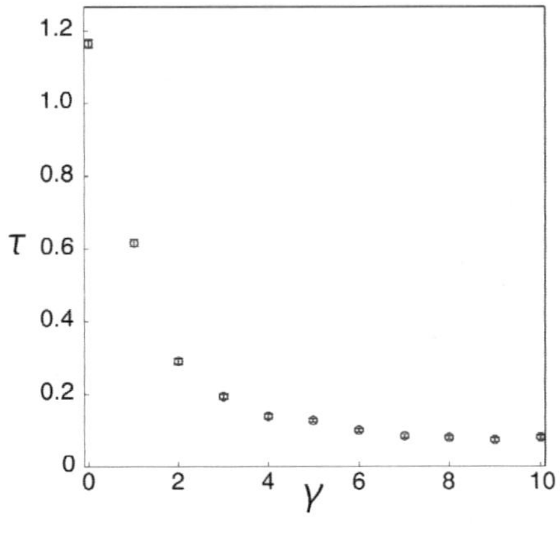

Figure 2. Integrated correlation time. The horizontal axis denotes the value of γ and the vertical one describes the average of the integrated correlation time. The statistical error falls within the size of the data points.

5.1. Operator form of the Fokker-Planck equation

First we rescale the instantaneous distribution as

$$\tilde{P}(\mathbf{x},t) = \frac{P(\mathbf{x},t)}{\sqrt{P^{(\mathrm{ss})}(\mathbf{x})}}. \tag{47}$$

Then the Fokker-Planck equation in the equilibrium system can be rewritten as

$$\frac{\partial \tilde{P}(\mathbf{x},t)}{\partial t} = \sum_{k=1}^{N} \left(\frac{1}{4T} \left(\frac{\partial E}{\partial [\mathbf{x}]_k} \right)^2 + \frac{1}{2} \frac{\partial^2 E}{\partial [\mathbf{x}]_k^2} - T \frac{\partial^2}{\partial [\mathbf{x}]_k^2} \right) \tilde{P}(\mathbf{x},t). \tag{48}$$

Let us here define the following operators as

$$[\mathbf{a}]_k = \frac{1}{2\sqrt{T}} \frac{\partial E}{\partial [\mathbf{x}]_k} + \sqrt{T} \frac{\partial}{\partial [\mathbf{x}]_k}, \tag{49}$$

$$\left[\mathbf{a}^\dagger\right]_k = \frac{1}{2\sqrt{T}} \frac{\partial E}{\partial [\mathbf{x}]_k} - \sqrt{T} \frac{\partial}{\partial [\mathbf{x}]_k}, \tag{50}$$

where $\mathbf{a}^\dagger$ denotes the adjoint of $\mathbf{a}$. These operators satisfy the following commutation relation:

$$\left[\left[\mathbf{a}^\dagger\right]_k, [\mathbf{a}]_l \right] = \frac{\partial^2 E}{\partial [\mathbf{x}]_k \partial [\mathbf{x}]_l} \tag{51}$$

and trivial ones as $\left[[\mathbf{a}]_k, [\mathbf{a}]_l \right] = 0$ and $\left[\left[\mathbf{a}^\dagger\right]_k, \left[\mathbf{a}^\dagger\right]_l \right] = 0$. The Fokker-Planck equation is simply expressed in terms of these operators as

$$\frac{\partial \tilde{P}(\mathbf{x},t)}{\partial t} = S(\mathbf{a}^\dagger, \mathbf{a}) \tilde{P}(\mathbf{x},t), \tag{52}$$

where

$$S(\mathbf{a}^\dagger, \mathbf{a}) = -{\mathbf{a}^\dagger}^{\mathrm{T}} \mathbf{a}.$$

Here we use the transpose of the vectors denoted by $(\cdot)^{\mathrm{T}}$. The operator on the right-hand side in the Fokker-Planck equation is Hermitian. The square root of the Gibbs-Boltzmann distribution can be an eigenfunction of this operator with zero eigenvalue. By using the same operator expression, we can rewrite the Fokker-Planck equation for the nontrivial solution satisfying the divergence-free condition (29). The additional force modifies the Fokker-Planck equation into

$$\begin{aligned}\frac{\partial \tilde{P}(\mathbf{x},t)}{\partial t} &= S(\mathbf{a}^{\dagger},\mathbf{a})\tilde{P}(\mathbf{x},t)\\ &+\gamma\sum_{k=1}^{N}\left(\frac{1}{2T}[\mathbf{B}]_k\frac{\partial E}{\partial[\mathbf{x}]_k}-\frac{\partial[\mathbf{B}]_k}{\partial[\mathbf{x}]_k}-[\mathbf{B}]_k\frac{\partial}{\partial[\mathbf{x}]_k}\right)\tilde{P}(\mathbf{x},t). \end{aligned}\tag{53}$$

From this observation, we find that the effect of the additional force in the Fokker-Planck equation can be given by $\gamma\mathbf{a}^{\dagger\mathrm{T}}\mathbf{B}(\mathbf{x})\tilde{P}(\mathbf{x},t)/\sqrt{T}$. The nontrivial solution of the vector field (28) can be expressed by the operators as

$$[\mathbf{B}(\mathbf{x})]_k=\sqrt{T}\left([\mathbf{a}^{\dagger}]_{k-1}+[\mathbf{a}]_{k-1}-[\mathbf{a}^{\dagger}]_{k+1}-[\mathbf{a}]_{k+1}\right). \tag{54}$$

Thus we reach

$$\frac{\partial \tilde{P}(\mathbf{x},t)}{\partial t}=W(\mathbf{a}^{\dagger},\mathbf{a})\tilde{P}(\mathbf{x},t), \tag{55}$$

where $\tilde{P}(\mathbf{x},t)=P(\mathbf{x},t)/\sqrt{P^{(\mathrm{ss})}(\mathbf{x})}$ and

$$W(\mathbf{a}^{\dagger},\mathbf{a}) = S(\mathbf{a}^{\dagger},\mathbf{a})-\gamma\sum_{k=1}^{N}[\mathbf{a}^{\dagger}]_k([\mathbf{a}]_{k-1}-[\mathbf{a}]_{k+1}).$$

We find the anti-Hermitian part on the right-hand side of the Fokker-Planck equation (55). The anti-Hermitian part does not change the steady state because the eigenfunction of the original Fokker-Planck equation is that for the annihilation operator $\mathbf{a}$. Thus we expect that the existence of the anti-Hermitian part of the operators on the right-hand side is closely related to the acceleration of the relaxation. In addition, the trivial solution with the exponentially divergent quantity also yields the anti-Hermitian part in operator form.

Furthermore, the above observation is supported from the Fokker-Planck equation for the divergence-free duplicate system (30). The Fokker-Planck equation for the duplicate system with nontrivial additional force can be rewritten as

$$\begin{aligned}\frac{\partial \tilde{P}(\mathbf{x}_1,\mathbf{x}_2,t)}{\partial t} &= \sum_{i=1}^{2}S(\mathbf{a}_i^{\dagger},\mathbf{a}_i)\tilde{P}(\mathbf{x}_1,\mathbf{x}_2,t)\\ &-\gamma\sum_{k=1}^{N}\left(\frac{\partial E}{\partial[\mathbf{x}_2]_k}\frac{\partial}{\partial[\mathbf{x}_1]_k}-\frac{\partial E}{\partial[\mathbf{x}_1]_k}\frac{\partial}{\partial[\mathbf{x}_2]_k}\right)\tilde{P}(\mathbf{x}_1,\mathbf{x}_2,t), \end{aligned}\tag{56}$$

where $\tilde{P}(\mathbf{x}_1,\mathbf{x}_2,t)=P(\mathbf{x}_1,\mathbf{x}_2,t)/\sqrt{P^{(\mathrm{ss})}(\mathbf{x}_1,\mathbf{x}_2)}$. Here we define the operators $\mathbf{a}_i$ and $\mathbf{a}_i^{\dagger}$ for each system. The commutation relation can be replaced with

$$\left[\left[\mathbf{a}_i^{\dagger}\right]_k,[\mathbf{a}_j]_l\right]=\delta_{ij}\frac{\partial^2 E}{\partial[\mathbf{x}_i]_k\partial[\mathbf{x}_i]_l} \tag{57}$$

and trivial ones as $\left[[\mathbf{a}_i]_k,[\mathbf{a}_j]_l\right]=0$ and $\left[\left[\mathbf{a}_i^{\dagger}\right]_k,\left[\mathbf{a}_j^{\dagger}\right]_l\right]=0$. We here use the Kronecker delta δ_{ij}, which takes unity when $i=j$ and otherwise vanishes. Thus we reach

$$\frac{\partial \tilde{P}(\mathbf{x}_1,\mathbf{x}_2,t)}{\partial t}=W(\{\mathbf{a}_i^{\dagger}\},\{\mathbf{a}_i\})\tilde{P}(\mathbf{x}_1,\mathbf{x}_2,t), \tag{58}$$

Journal of Physics: Conference Series **638** (2015) 012003 doi:10.1088/1742-6596/638/1/012003

where

$$W(\{\mathbf{a}_i^\dagger\},\{\mathbf{a}_i\}) = \sum_{i=1}^{2} S(\mathbf{a}_i^\dagger,\mathbf{a}_i) - \gamma\left(\mathbf{a}_2^{\dagger\mathrm{T}}\mathbf{a}_1 - \mathbf{a}_1^{\dagger\mathrm{T}}\mathbf{a}_2\right). \tag{59}$$

The artificial force satisfying the divergence-free condition involves the anti-Hermitian part also in the case for the duplicate system. From this common property, the anti-Hermitian part of the operators in the Fokker-Planck equation is expected to accelerate the relaxation. In addition, when we violate the DBC in the master equation, namely, $W(\mathbf{x}|\mathbf{y}) \neq W(\mathbf{y}|\mathbf{x})$ in the expression in Equation (13), we find that the transition matrix takes the asymmetric form. The key point comes from the asymmetry of the transition matrix in the master equation or operators in the Fokker-Planck equation. The relaxation time is characterized by the gap between the first and second largest eigenvalues of the transition matrix and operators. Below let us analyze the eigenvalues for asymmetric operators by identifying both of them for simplicity. Then we prove that the gap of the cases with the additional force is always opened in comparison to that without it.

5.2. *Shift of the second eigenvalues*

Let us consider the eigenvalues for the asymmetric matrix and operators. We restrict ourselves to the case of the nontrivial solution of the vector field in the single system. The following analysis can be straightforwardly generalized to the duplicate system case. The trivial eigenfunction for the operator $W(\mathbf{a}^\dagger,\mathbf{a})$ is $\sqrt{P^{(\mathrm{ss})}(\mathbf{x})}$ with zero eigenvalue that is the largest. Thus relaxation to steady state $P^{(\mathrm{ss})}(\mathbf{x})$ is assured [18]. The relaxation speed is dominated by the gap between the first and the second largest eigenvalues. As seen above, introduction of the artificial force induces the anti-Hermitian part of the operator $W(\mathbf{a}^\dagger,\mathbf{a})$, while the Hermitian part is fixed. We here utilize a mathematical tool from linear algebra known as the Ostrowski-Taussky inequality [19]

$$|\det A| \geq \det\frac{A+A^\dagger}{2}, \tag{60}$$

where A is an arbitrary operator with $(A+A^\dagger)/2$ being positive definite. We can show that the anti-Hermitian part opens the gap larger by applying the Ostrowski-Taussky inequality to the characteristic polynomial $\det\left(\lambda I - W(\mathbf{a}^\dagger,\mathbf{a})\right)$, where I is an identity operator. Since the largest eigenvalues for both $W(\mathbf{a}^\dagger,\mathbf{a})$ and $\left(W(\mathbf{a}^\dagger,\mathbf{a}) + W^\dagger(\mathbf{a}^\dagger,\mathbf{a})\right)/2$ are $\lambda_1^{(\mathrm{asym})} = \lambda_1^{(\mathrm{sym})} = 0$, the Ostrowski-Taussky inequality yields

$$\left|\det\left(\lambda I - W(\mathbf{a}^\dagger,\mathbf{a})\right)\right| \geq \det\left(\lambda I - S(\mathbf{a}^\dagger,\mathbf{a})\right) \tag{61}$$

for arbitrary real $\lambda > 0$. We show a schematic picture of the behavior of two characteristic functions in Figure 3. The root of both of the characteristic functions is located at $\lambda = 0$. In addition, Inequality (61) ensures that the slope of the characteristic function of the asymmetric operator at $\lambda = 0$ is steeper than that of the corresponding symmetric part. To prove the fact that the second largest eigenvalue, the nearest root to the origin, is located at the further point by induction of the anti-Hermitian operator, let us consider $\det\left(\lambda I - W(\mathbf{a}^\dagger,\mathbf{a})\right)/\lambda$, which has the same roots as those of the characteristic function except for $\lambda = 0$. We can prove the lemma that the Hermitian and anti-Hermitian operators $\tilde{S}_\lambda$ and $\tilde{G}$ exists satisfying

$$\frac{\det\left(\lambda I - W(\mathbf{a}^\dagger,\mathbf{a})\right)}{\lambda} = \det\left(\tilde{S}_\lambda(\mathbf{a}^\dagger,\mathbf{a}) - \tilde{G}(\mathbf{a}^\dagger,\mathbf{a})\right) \tag{62}$$

for arbitrary real λ using the appropriate unitary transformation. Since the unitary transformation does not change the symmetries and determinants of the operators, we reach

$$\det\left(\tilde{S}_\lambda(\mathbf{a}^\dagger,\mathbf{a})\right) = \frac{\det\left(\lambda I - S(\mathbf{a}^\dagger,\mathbf{a})\right)}{\lambda}. \tag{63}$$

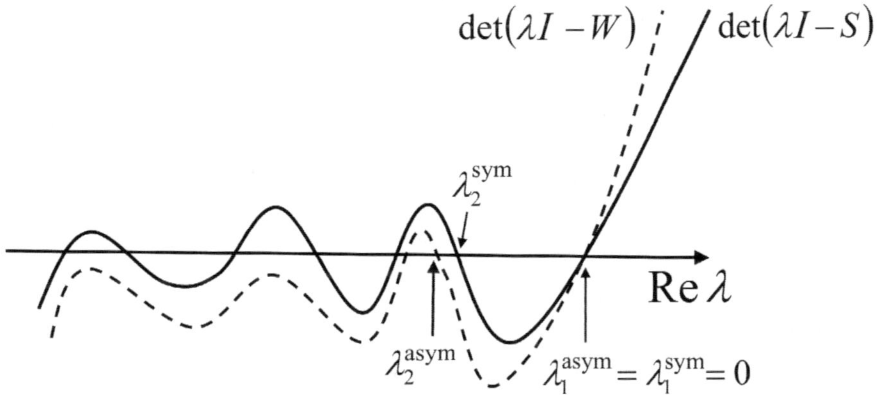

Figure 3. Real parts of $\det\left(\lambda I - W(\mathbf{a}^\dagger, \mathbf{a})\right)$ and $\det\left(\lambda I - S(\mathbf{a}^\dagger, \mathbf{a})\right)$. Inequality (64) claims $\det\left(\lambda I - W(\mathbf{a}^\dagger, \mathbf{a})\right) \leq \det\left(\lambda I - S(\mathbf{a}^\dagger, \mathbf{a})\right)$ for $\lambda_2^{(\mathrm{sym})} < \lambda < 0$.

For λ satisfying $\lambda_2^{(\mathrm{sym})} < \lambda < 0$, where $\lambda_2^{(\mathrm{sym})}$ denotes the second largest eigenvalue for the symmetric operator, $\tilde{S}_\lambda(\mathbf{a}^\dagger, \mathbf{a})$ is positive definite. Resorting the Ostrowski-Taussky inequality, we find

$$\left|\frac{\det\left(\lambda I - W(\mathbf{a}^\dagger, \mathbf{a})\right)}{\lambda}\right| \geq \frac{\det\left(\lambda I - S(\mathbf{a}^\dagger, \mathbf{a})\right)}{\lambda}. \tag{64}$$

This inequality shows that the second largest eigenvalue of the asymmetric operator is further from the origin than that of the corresponding symmetric operator. All of the above ingredients shows that

$$\mathrm{Re}\lambda_2^{(\mathrm{asym})} \leq \lambda_2^{(\mathrm{sym})}. \tag{65}$$

Therefore, the acceleration of convergence by the introduction of asymmetry in the transition dynamics is guaranteed in terms of the shift of eigenvalues.

6. Conclusion

We investigated several solutions of nontrivial forces to accelerate convergence to the desired distribution in Langevin systems. The solutions are roughly categorized into two types. The first is the addition of an exponentially divergent force to the gradient of the potential, i.e., natural force. However, this type of force involves numerical instabilities such as overflow. In contrast, the second solution, which takes advantage of rotation of the vector field, is free from such numerical instabilities. The latter solution can be simplified by employing the idea in the duplicated system inspired by the SDBC. Our proposed method has been indeed confirmed by numerical tests to accelerate convergence. This aspect of violation of the DBC can give mathematical assurance in terms of the shift of eigenvalues for asymmetric operators. In numerical experiments, we also confirm that our method shortens the correlation time compared to that in the ordinary implementation of Langevin dynamics. Previous studies on using MCMC method without the DBC also revealed the reduction of the correlation time, and we may thus regard this property as a beneficial aspect of the DBC violation. This advantage might be explained by some mathematical property on the violation of the DBC, but we have not reached any conclusive result in this regard. We hope that a future study will clarify the reduction of the correlation time for the case violating the DBC.

In the present study, we utilize a mathematical tool in linear algebra to prove the acceleration of convergence to the steady state. However, recent developments in nonequilibrium statistical mechanics can explain two of the beneficial points of the violation of the DBC. The acceleration of convergence and reduction of the correlation time rely on the change of frequency of state transitions. Let us consider the case with local minima of potential energy. The existence of local minima can be a bottleneck in the convergence toward the equilibrium state. This is because the typical events are to be trapped in the local minima. As shown in our present study, violation of the DBC can help the particles escape from local minima and accelerates the relaxation to the steady state. This implies that the particles pass through the modified path from the typical transitions in the equilibrium case. The modification of the path probability between different microscopic states by additional forces has been recently discussed in the context of biased sampling in nonequilibrium statistical mechanics [20, 21]. The detailed analysis on the effect of the additional force is reported from the viewpoint of biased sampling and its optimality by considering a kind of variational principle [22]. Such a viewpoint will give a deeper understanding on the faster convergence to the steady state resulting from the violation of the DBC and fundamental aspects on nonequilibrium statistical mechanics through concrete instances near the equilibrium system.

Acknowledgments

One of the authors (M.O.) is grateful for fruitful discussions with M. Kikuchi, H. Touchette, and K. Hukushima. This work was supported by MEXT in Japan, KAKENHI No. 15H03699, 24740263 and No. 25120008, and the Kayamori Foundation of Informational Science Advancement.

References

[1] Swendsen R H and Wang J S 1987 *Phys. Rev. Lett.* **58**(2) 86–88
[2] Hukushima K and Nemoto K 1996 *Journal of the Physical Society of Japan* **65** 1604–1608
[3] Neal R 2001 *Statistics and Computing* **11** 125–139 ISSN 0960-3174
[4] Ohzeki M 2010 *Phys. Rev. Lett.* **105**(5) 050401
[5] Ohzeki M and Nishimori H 2010 *J. Phys. Soc. Jpn.* **79**(8) 084003
[6] Ohzeki M and Nishimori H 2011 *Journal of Physics: Conference Series* **302** 012047
[7] Ohzeki M and Nishimori H 2011 *Physica E: Low-dimensional Systems and Nanostructures* **43** 782 – 785
[8] Ohzeki M 2012 *Phys. Rev. E* **86**(6) 061110
[9] Suwa H and Todo S 2010 *Phys. Rev. Lett.* **105**(12) 120603
[10] Turitsyn K S, Chertkov M and Vucelja M 2011 *Physica D: Nonlinear Phenomena* **240** 410 – 414
[11] Fernandes H C and Weigel M 2011 *Computer Physics Communications* **182** 1856 – 1859
[12] Ichiki A and Ohzeki M 2013 *Phys. Rev. E* **88**(2) 020101
[13] Nishimori H and Ortiz G 2011 *Elements of Phase Transitions and Critical Phenomena (Oxford Graduate Texts)* (Oxford University Press, USA)
[14] Hatano T and Sasa S i 2001 *Phys. Rev. Lett.* **86**(16) 3463–3466
[15] Sughiyama Y and Ohzeki M 2011 *Physica E: Low-dimensional Systems and Nanostructures* **43** 790 – 793 ISSN 1386-9477
[16] Kloeden P E and Platen E 2011 *Numerical Solution of Stochastic Differential Equations* corrected ed (New York: Springer)
[17] Ohzeki M and Ichiki A 2013 *cond-mat* **abs/1307.0434**
[18] Risken H 1996 *The Fokker-Planck Equation: Methods of Solutions and Applications* 2nd ed Springer Series in Synergetics (Springer)
[19] Horn R and Johnson C 2012 *Matrix Analysis* Matrix Analysis (Cambridge University Press)
[20] Nemoto T and Sasa S 2011 *Phys. Rev. E* **83**(3) 030105
[21] Sughiyama Y and Ohzeki M 2013 *Interdisciplinary Information Sciences* **19** 93–99
[22] Ichiki A and Ohzeki M 2015 *Phys. Rev. E* **91**(6) 062105

Journal of Physics: Conference Series **638** (2015) 012004 doi:10.1088/1742-6596/638/1/012004

Spatiotemporal pattern formation in a prey-predator model under environmental driving forces

Anuj Kumar Sirohi

School of Computational and Integrative Sciences, Jawaharlal Nehru University, New Delhi-110067, INDIA

E-mail: anujsirohi.amu@gmail.com

Malay Banerjee

Department of Mathematics and Statistics, Indian Institute of Technology, Kanpur-208016, INDIA

E-mail: malayb@iitk.ac.in

Anirban Chakraborti

School of Computational and Integrative Sciences, Jawaharlal Nehru University, New Delhi-110067, INDIA

E-mail: anirban@jnu.ac.in

Abstract. Many existing studies on pattern formation in the reaction-diffusion systems rely on deterministic models. However, environmental noise is often a major factor which leads to significant changes in the spatiotemporal dynamics. In this paper, we focus on the spatiotemporal patterns produced by the predator-prey model with ratio-dependent functional response and density dependent death rate of predator. We get the reaction-diffusion equations incorporating the self-diffusion terms, corresponding to random movement of the individuals within two dimensional habitats, into the growth equations for the prey and predator population. In order to have the noise added model, small amplitude heterogeneous perturbations to the linear intrinsic growth rates are introduced using uncorrelated Gaussian white noise terms. For the noise added system, we then observe spatial patterns for the parameter values lying outside the Turing instability region. With thorough numerical simulations we characterize the patterns corresponding to Turing and Turing-Hopf domain and study their dependence on different system parameters like noise-intensity, etc.

1. Introduction

Self-organizing spatial pattern formation in interacting population models is an important area of research to understand the distribution of the species within their habitats. Mainly reaction-diffusion equation models are studied analytically with supportive numerical simulations to understand the nature of population patches and whether they are changing with respect to time or not. Investigations for various types of Turing and non-Turing pattern formations in two dimensional prey-predator models have received significant attention [1, 2, 3, 4, 5]. Spatio-temporal prey-predator models exhibit various types of self-organizing spatial patterns; either

Journal of Physics: Conference Series **638** (2015) 012004 doi:10.1088/1742-6596/638/1/012004

they are stationary or non-stationary. Stationary patterns are characterized by heterogeneous distribution of populations over space which do not change with time once they reach the non-constant steady-state. On the other hand, the non-stationary patterns change with time continuously and never reach any non-constant steady-state. The stationary patterns produced by the spatio-temporal prey-predator models for two dimensional spatial domains can be classified as: (i) spot patterns (cold-spots and hot-spots), (ii) labyrinthine/stripe patterns, or (iii) the mixture of spot and stripe patterns. These three types of patterns are mostly observed for parameter values taken from Turing-domain or Turing-Hopf-domain, when the diffusivity of two species differ by a significant magnitude. Sometimes one does find labyrinthine patterns as a non-stationary pattern but for parameter values well inside the Turing-Hopf domain. The non-stationary patterns may include periodic patterns (sometimes as periodic traveling waves), interacting spiral patterns and chaotic patterns. The occurrences of spatio-temporal chaos for parameter values lying within the Turing domain or outside the Turing-domain remains a debatable and controversial issue [6]. There is no unified approach in determining the parameter values in a systematic way or in identifying the mechanism which leads to the spatio-temporal chaotic patterns [2, 7]. Derivation of necessary-sufficient condition(s), either in terms of parametric restrictions or system characterization for the onset of spatio-temporal chaotic patterns for prey-predator type interacting models thus remains an open problem.

Recently, the role of environmental noise on spatio-temporal pattern formation has been studied by the researchers for the noise added spatio-temporal models of interacting populations [8, 9, 10, 11]. In most of the cases, the authors have been used temporally correlated and spatially uncorrelated coloured noise terms. Although in some investigations, the alteration of spatial patterns due to the presence of additive coloured noise terms have been reported, this type of formulation is not ecologically viable as they correspond to spontaneous generation or inhibition of populations at the places with no population or very low concentration of population. On contrary, the consideration of multiplicative noise terms signify the noisy variation in some system parameters involved with the reaction kinetics. It is evident that due to the environmental fluctuations, the carrying capacity, birth and death rates, intensity of intra- and inter-species competition rates, predation rates exhibit random fluctuations to a modular extent around some average value [12]. Perturbation of one or more rate constants by the noise terms leads to spatio-temporal models with multiplicative noise terms. Several noise induced phenomena like noise-induced phase transitions, noise enhanced stability/instability, noise delayed extinction, stochastic resonance are investigated for the spatio-temporal ecological models considered within fluctuating environments. These analytical/numerical findings provide important information towards how the species establish within their respective habitats or help us to identify the important risk factors leading towards the extinction of species.

The main objective of this paper is to study a simple model of predator-prey with ratio-dependent functional response and density dependent death rate of predator, along with the self-diffusion terms corresponding to the random movement of the individuals within two dimensional habitats, and study the influence of small amplitude heterogeneous perturbations to the linear intrinsic growth rates.

2. The spatiotemporal model

Consider that $u \equiv u(x, y, t)$ and $v \equiv v(x, y, t)$ represent the prey and predator population densities at any time t and at the spatial location $(x, y) \in \Omega$, where $\Omega \in R^2$ is a square domain in $\mathbb{R}$ with boundary $\partial\Omega$. The spatiotemporal dynamics of prey-predator interaction with logistic growth for prey, ratio-dependent functional response and density dependent death rate for predators is governed by the following system of nonlinear coupled partial differential equations (PDE), in terms of dimensionless variables and parameters:

$$\frac{\partial u}{\partial t} = u(1-u) - \frac{\alpha uv}{u+v} + \nabla^2 u \equiv f(u,v) + \nabla^2 u, \quad (1)$$

$$\frac{\partial v}{\partial t} = \frac{\beta uv}{u+v} - \gamma v - \delta v^2 + d\nabla^2 v \equiv g(u,v) + \nabla^2 v \quad (2)$$

subjected to positive initial conditions:

$$u(x,y,0) \equiv u_0(x,y) > 0, v(x,y,0) \equiv v_0(x,y) > 0, \text{for all}(x,y) \in \Omega \quad (3)$$

and zero-flux boundary conditions:

$$\frac{\partial u}{\partial \nu} = \frac{\partial v}{\partial \nu} = 0, \text{ for all}(x,y) \in \partial\Omega, t > 0 \quad (4)$$

where ν is the outward drawn unit normal vector on the boundary, and α, β, γ and δ are positive dimensionless parameters. The ratio of diffusivity of two species is denoted as d. The diffusion terms in this reaction-diffusion system for the prey and predator population are introduced to take care of the random movements of the individuals within two dimensional habitats. The zero-flux boundary condition indicates that predator-prey system is self-contained within the two dimensional habitat with no population flux across the boundary.

2.1. Turing Bifurcation

The ecologically feasible and coexisting equlibrium point for the correspoding temporal model are spatially homogeneous steady state for the reaction diffusion model (1-2) and can be obtained by solving the following two algebric equations, with $u, v > 0$:

$$f(u,v) = u(1-u) - \frac{\alpha uv}{u+v} = 0, \quad (5)$$

$$g(u,v) = \frac{\beta uv}{u+v} - \gamma v - \delta v^2 = 0. \quad (6)$$

If we denote the coexisting equilibrium point by $E_*\,(u_*, v_*)$, then u_* and v_* satisfy the equtions $1-u-\frac{\alpha v}{u+v} = 0 = \frac{\beta u}{u+v}-\gamma-\delta v$. The explicit expression for the components of E_*, conditions for their feasible existence and possible number of coexisting steady-states are discussed in Ref. [7] extensively. Further we assume that E_* is coexisting homogeneous steady-state which is locally asymptotically stable for the temporal counterpart of the system (1-2), i.e., the conditions $a_{11} + a_{22} < 0$ and $a_{11}a_{22} - a_{12}a_{21} > 0$ are simultaneously satisfied, where a_{11}, a_{12}, a_{21} and a_{22} are given by

$$a_{11} = 1 - 2u_* - \frac{\alpha v_*^2}{(u_*+v_*)^2}, \quad a_{12} = -\frac{\alpha u_*^2}{(u_*+v_*)^2}, \quad a_{21} = \frac{\beta v_*^2}{(u_*+v_*)^2},$$

$$a_{22} = -\gamma - 2\delta v_* + \frac{\beta u_*^2}{(u_*+v_*)^2}.$$

The basic criteria for the diffusion-driven instability or Turing instability is that the stable homogeneous steady-state becomes unstable due to small amplitude heterogeneous perturbation around the homogeneous steady-state. Hence for the occurrence of Turing instability, first we need a locally stable coexisting steady-state for the corresponding temporal model. The linear stability analysis for spatiotemporal model (1-2) around the homogeneous steady-state E_*

and the conditions required for the onset of Turing-instability are well established and used by several researchers. Here, we just mention the three basic mathematical criteria required for Turing-instability of the model (1-2) around $E_*(u_*, v_*)$, as follows:

$$a_{11} + a_{22} < 0, \tag{7}$$
$$a_{11}a_{22} - a_{12}a_{21} > 0, \tag{8}$$
$$da_{11} + a_{22} > 2\sqrt{d}\sqrt{a_{11}a_{22} - a_{12}a_{21}}, \tag{9}$$

and Turing-instability sets in at the critical wave number

$$k_{cr}^2 = \frac{da_{11} + a_{22}}{2\sqrt{d}} > 0. \tag{10}$$

Satisfaction of the condition (9) implies homogeneous steady-state becomes unstable under small amplitude heterogeneous perturbation for wavenumbers $k > k_{cr}$ and Turing pattern emerges. Type(s) of emerging spatial pattern(s) (as mentioned earlier in the introduction), depend upon the specific choice of parametric values and magnitude of d.

The Turing-bifurcation curve is defined by the following equation

$$da_{11} + a_{22} = 2\sqrt{d}\sqrt{a_{11}a_{22} - a_{12}a_{21}}.$$

Solving this equation for d, we can find the critical threshold for the ratio of diffusivity d_{cr}, above which the Turing patterns emerge. In the next section we are going to report the varieties of spatial patterns generated by the system (1-2) for suitable choice of the parameter values satisfying the Turing bifurcation conditions.

3. Self-organizing spatial patterns

In this section, we present the numerical simulation results and varieties of spatial patterns that are exhibited by the spatiotemporal model (1-2) for parameter values within the Turing and Turing-Hopf domain. If we take the parameter values $\alpha = 2.0$, $\beta = 1$, $\gamma = 0.6$ and $\delta = 0.1$ then we find the unique coexisting equilibrium point $E_*(0.2287, 0.1436)$ which is a homogeneous steady-state for the system (1-2). E_* is locally asymptotically stable for the temporal counterpart as the conditions (7) and (8) are satisfied. In order to understand the role of diffusivity of the prey and the predator towards the destabilization of the homogeneous steady-state we need to find the quantity d_{cr} as mentioned in the previous section. For the chosen set of parameter values, we find $d_{cr} = 4.9157$ and hence the heterogeneous perturbations lead to Turing patterns for $d > d_{cr}$.

The existence and non-existence of Turing patterns are solely dependent upon the magnitude of the parameters involved with the model under consideration. Here we consider α and d as the bifurcation parameters to construct the Turing bifurcation diagram. We have presented the Turing bifurcation diagram in αd-parametric plane (see Fig. 1). The bifurcation diagram is prepared keeping $\beta = 1$, $\gamma = 0.6$ and $\delta = 0.1$ fixed. α and d are considered as the controlling parameters to obtain different spatiotemporal patterns. We have presented three bifurcation curves in the bifurcation diagram, namely Turing-bifurcation curve (blue curve), temporal Hopf-bifurcation curve (red dashed line) and temporal homoclinic bifurcation curve (black dotted curve). The coexisting equilibrium point E_* for the temporal model, corresponding to the system (1-2), is stable for $\alpha < \alpha_h$ and it looses stability through the Hopf-bifurcation at $\alpha_h = 2.01$ (approx). The Turing instability region is the region lying above the Turing bifurcation curve and which is divided into two parts by the vertical line $\alpha = \alpha_h$. Part of

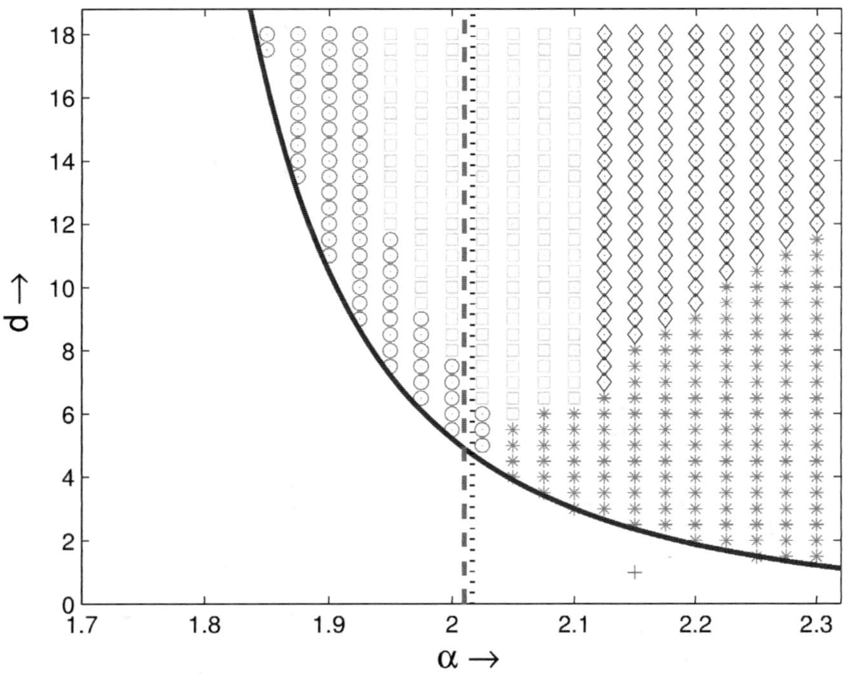

Figure 1. Five different types of spatial patterns observed for the parameter values lying within the Turing instability region (above the blue curve which is Turing bifurcation curve) and below Turing instability region. Choices of parameter values are marked with four different symbols based upon the resulting patterns: ○ cold-spot, □ mixture of spot-stripe, ◇ labyrinthine, ∗ chaotic and + interacting spiral. Hopf-bifurcation curve (shown by red dashed line) and temporal homoclinic bifurcation curve (shown by black dotted curve).

the Turing domain lying in the region $\alpha > \alpha_h$ is the Turing-Hopf domain where temporal and spatiotemporal perturbations are both unstable.

For the chosen set of parameter values we find only one feasible coexisting equilibrium point E_* therefore we will discuss only the spatial-pattern formation around this homogeneous steady-state. We have obtained various patterns generated by the model under consideration through exhaustive numerical simulations. All numerical simulations are carried out over a 200×200 lattice with time-step $\Delta t = 0.01$ and spatial steps $\Delta x = \Delta y = 2$. All numerical simulations are performed using the forward Euler method for the temporal part and five point finite difference scheme for the diffusion part. A small amplitude heterogeneous perturbations to the homogeneous steady-state are introduced by $u_0(x_i, y_j) = u_* + \epsilon\xi_{ij}$, $v_0(x_i, y_j) = v_* + \epsilon\eta_{ij}$ where $\epsilon = 0.001$, ξ_{ij} and η_{ij} $(i, j = 1, 2)$ are spatially uncorrelated Gaussian white noise terms. We have observed four different types of patterns within the Turing domain, cold-spot, labyrinthine, mixture of spot and stripe and spatiotemporal chaotic patterns.

We have observed only cold-spot and mixture of spot-stripe patterns for parameter values within the Turing domain, whereas mixture of spot-stripe, labyrinthine and chaotic patterns are observed for parameters lying in the Turing-Hopf domain. Numerical simulations have been performed for systematic choices of α and d within the Turing domain in order to understand the entire variety of spatiotemporal patterns for parameter values within the Turing domain. Obtained patterns for different combinations of α and d are marked with four different symbols.

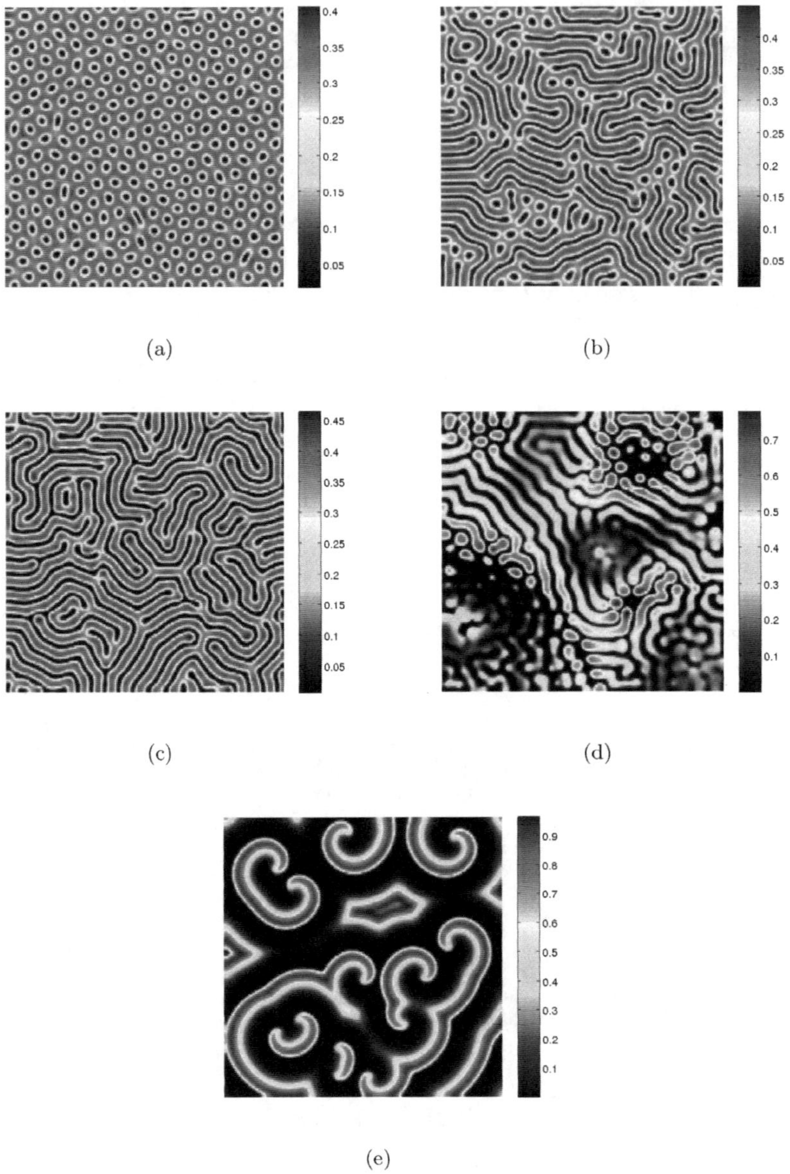

Figure 2. Spatial distribution of prey (u) obtained at $t = 1000$ for $\beta = 1$, $\gamma = 0.6$, $\delta = 0.1$, $d = 10.0$ and (a) $\alpha = 1.95$; (b) $\alpha = 2.05$; (c) $\alpha = 2.15$; (d) $\alpha = 2.30$; and (e) $d = 1.0$, $\alpha = 2.15$. The color bar in each panel represents the magnitude of u.

The whole regime of patterns in the Turing domain is shown in the Fig. 1.

A sample of each type of patterns observed within the Turing domain are shown in Fig. 2. At the lower panel we have presented the interacting spiral pattern which is observed for parameter values just outside the Turing-Hopf domain and for $d = 1.0$. As the Turing instability can not occur for $d = 1.0$, we can label this pattern as non-Turing pattern. To understand the stationary nature of the observed patterns, we have calculated the time evolution of spatial average for the prey and predator densities against time corresponding to the patterns reported in Fig. 2 and the results are presented in Fig. 3. The time evolution of $\langle u\rangle$ and $\langle v\rangle$ are considered over a longer period of time to ensure that the final observations are far away from the transient behavior. Looking at the time evolution of spatial averages presented at Fig. 3, one can easily understand that cold-spot, mixture of spot-stripe and labyrinthine patterns are stationary patterns whereas the interacting spiral pattern is a non-stationary pattern. The time evolution of the spatial averages are continuously oscillating and never settle down to any stationary value. Corresponding to the spatiotemporal chaotic pattern presented in Fig. 2(d), the time evolution of spatial averages exhibit irregular oscillations for $t > 650$ and the irregularity sustains at all future time. Here we have presented the results up to $t = 1000$ from the sake of brevity.

4. Noise Added Model

In this section we are interested to study the effect of noise on the spatiotemporal pattern formation for the model (1-2). For this purpose we introduce uncorrelated multiplicative white noise terms to the growth equations of both the prey and the predator populations, accordingly the noise added model is given by:

$$\frac{\partial u}{\partial t} = u(1-u) - \tfrac{\alpha uv}{u+v} + \nabla^2 u + \sigma_1 u \xi_1(t,x,y), \tag{11}$$

$$\frac{\partial v}{\partial t} = \tfrac{\beta uv}{u+v} - \gamma v - \delta v^2 + \nabla^2 v + \sigma_2 u \xi_2(t,x,y), \tag{12}$$

where $\xi_1(t,x,y)$ and $\xi_2(t,x,y)$ are two temporally as well as spatially uncorrelated Gaussian white noise terms. The averages and correlations of the noise functions are given by,

$$\langle \xi_1(t,x,y)\rangle = \langle \xi_2(t,x,y)\rangle = 0$$

$$\langle \xi_j(t,x,y)\,\xi_j(t_1,x_1,y_1)\rangle = 2\sigma_j \delta(t-t_1)\,\delta(x-x_1)\,\delta(y-y_1)\,, j=1,2,$$

where σ_1 and σ_2 are the intensities of the environmental driving forces. The introduction of the noise terms can be justified from ecological point of view also. Here we are considering the dimensionless model and hence we can see from (1-2) that the intrinsic growth rate of the prey is '1' and the intrinsic death rate for the predators is 'γ'. Now we assume that the intrinsic growth rate of the prey and the death rate of the predator population are subjected to environmental fluctuations and hence we can perturb them by spatiotemporal noise terms. As a result, if we consider the introduction of noise terms into the said rate constants, that is, $1 \to 1 + \sigma_1\xi_1(t,x,y)$ and $\gamma \to \gamma + \sigma_2\xi_2(t,x,y)$ then we find the noise added model (11) - (12) from the original spatiotemporal model (1-2).

In order to understand the effect of noise on the spatiotemporal pattern formation, we have performed numerical simulations with the changing magnitude of noise intensities. Here we consider the parameter set $\alpha = 2.15$, $\beta = 1$, $\gamma = 0.6$, $\delta = 0.1$, $d = 8.5$ and assume that both the noise intensities are same, that is $\sigma_1 = \sigma_2 = \sigma$. The numerical simulation is performed for $\sigma = 0.01$ and the resulting distribution of the prey population is shown at Fig. 4(b) along with the pattern observed for $\sigma = 0$ (see Fig. 4(a)). There is no significant difference in the reported patterns as the noise intensity is small. We have also calculated the spatial averages for the prey and predator populations up to $t = 3000$ for $\sigma = 0$ and $\sigma = 0.0010$ and the results are presented at Fig. 4(c)-(d). It is clear from these figures that the small noise does not change the system characteristics when the spatiotemporal patterns are oscillatory in nature. One can observe similar results for the parameter values corresponding to the stationary patterns and when noise intensities are not very high.

In order to support our claim, we have prepared a bifurcation diagram for the same set of parameter values, as mentioned above, and considering the ratio of diffusivity as the bifurcation

Journal of Physics: Conference Series **638** (2015) 012004 doi:10.1088/1742-6596/638/1/012004

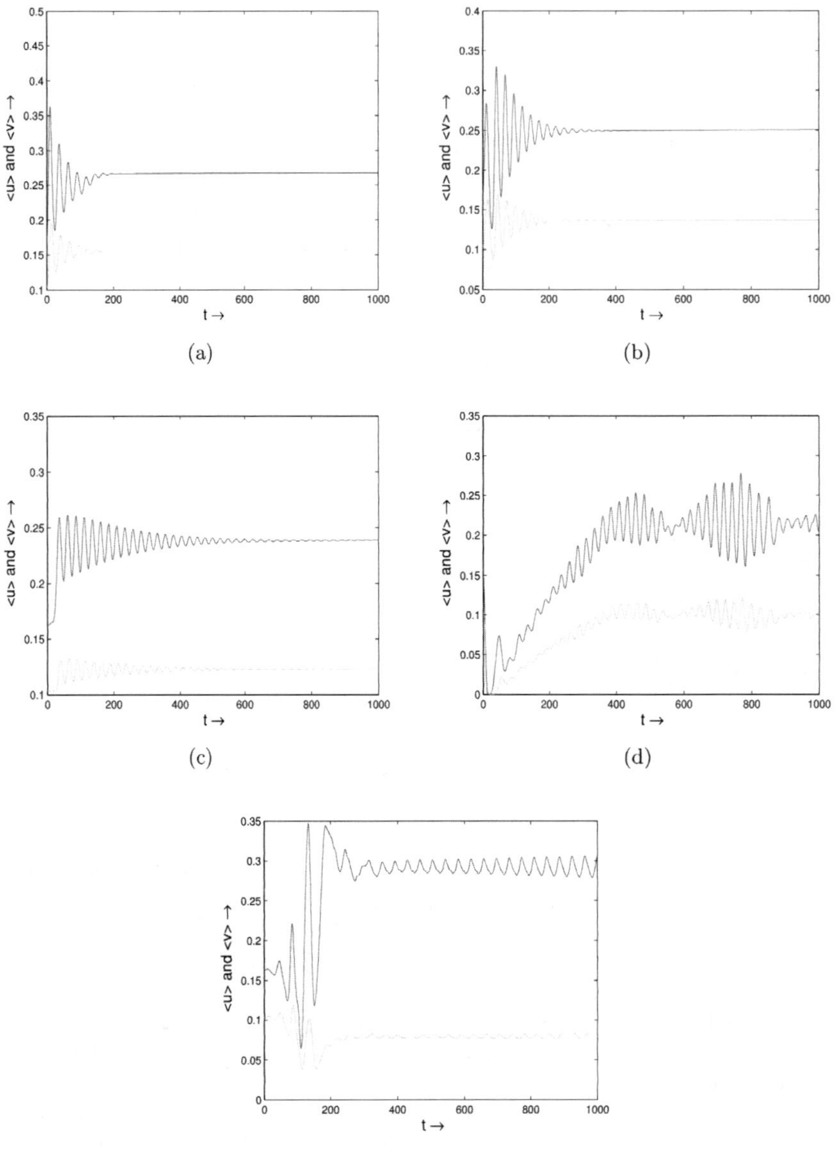

Figure 3. Time evolution of spatial averages of the prey $\langle u \rangle$ and predator $\langle v \rangle$ populations, presented in (dark) blue and and (light) green colors, respectively, for $t \in [0, 1000]$ corresponding to the patterns presented in Fig. 2.

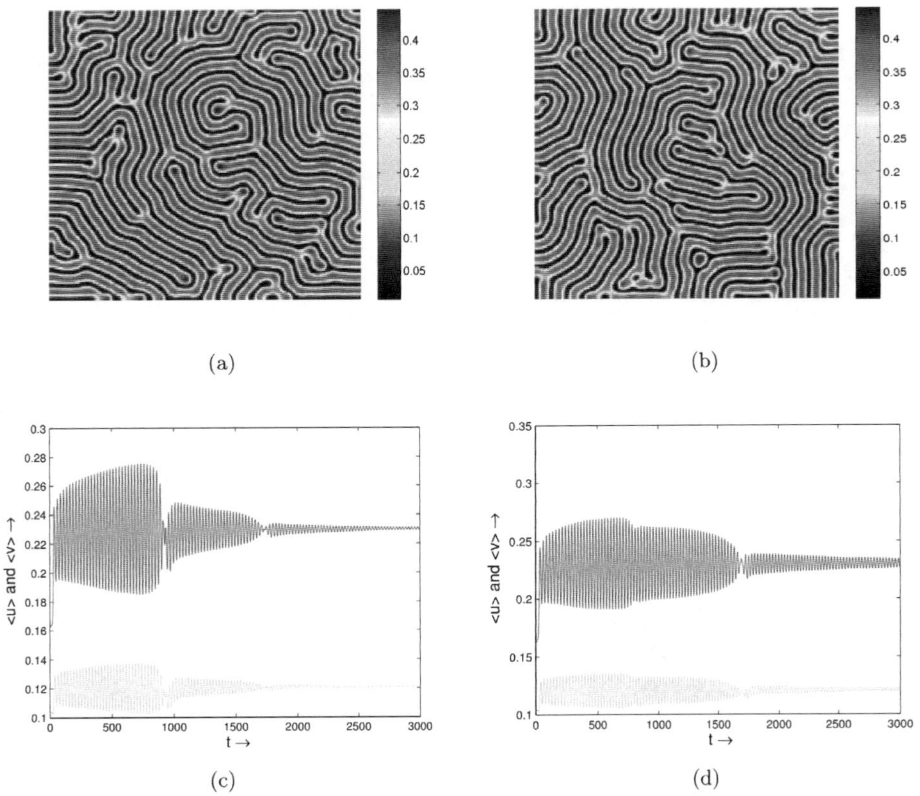

Figure 4. Spatial distribution of prey obtained at $t = 3000$ for $\alpha = 2.15, \beta = 1, \gamma = 0.6, \delta = 0.1$ and $d = 8.5$ (a) $\sigma = 0$ (b) $\sigma = 0.01$; the color bar in each panel represents the magnitude of u. Time evolution of spatial averages of prey $\langle u \rangle$ and predator $\langle v \rangle$ populations in (dark) blue and and (light) green colors, respectively, for (c) $\sigma = 0$, and (d) $\sigma = 0.01$.

parameter. The bifurcation diagrams are prepared for the spatiotemporal model (1-2) as well as for the noise added model (11) - (12) with $\sigma = 0.01$, see Fig. 5. We have performed the numerical simulations for $d = 1.0$ to $d = 11.0$ with increments of 0.5 and spatial averages are calculated for the prey population up to $t = 3000$. After discarding the initial transients, all local maxima and minima of the spatial average of the prey population are plotted against each values of 'd'. If we compare the bifurcation diagrams, then it is evident that the number of local maxima and minima increases with the introduction of small amplitude noise terms mainly within the parameter regime where we find non-stationary spatial patterns in the absence of noise. One natural question arises whether the bifurcation diagram presented at the right panel of Fig. 5 changes from one simulation to other. It does not change as we have taken the spatial average at each time step and the spatial domain is sufficiently large to avoid such kind of variation in the bifurcation diagram.

After understanding the effect of small intensity noise on the resulting patterns, it is important to understand what happens with the increasing magnitude of noise intensities. Of course, we can find some noisy patterns when the magnitude of σ is significantly large and some times it is difficult to identify the particular type of patterns they are representing. In order to avoid this type of confusion and appropriate understanding of the underlying phenomena, we have

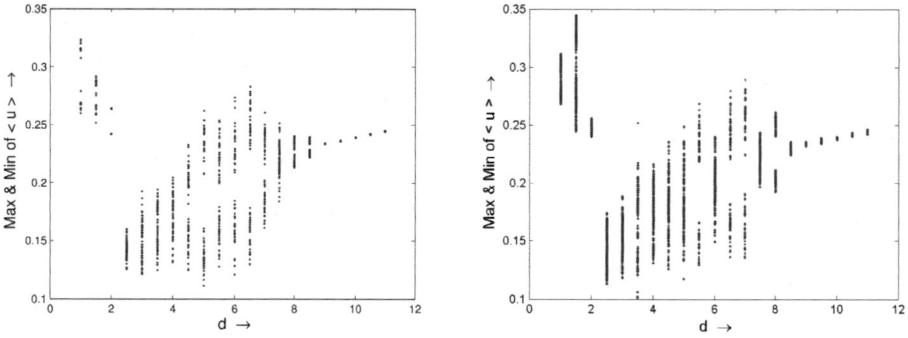

(a) (b)

Figure 5. Bifurcation diagram with respect to bifurcation parameter d with other parameters remaining same as in the previous figure for (a) $\sigma = 0$ and (b) $\sigma = 0.01$.

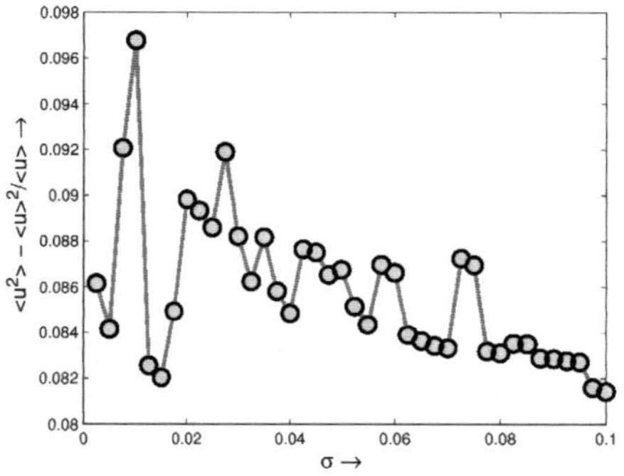

Figure 6. Signal to Noise ratio measure for $\alpha = 2.15$, $\beta = 1$, $\gamma = 0.6$, $\delta = 0.1$, $d = 8.5$.

calculated the signal-to-noise-ratio (SNR) against a range of noise intensities. Keeping $\alpha = 2.15$, $\beta = 1$, $\gamma = 0.6$, $\delta = 0.1$, $d = 8.5$ fixed, we have performed the numerical simulations for a range of values of σ. After discarding the initial transients, we have calculated the signal-to-noise-ratio measure $\frac{<u^2>-<u>^2}{<u>}$ for $u(t, x, y)$ obtained at $t = 3000$ and the average is taken over the entire spatial domain [13]. The plot of the SNR measure is shown at Fig. 6 when σ varies within the range $[0.0025, 0.1]$. From the figure, it is evident that the relative variance of the population is not a monotonic increasing function of the noise intensities. The environmental driving force can cause the dispersion of the populations from their average value but the internal mechanism can compensate to maintain the size of the population patches and as a result we find that the relative variance is an increasing function of the noise intensity.

5. Conclusion

In this paper, we have studied a predator-prey model with ratio-dependent functional response and density dependent death rate of predator, incorporated with the self-diffusion terms corresponding to the random movement of the individuals within two dimensional habitats. Extensive numerical simulations were performed to understand the regimes of patterns that the model exhibited within and outside of the Turing boundary. Typically, four types of patterns were observed within the Turing domain and interacting spiral pattern as non-Turing pattern.

Journal of Physics: Conference Series **638** (2015) 012004
doi:10.1088/1742-6596/638/1/012004

The effect of environmental driving forces (noise) were also studied. Interestingly, we observed that the small amplitude noise does not alter the stationary pattern; only the time taken to reach the stationary pattern increases. However, irregularity increases even with small noise intensities within the non-stationary regime. We observed that dispersion of population from their spatial average does not increase gradually with the noise intensities; rather noise and demographic interaction play crucial roles to determine the distribution of populations within their habitats, even during the adverse environmental conditions. Our next attempt will be to address the issues of parameter values within the spatiotemporal chaotic domain, and also investigate different noise scenarios such as $\sigma_1 \neq \sigma_2$.

Acknowledgments

AC acknowledges the financial support by institutional research funding IUT (IUT39-1) of the Estonian Ministry of Education and Research.

References

[1] Alonso D, Bartumeus F and Catalan J 2002 *Ecology* **83** 28–34
[2] Banerjee M and Petrovskii S 2011 *Theoretical Ecology* **4** 37–53
[3] Baurmann M, Gross T and Feudel U 2007 *Journal of Theoretical Biology* **245** 220–229
[4] Medvinsky A B, Petrovskii S V, Tikhonova I A, Malchow H and Li B L 2002 *SIAM Review* **44** 311–370
[5] Wang W, Liu Q X and Jin Z 2007 *Physical Review E* **75** 051913
[6] Malchow H, Petrovskii S V and Venturino E 2007 *Spatiotemporal Patterns In Ecology And Epidemiology: Theory, Models, And Simulation* (Chapman and Hall/CRC)
[7] Banerjee M and Banerjee S 2012 *Mathematical Biosciences* **236** 64–76
[8] Malchow H, Hilker F M, Petrovskii S V and Brauer K 2004 *Ecological Complexity* **1** 211–223
[9] Sun G Q, Jin Z, Li L and Liu Q X 2009 *Journal of Biological Physics* **35** 185–196
[10] Wang W, Li W, Li Z and Zhang H 2011 *Biosystems* **104** 48 – 56
[11] Li L and Jin Z 2012 *Nonlinear Dynamics* **67** 1737–1744
[12] May R M 1972 *Science* **177** 900–902
[13] Pikovsky A S and Kurths J 1997 *Physical Review Letters* **78** 775–778

Journal of Physics: Conference Series **638** (2015) 012005 doi:10.1088/1742-6596/638/1/012005

A finite variant of the Toom Model

Arvind Ayyer

Department of Mathematics,
Indian Institute of Science,
Bangalore - 560012, India

E-mail: arvind@math.iisc.ernet.in

Abstract. We present results for a finite variant of the one-dimensional Toom model with closed boundaries. We show that the steady state distribution is not of product form, but is nonetheless simple. In particular, we give explicit formulas for the densities and some nearest neighbour correlation functions. We also give exact results for eigenvalues and multiplicities of the transition matrix using the theory of $\mathscr{R}$-trivial monoids in joint work with A. Schilling, B. Steinberg and N. M. Thiéry.

1. Introduction

Dynamical variants of the Ising model have been of interest since Glauber's pioneering work [1] on the subject. Although there are many dynamical rules for the Ising model whose steady state distribution is Gibbsian, there are many other dynamical rules whose steady state distribution has very different structure, see for example [2, 3].

The model that we will be interested in is a probabilistic cellular automaton of Ising spins on $\mathbb{Z}^2$ due to A. Toom [4]. The dynamics is given by the so-called NEC (north-east-center) rule. Recall that cellular automata are discrete-time dynamical models, where the entire configuration is updated simultaneously. Denote the spin at the site (i,j) at integer time $t \in \mathbb{Z}_+$ as $\sigma_{i,j}(t)$, where $\sigma_{i,j}(t)$ can be $\pm$. The dynamics then evolves as follows.

$$\sigma_{i,j}(t+1) = \begin{cases} + & \text{with prob. p} \\ - & \text{with prob. q} \\ \text{sign}(\sigma_{i,j}(t) + \sigma_{i+1,j}(t) + \sigma_{i,j+1}(t)) & \text{with prob. 1-p-q} \end{cases}$$

To be more explicit, the spin at a given site at time $t+1$ evolves according to three spins at time t; itself, its north neighbour and its east neighbour. We illustrate the rule below, where we only show the case where $\sigma_{i,j}(t) = +$. The other case is similar. For clarity, we have circled the changing spin.

$$\begin{array}{cc} + & \\ \oplus & + \end{array} \xrightarrow{q} \begin{array}{cc} + & \\ \ominus & + \end{array}, \quad \begin{array}{cc} + & \\ \oplus & - \end{array} \xrightarrow{q} \begin{array}{cc} + & \\ \ominus & - \end{array}, \quad \begin{array}{cc} - & \\ \oplus & - \end{array} \xrightarrow{1-p} \begin{array}{cc} - & \\ \ominus & - \end{array}$$

Notice that the model becomes a deterministic NEC majority rule when $p = q = 0$. If $p \neq q$, the rule favours one spin over the other, which can be thought of as the effect of a magnetic field.

It turns out that the model exhibits a noise-dependent phase transition, which was demonstrated rigorously by Toom [4]. He showed that for sufficiently high p, q, there is a unique steady state, whereas for low enough p, q, there are at least two steady states.

Derrida, Lebowitz, Speer and Spohn [5, 6] considered this model on the third quadrant in $\mathbb{Z}^2$, with the boundary condition that all spins on the negative x-axis are $-$ and all on the negative y-axis are $+$. When $p = q = 0$, any configuration in which $+$'s and $-$'s are separated by a single interface of staircase shape is stationary, see Figure 1. They were interested in the dynamics of the interface for very small p, q. First, notice that in that case, all defects in the bulk of these phases are quickly cancelled out. However, the interface can and does fluctuate because of spontaneous sign changes of the $+$'s below and the $-$'s to the left of the interface. Figure 1 shows how this happens for a $+$ spin below the interface. Notice that the new interface is still of staircase shape.

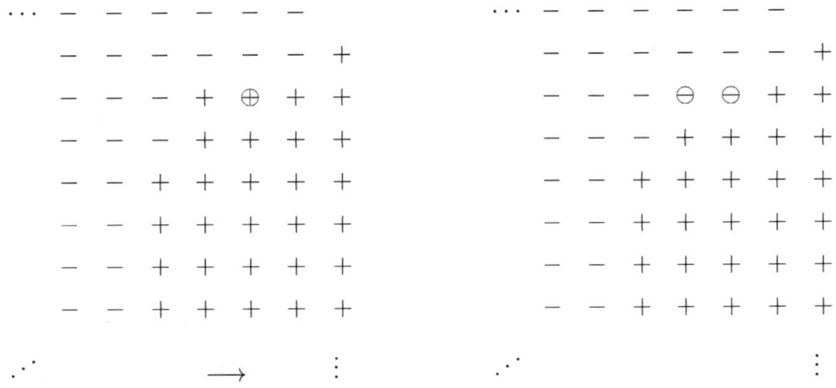

Figure 1. Dynamics in the PCA at low temperature in the third quadrant. The $+$ spin which will spontaneously flip is shown on the left as $\oplus$ and the spins which flipped to $-$ are shown in the right as $\ominus$.

Derrida, Lebowitz, Speer and Spohn [5, 6] understood the interface dynamics by thinking of the interface as itself an Ising spin configuration S on $\mathbb{Z}_+$ as follows. Look at the steps taken by the interface starting from the north-east corner. If the n'th step is vertical, set $S_n = +$, otherwise set $S_n = -$.

The transition shown in Figure 1 corresponds to the transition

$$+ - \underline{- - +} + - + + + + \cdots \quad \longrightarrow \quad + - \underline{+ - -} + - + + + + \cdots, \tag{1.1}$$

where the part undergoing the change has been underlined. In this one-dimensional model, each $\pm$ exchanges with the first $\mp$ on its right with rate $\lambda_\pm$. They also proposed to call this one-dimensional version the "Toom model".

However, there turns out to be a simplification of this model due to Lebowitz, Neuhauser and Ravishankar (LNR) [7]. Rather than allowing all spins to exchange, they only allow the leftmost spin in a block to exchange with the first opposite spin to its right. The transition in (1.1) would not be allowed in this model. An example of an allowed transition is

$$+\underline{---+}+-++++\cdots \quad\longrightarrow\quad +\underline{+---}+-++++\cdots. \tag{1.2}$$

Further, the first spin flips independently with a prescribed rate α. They studied the steady state distribution of this model when $\lambda_+ = \lambda_- = \alpha = 1$. The most interesting result there from our point of view is [7, Theorem 2] that the density at site n goes like $\dfrac{1}{2\sqrt{\pi n}}$.

It is this variant that we will refine. We consider the model on an interval of L sites with closed boundaries. We modify the LNR model by introducing two parameters α and β. In the original model [7], α and β were taken to be 1. We will also make the connection to exclusion processes explicit by replacing the spins $\pm$ by $0, 1$. Each site is then occupied either by a particle of type 0 or that of type 1. The dynamics is as follows: the leftmost particle in a block of particles of the same type can exchange with the opposite type particle to the right of the block. Clearly, the site preceding the block must be of opposite character if there is such a site. This exchange occurs with rate α if the block contains 0's and β if the block contains 1's. Note that the block size can be one. To recap,

$$\begin{array}{lll}
\underbrace{0\ldots0}_{k}1 \longrightarrow 1\underbrace{0\ldots0}_{k}, & \text{with rate } \alpha, & \multirow{2}{*}{\text{if the block begins at the first site, and}}\\
\underbrace{1\ldots1}_{k}0 \longrightarrow 0\underbrace{1\ldots1}_{k}, & \text{with rate } \beta, & \\
1\underbrace{0\ldots0}_{k}1 \longrightarrow 11\underbrace{0\ldots0}_{k}, & \text{with rate } \alpha, & \multirow{2}{*}{\text{if the block does not begin at the first site.}}\\
0\underbrace{1\ldots1}_{k}0 \longrightarrow 00\underbrace{1\ldots1}_{k}, & \text{with rate } \beta, &
\end{array} \tag{1.3}$$

An example of this dynamics is given with different notation in (1.2). The model is particle conserving and one defines it using integers (n_0, n_1), with $n_0 + n_1 = L$, to denote the number of holes and particles respectively. There are thus $\binom{n_0+n_1}{n_0}$ configurations. We will call this the (n_0, n_1)-system. The rules then are just given by (1.3) for any positive value of k. In particular, the rightmost block cannot by itself make any transitions. It is of course affected by the transitions involving the block immediately to its left.

Note that the presence of the boundary is crucial to having a nonequilibrium steady state. If we consider the model on a ring of L sites, then there is a unique absorbing set of configurations with all n_0 holes in one block and all n_1 particles in another block. These configurations are translations of one another, and the steady state is uniformly distributed on this small set of configurations. On the interval model, however, the boundary acts to break up the size of blocks. For example, in the $(3, 3)$ system, if one starts with the configuration 001110, then one can have a transition to the configuration 100110 with rate α. Then one has split the 1's from a block of size 3 to two blocks.

We will present results for this model in the rest of the paper. The proofs will be delegated to a subsequent publication [8]. In Section 2, we demonstrate the structure of the Markov matrices for the model. We will first give a recurrence for constructing the matrices, and give formulas for their eigenvalues. In Section 3, we present formulas for the density and other nearest neighbour correlations. We will also present a conjecture for the partition function.

2. The Markov Matrices

Here we will show that the Markov matrices for the model have a rich structure. Recall that the Markov matrix of the (n_0, n_1)-system is a matrix of size $\binom{n_0+n_1}{n_0}$ indexed by configurations of the system whose (i,j)'th entry is the rate of going from state j to state i. We will denote the matrix by M_{n_0,n_1}.

As an illustration, consider the example with $n_0 = n_1 = 2$. The configurations, in lexicographic order are: $(0011, 0101, 0110, 1001, 1010, 1100)$. The Markov matrix in this ordered basis is given by

$$M_{2,2} = \begin{pmatrix} -\alpha & \beta & \beta & 0 & 0 & 0 \\ 0 & -\beta-2\alpha & 0 & \beta & 0 & 0 \\ 0 & \alpha & -\beta-\alpha & 0 & \beta & \beta \\ \alpha & \alpha & 0 & -\beta-\alpha & \beta & 0 \\ 0 & 0 & \alpha & 0 & -2\beta-\alpha & 0 \\ 0 & 0 & 0 & \alpha & \alpha & -\beta \end{pmatrix} \tag{2.1}$$

The steady state distribution is the column (or right) eigenvector with eigenvalue 0. We will normalise it so that the sum of the entries is 1, i.e. it represents a probability distribution. We will denote the vector by π_{n_0,n_1}. For this case,

$$\pi_{2,2} = \left(\frac{\beta^2}{(\alpha+\beta)^2}, \frac{\alpha\beta^3}{(\alpha+\beta)^4}, \frac{\alpha^2\beta\,(\alpha+2\,\beta)}{(\alpha+\beta)^4}, \frac{\alpha\beta^2\,(2\,\alpha+\beta)}{(\alpha+\beta)^4}, \frac{\alpha^3\beta}{(\alpha+\beta)^4}, \frac{\alpha^2}{(\beta+\alpha)^2} \right) \tag{2.2}$$

Note the simple probabilities of configurations with just two clusters. We will also see later that M_{n_0,n_1} has simple formulas for eigenvalues and their multiplicities. The eigenvalues of $M_{2,2}$ are

$$\begin{array}{cl} 0 & \text{with multiplicity 1,} \\ -\alpha-\beta & \text{with multiplicity 4,} \\ -2\alpha-2\beta & \text{with multiplicity 1.} \end{array} \tag{2.3}$$

We will begin by explaining the recursive structure of the Markov matrices. To that end, we will first need to define some auxiliary matrices. Let $\mathbb{1}_k$ denote the $k \times k$ identity matrix and let A_{n_0,n_1} be a matrix of size $\binom{n_0+n_1-2}{n_1-1} \times \binom{n_0+n_1-1}{n_1}$ defined as a concatenation of identity matrices of increasing sizes which are justified at the top row as,

$$A_{n_0,n_1} = \begin{pmatrix} \mathbb{1}_{\binom{n_1-1}{n_1-1}} & \mathbb{1}_{\binom{n_1}{n_1-1}} & \cdots & \mathbb{1}_{\binom{n_0+n_1-2}{n_1-1}} \end{pmatrix}, \tag{2.4}$$

and the rest of whose entries are zero. For example,

$$A_{3,3} = \begin{pmatrix} 1 & 1 & 0 & 0 & 1 & 0 & 0 & 0 & 0 & 0 \\ 0 & 0 & 1 & 0 & 0 & 1 & 0 & 0 & 0 & 0 \\ 0 & 0 & 0 & 1 & 0 & 0 & 1 & 0 & 0 & 0 \\ 0 & 0 & 0 & 0 & 0 & 0 & 0 & 1 & 0 & 0 \\ 0 & 0 & 0 & 0 & 0 & 0 & 0 & 0 & 1 & 0 \\ 0 & 0 & 0 & 0 & 0 & 0 & 0 & 0 & 0 & 1 \end{pmatrix}.$$

Similarly, let B_{n_0,n_1} be a matrix of size $\binom{n_0+n_1-2}{n_0-1} \times \binom{n_0+n_1-1}{n_0}$ defined as a concatenation of identity matrices of increasing sizes which are justified at the bottom row,

$$B_{n_0,n_1} = \begin{pmatrix} \mathbb{1}_{\binom{n_0+n_1-2}{n_0-1}} & \cdots & \mathbb{1}_{\binom{n_0}{n_0-1}} & \mathbb{1}_{\binom{n_0-1}{n_0-1}} \end{pmatrix}, \tag{2.5}$$

and the rest of whose entries are zero. For example,

$$B_{3,3} = \begin{pmatrix} 1 & 0 & 0 & 0 & 0 & 0 & 0 & 0 & 0 & 0 \\ 0 & 1 & 0 & 0 & 0 & 0 & 0 & 0 & 0 & 0 \\ 0 & 0 & 1 & 0 & 0 & 0 & 0 & 0 & 0 & 0 \\ 0 & 0 & 0 & 1 & 0 & 0 & 1 & 0 & 0 & 0 \\ 0 & 0 & 0 & 0 & 1 & 0 & 0 & 1 & 0 & 0 \\ 0 & 0 & 0 & 0 & 0 & 1 & 0 & 0 & 1 & 1 \end{pmatrix}.$$

We will write down the Markov matrix in the lexicographically ordered basis of configurations, just as in the example (2.1). The transition matrix M_{n_0,n_1} can be expressed in terms of M_{n_0-1,n_1} and M_{n_0,n_1-1} using 2×2 blocks as,

$$M_{n_0,n_1} = \left(\begin{array}{c|c} M^{(0,0)} & M^{(0,1)} \\ \hline M^{(1,0)} & M^{(1,1)} \end{array} \right),$$

where each block can be expressed in terms of smaller blocks as

$$\begin{aligned} M^{(0,0)} &= M_{n_0-1,n_1} - \alpha \left(\frac{0}{A_{n_0,n_1}} \right), \\ M^{(0,1)} &= \beta \left(\frac{0}{B_{n_0,n_1}} \right), \\ M^{(1,0)} &= \alpha \left(\frac{A_{n_0,n_1}}{0} \right), \\ M^{(1,1)} &= M_{n_0,n_1-1} - \beta \left(\frac{B_{n_0,n_1}}{0} \right). \end{aligned}$$

Here 0 represents a zero matrix, A_{n_0,n_1} and B_{n_0,n_1} are defined in (2.4) and (2.5) respectively and with the initial condition $M_{0,n_1} = M_{n_0,0} = (0)$, the 1×1 zero matrix. The sizes of the blocks of zero matrices is fixed by the fact that $M^{(0,0)}$ and $M^{(1,1)}$ are square matrices of size $\binom{n_0+n_1}{n_1}$ and $\binom{n_0+n_1}{n_0}$ respectively. One can check that the matrix of the example at the beginning of this section in (2.1) can be recovered from the recursion above. The proof of this recursion follows from an elementary analysis of transitions in the ordered basis. For instance, $M^{(0,1)}$ represents all transitions from configurations that begin with a 1 to those that begin with a 0.

The deepest results that have been obtained about this model are on the eigenvalues and their multiplicities of the Markov matrices in joint work with A. Schilling, B. Steinberg and N. M. Thiéry [9]. The result there is more general, but it reduces to the following in our case. The characteristic polynomial of the transition matrix M_{n_0,n_1} is given by the explicit product formula

$$|\lambda \mathbb{1} - M_{n_0,n_1}| = \prod_{k=0}^{\min(n_0,n_1)} (\lambda + k(\alpha+\beta))^{\binom{n_0}{k}\binom{n_1}{k}}, \tag{2.6}$$

where the factor for $k = 0$ corresponds to the zero eigenvalue, which naturally occurs with multiplicity one. One can check that the eigenvalues for $n_0 = n_1 = 2$ in (2.6) match those given in (2.3).

The proof of (2.6) is highly nontrivial and has been obtained using techniques from the representation theory of $\mathscr{R}$-trivial monoids; another application of this theory is to a nonabelian model of sandpiles [10]. No elementary proof of (2.6) has been found so far.

Journal of Physics: Conference Series **638** (2015) 012005 doi:10.1088/1742-6596/638/1/012005

3. Correlation Functions

We will now state exact results for some correlation functions including the density. In some cases, the analysis is straightforward and will be explained, but the more complicated cases will be stated without proof. Detailed proofs will appear elsewhere [8].

We will use η_i to denote the occupation variable of the particle at site i and angle brackets $\langle\cdot\rangle$ to denote averages in the steady state distribution. As before, we will set $n_0 + n_1 = L$. First of all, note that the system satisfies an obvious particle-hole symmetry. Namely $\langle\eta_1, \dots, \eta_L\rangle$ in the (n_0, n_1)-system is equal to $\langle 1-\eta_1, \dots, 1-\eta_L\rangle$ in the (n_1, n_0)-system with α and β interchanged.

It turns out that the nature of explicit formulas for correlation functions is different for the sites less than both n_0 and n_1. Consequently, it will be useful to denote $n^* = \min(n_0, n_1)$. This is because looking at the conditional process on the first n^* sites is equivalent to looking at the first n^* sites in the semi-infinite system [7] because no matter how large a block of particles (holes) is within this subsystem, there is always a hole (particle) far enough to the right.

First of all, the density at the first site is given by

$$\langle\eta_1\rangle = \frac{\alpha}{\alpha+\beta}. \tag{3.1}$$

This is clear since there is a hole somewhere in the system for an exchange with the leftmost site (which is of course the leftmost site for the first block),

$$0 = \frac{d}{dt}\langle\eta_1\rangle = \alpha\langle 1-\eta_1\rangle - \beta\langle\eta_1\rangle.$$

With just a little more work, we can also show that the density in the first n^* sites is constant, i.e.,

$$\langle\eta_k\rangle = \frac{\alpha}{\alpha+\beta}, \tag{3.2}$$

for $1 \le k \le n^*$. To see this, suppose that $2 \le k \le n^*$, and consider the master equation for $\langle\eta_k\rangle$. First, we look at the outgoing contribution. If η is a configuration with $\eta_k = 1$ then, if $\eta_{k-1} = 1$ then η_k cannot change instantaneously. If $\eta_{k-1} = 0$, then η_k can become 0 in two ways: at rate α due to an exchange with η_{k-j} for some $j \ge 1$ and at rate β due to an exchange with η_{k+j} for some $j \ge 1$. Thus the total negative contribution to $\frac{d}{dt}\langle\eta_k\rangle$ is

$$-(\alpha+\beta)\langle(1-\eta_{k-1})\eta_k\rangle.$$

Similarly, if η is a configuration with $\eta_k = 0$ then, if $\eta_{k-1} = 0$ then η_k cannot change instantaneously. If $\eta_{k-1} = 1$ then η_k can become 1 in two ways: at rate β due to an exchange with η_{k-j} for some $j \ge 1$ and at rate α due to an exchange with η_{k+j} for some $j \ge 1$. Thus the total positive contribution to $\frac{d}{dt}\langle\eta_k\rangle$ is

$$(\alpha+\beta)\langle\eta_{k-1}(1-\eta_k)\rangle.$$

So

$$\begin{aligned}\frac{d}{dt}\langle\eta_k\rangle &= (\alpha+\beta)[-\langle(1-\eta_{k-1})\eta_k\rangle + \langle\eta_{k-1}(1-\eta_k)\rangle] \\ &= (\alpha+\beta)[\langle\eta_{k-1}\rangle - \langle\eta_k\rangle],\end{aligned} \tag{3.3}$$

and this must vanish in the steady state. Using the initial condition in (3.1) proves (3.2).

The reason the above proof does not go through if $k > n^*$ is that it assumes there is a particle of the opposite type far enough to the right of the block starting at k. When k is too large, such an assumption does not hold. The density at a general site is given by the more complicated formula,

$$\langle \eta_k \rangle = \frac{\beta^{k-n_1}}{(\alpha+\beta)^k} \left(\sum_{j=0}^{L-k} \binom{k-1}{n_1-j-1} \alpha^{n_1-j}\beta^j + \sum_{j=0}^{n_1-k-1} \binom{k}{j} \beta^{n_1-j}\alpha^j \right). \tag{3.4}$$

A little bit of manipulation shows that (3.4) reduces to (3.2) when $k \le n^*$. The proof of this result is more complicated and uses results for other correlations stated in (3.6) and (3.7).

For more complicated correlation functions such as block size distributions, we have some preliminary results. For $k \le n_1$,

$$\langle \eta_1 \dots \eta_k \rangle = \left(\frac{\alpha}{\alpha+\beta} \right)^k. \tag{3.5}$$

To see this, first consider the outgoing transitions for a configuration η with $\eta_1 = \cdots = \eta_k = 1$. Note that $\eta_2, \dots, \eta_k$ cannot change instantaneously. The only position which can effect a transition is the first one, where a hole to the right of k exchanges with the particle at the first site with rate β. The only incoming transitions to η are from configurations η' in which $\eta'_1 = \cdots = \eta'_{k-1} = 1$ and $\eta'_k = 0$ is nonzero. There is a transition to η at rate α which exchanges a particle to the right of k with the hole at the kth site. Therefore

$$0 = \frac{d}{dt}\langle \eta_1 \dots \eta_k \rangle = \alpha \langle \eta_1 \dots \eta_{k-1}(1-\eta_k) \rangle - \beta \langle \eta_1 \dots \eta_k \rangle,$$

which implies the recursion,

$$\langle \eta_1 \dots \eta_k \rangle = \frac{\alpha \langle \eta_1 \dots \eta_{k-1} \rangle}{\alpha+\beta}.$$

Using (3.1) as the initial condition gives the desired result.

Our final results are about correlations at a block of sites at the end of the system. The probability of having a block of $L+1-k$ 1's starting at position k is given by

$$\langle \eta_k \cdots \eta_L \rangle = \left(\frac{\beta}{\alpha+\beta} \right)^{k-1} \sum_{i=0}^{k-1-n_0} \binom{k-1}{i} \left(\frac{\alpha}{\beta} \right)^i. \tag{3.6}$$

The formula becomes much simpler if we also look at the joint correlation with a hole at site $k-1$.

$$\langle (1-\eta_{k-1})\eta_k \cdots \eta_L \rangle = \binom{k-2}{n_0-1} \frac{\alpha^{k-1-n_0}\beta^{n_0}}{(\alpha+\beta)^{k-1}}. \tag{3.7}$$

Note that both formulas (3.6) and (3.7) give 0 when we set $k = n_0$ as expected. These will be proved in a subsequent publication [8]. One can get the joint correlation of blocks of holes in the first and the last sites by interchanging α and β in (3.5), (3.6) and (3.7). We do not yet have exact formulas for block correlations in the bulk.

Steady state probabilities are rational functions of α and β, and hence there is no unique way to write them. Suppose we write the steady state probabilities such that the greatest common factor of the numerators is 1. This forces them to be written in a unique way. See for example, (2.2). We will call the *normalization factor*, denoted $Z(n_0, n_1)$, in the (n_0, n_1)-system to be the least common multiple of the denominators of the steady state probabilities when they are written in this way. The formula for the normalization factor $Z(n_0, n_1)$ seems to be tantalisingly clean,

$$Z(n_0, n_1) = (\alpha+\beta)^{n_0 n_1}. \tag{3.8}$$

When $n_0 = n_1 = 2$, this matches with (2.2). Although this formula seems quite simple, the proof has resisted the best of our efforts, and we leave it as a conjecture for now.

To conclude, let us compare some of our results with those of Lebowitz, Neuhauser and Ravishankar [7]. First of all, our model is on the finite one-dimensional lattice, whereas theirs is on $\mathbb{Z}_{\geq 0}$. We have a more general model in the sense that we have more parameters. Note that the ratio of α and β is the only extra parameter here, but in our generalised model [9], we have many more. We have explicit formulas for many correlation functions in (3.4),(3.5),(3.7),(3.6), and a conjecture for the partition function (3.8).

One can compare the densities in both models by performing asymptotics of (3.4). Of course, when $\alpha = \beta$ and k is large, the density at site k goes exactly like $\frac{1}{2\sqrt{\pi k}}$, exactly as in [7]. A preliminary analysis seems to indicate that the density decays like $1/k$ for all values of α/β.

Acknowledgements
This work is partially supported by UGC Centre for Advanced Studies. We thank J. L. Lebowitz and E. R. Speer for discussions during the initial stages of the project and an anonymous referee for helpful comments.

References

[1] Glauber R J 1963 *J. Mathematical Phys.* **4** 294–307
[2] Grinstein G, Lee D H and Sachdev S 1990 *Phys. Rev. Lett.* **64**(16) 1927–1930
[3] Garrido P L, Lebowitz J L, Maes C and Spohn H 1990 *Phys. Rev. A* **42**(4) 1954–1968
[4] Toom A L 1980 *Advances in Probability* **6** 12
[5] Derrida B, Lebowitz J L, Speer E R and Spohn H 1991 *Phys. Rev. Lett.* **67** 165–168
[6] Derrida B, Lebowitz J L, Speer E R and Spohn H 1991 *J. Phys. A* **24** 4805–4834
[7] Lebowitz J L, Neuhauser C and Ravishankar K 1996 *Stochastic Process. Appl.* **64** 187–208
[8] Ayyer A 2015 Exact results for a finite variant of the Toom model, in preparation
[9] Ayyer A, Schilling A, Steinberg B and Thiéry N M 2015 *International Journal of Algebra and Computation* pp. 1–63 arXiv:1401.4250
[10] Ayyer A, Schilling A, Steinberg B and Thiery N M 2015 *Communications in Mathematical Physics* arXiv:1305.1697

Journal of Physics: Conference Series **638** (2015) 012006 doi:10.1088/1742-6596/638/1/012006

Branching random walk with step size coming from a power law

Ayan Bhattacharya

E-mail: ayanbhattacharya.isi@gmail.com
Statistics and Mathematics Unit, Indian Statistical Institute, 203 B. T. Road, Kolkata 700108.

Rajat Subhra Hazra

E-mail: rajatmaths@gmail.com
Statistics and Mathematics Unit, Indian Statistical Institute, 203 B. T. Road, Kolkata 700108.

Parthanil Roy

E-mail: parthanil.roy@gmail.com
Statistics and Mathematics Unit, Indian Statistical Institute, 203 B. T. Road, Kolkata 700108.

Abstract. In their seminal work [15], Brunet and Derrida made predictions on the random point configurations associated with branching random walks. We shall discuss the limiting behavior of such point configurations when the displacement random variables come from a power law. In particular, we establish that two prediction of [15] remains valid in this setup and investigate various other issues mentioned in their paper.

1. Introduction

Branching random walk is a collection of particles that starts from a single particle situated at the origin, and branch and disperse independently of each other. It can be applied to model how a growing population (of bacteria, animals, particles, etc.) occupies an environment and hence is useful in biology, ecology, statistical physics, etc. In this article, we shall assume that the particles will branch in a supercritical fashion statisfying the Kesten-Stigum condition (see (2.1) below) and their displacements will follow a power law (see (2.2) below).

Branching random walk is of interest starting from [19], [20], [8]. Recently, its extreme value theory has gained prominence due to the connection to tree indexed random walk and Gaussian free field; see [12], [18, 1], [3], [22], [9, 10], [11]. See also [14, 13], [21], [6] for related results on extremes of branching Brownian motion. [15] predicted various properties of the limiting point configurations of branching random walks and branching Brownian motion. These predictions have been verified mathematically by [4, 5] and [2] for branching Brownian motion, and by [22] for branching random walks with light-tailed displacements. In this paper, we go beyond the usual setup of light-tailed displacements and announce the results when the step sizes follow a power law. For detailed discussions, technical background and proofs of the stated results in a slightly general setup, the readers are referred to [7].

Journal of Physics: Conference Series **638** (2015) 012006 doi:10.1088/1742-6596/638/1/012006

2. Notation and Model

We first give the description of a Galton-Watson tree which will form a building block for our model. Such a tree is obtained by the following random mechanism. Start with one particle and call it the root of the tree (denoted by o). This particle produces a random number of offsprings according to a discrete probability distribution and dies immediately. These particles form the first generation. Now each of these particle undergoes the same mechanism independently of each other and the past. The new particles form the second generation and this goes on. Let Z_i be the number of particles in the ith generation. We assume that $Z_1 \geq 1$, $\mu := \langle Z_1 \rangle \in (1, \infty)$ and the *Kesten-Stigum condition* holds, i.e.,

$$\langle Z_1 \log Z_1 \rangle < \infty. \tag{2.1}$$

It is well-known that under the above assumptions Z_n/μ^n is a martingale sequence that converges almost surely to a positive random variable W.

We denote this Galton-Watson tree by $\mathbb{T} = (V, E)$, where V and E denote the collection of vertices and edges, respectively. After obtaining this infinite tree, we assign independent and identically distributed edge random variables $\{X_{\mathbf{e}} : \mathbf{e} \in E\}$ (that are also independent of the Galton-Watson process $\{Z_i\}_{i \geq 0}$) on the edges satisfying the *power law*

$$\mathbf{P}(X_{\mathbf{e}} > x) \sim cx^{-\alpha} \tag{2.2}$$

as $x \to \infty$ for some constant $c > 0$ and $\alpha > 0$. To each vertex $\mathbf{v} \in V$, we attach a displacement random variable $S_{\mathbf{v}}$, which is the sum of all edge random variables on the unique geodesic path from the root o to the vertex $\mathbf{v}$. The size of this path is called the generation of $\mathbf{v}$ and is denoted by $|\mathbf{v}|$. Our goal is to find the limit of the random point configuration $\{b_n^{-1} S_{\mathbf{v}} : |\mathbf{v}| = n\}$ denoted by N_n, where $b_n = c^{1/\alpha} \mu^{n/\alpha}$ and verify that this limit honours the prediction of [15].

It is important to note that $\mu^{n/\alpha}$ is the right scaling for the weak convergence of the random point configuration of the nth generation displacement random variables because according to [17], it is the correct scaling for the maxima. If we scale by a sequence $b_n \gg \mu^{n/\alpha}$, then all the points will be killed by this normalization. On the other hand, if we scale by $b_n \gg \mu^{n/\alpha}$, then many of the points will diverge to infinity.

3. Main Results

Let $\{T_l\}_{l \geq 1}$ be a sequence of independently and identically distributed random variables with probability mass function

$$\gamma(y) = r \sum_{i=0}^{\infty} \frac{1}{\mu^i} \mathbf{P}(Z_i = y) \tag{3.1}$$

with $r = \mu/(\mu - 1)$, for $y \in \mathbb{N}$ independent of W. Let $\{j_l\}_{l \geq 1}$ be the collection of all Poisson points of the Poisson process on $(0, \infty]$ with intensity $\nu_\alpha(dx) = \alpha x^{-\alpha-1} dx$ independent of $\{T_l\}_{l \geq 1}$ and W. The point 0 is removed from the state space of the Poisson process because the intensity blows up near 0. Our main result says the limiting random point configuration is a Cox cluster process. The clusters appear in the limit because of the strong dependence among the nth generation displacement random variables.

Theorem 3.1. *N_n converges in distribution to a random point configuration N_* with $\{(rW)^{1/\alpha} j_l\}_{l=1}^{\infty}$ as the collection of points with the lth point repeating T_l many times.*

Clearly, the points of the limiting random point configuration arise as a combined effect of two kinds of randomness: one randomness is due to the displacement random variables and the other is coming from the underlying branching process. The clusters appear here due to the strong dependence structure of the displacement random variables $\{S_{\mathbf{v}} : |\mathbf{v}| = n\}$. The randomness in the intensity measure arises from the martingale limit W in contrast to the light-tailed case, where similar randomness comes from the derivative martingale limit. Note also that a W-mixture was already present in Theorem 1 of [17].

Recall that according to [15], the limiting random point configuration is a "decorated Poisson point process" and satisfies "superposability" property. In our setup, we can show that the limiting random point configuration is a *randomly scaled scale-decorated Poisson point process* and it satisfies the an analogue of superposability.

First we would like to explain what is meant by a scale-decorated Poisson process. Let $\mathcal{Q}$ (will be referred as scale-decoration) denotes a random point configuration on $[-\infty,\infty]\setminus\{0\}$. Let Λ (will be referred as underlying Poisson process) be a Poisson process on $(0,\infty)$ with intensity measure $m(dx)$ and $\{\lambda_i\}_{i\geq 1}$ be the collection of Poisson points of Λ. Consider independent copies $\{\mathcal{Q}_i\}_{i\geq 1}$ of $\mathcal{Q}$ such that the entire sequence is independent of Λ. We multiply each point of the random point configuration $\mathcal{Q}_i$ by λ_i and denote it by $\mathbf{S}_{\lambda_i}\mathcal{Q}_i$. The superposed point configuration $\cup\mathbf{S}_{\lambda_i}\mathcal{Q}_i$ is called a scale-decorated Poisson point process and we denote it by $ScDPPP(m,\mathcal{Q})$. We further multiply each point of this superposed point configuration by a positive random variable U (will be referred as scaling random variable) independent of both Λ and $\{\mathcal{Q}_i\}_{i\geq 1}$. The resultant random point configuration $\cup\mathbf{S}_{U\lambda_i}\mathcal{Q}_i$ is called a randomly scaled scale-decorated Poisson point process and is denoted by $SScDPPP(m,\mathcal{Q},U)$. The next result and the subsequent remark confirm that the predictions of [15] remains valid in our setup.

Theorem 3.2. *The limiting random point configuration N_* is a randomly scaled scale-decorated Poisson point process. The scale-decoration is the random point configuration consisting of* 1 *repeating T many times, where T has probability mass function* (3.1). *The intensity measure of the underlying Poisson process is $\nu_\alpha(dx)$. The scaling random variable is $(rW)^{1/\alpha}$, where $r=\mu/(\mu-1)$ and W is the martingale limit of the underlying branching process.*

Remark 3.3. If $Z_1\equiv d$, then the underlying tree is d-regular. In this case, we denote the limiting point configuration by $N_*^{(d)}$. Consider two independent copies $N_*^{(d,1)}$ and $N_*^{(d,2)}$ of $N_*^{(d)}$. Take two positive real numbers a_1 and a_2 satisfying $a_1^\alpha+a_2^\alpha=1$. Multiply each point $N_*^{(d,1)}$ by a_1, each point of $N_*^{(d,2)}$ by a_2, and superpose the scaled points. The new random point configuration will have the same distribution as that of $N_*^{(d)}$. This is the analogue of superposability enjoyed by our limiting random point configuration and it confirms a related prediction of [15] in our setup. In [16], this property has been investigated and referred to as α-stability of random point configurations.

3.1. Order statistics

The asymptotic distribution of order statistics of the displacement random variables attached to the vertices of the nth generation can be easily derived from Theorem 3.1. In particular, we rediscover a result of [17] on the limiting behaviour of the rightmost particle coming from the nth generation. Let $M_n^{(k)}$ denote the kth upper order statistic coming from the nth generation, and $G_n^{(k)}=M_n^{(k)}-M_n^{(k+1)}$ be the kth gap statistic. In order to study the asymptotic properties of these statistics, we need a few more notations as described below. We denote by π a partition of an integer l of the form $l=i_1y_1+i_2y_2+\cdots+i_{|\pi|}y_{|\pi|}$, where each i_j repeats y_j many times in the partition, and $i_1<i_2<\cdots<i_{|\pi|}$. Here $|\cdot|$ denotes the number of distinct elements in a partition. The set of all such partitions of the integer l is denoted by Π_l.

- **Property 1** (k^{th} upper order statistic) For all $x>0$, $\lim_{n\to\infty}\mathbf{P}\left(M_n^{(k)}\leq c^{1/\alpha}\mu^{n/\alpha}x\right)$

$$=\left\langle e^{-rWx^{-\alpha}}\right\rangle+\sum_{l=1}^{k-1}\sum_{\pi\in\Pi_l}\left\langle\prod_{j=1}^{|\pi|}\left(\left(rWx^{-\alpha}\gamma(i_j)\right)^{y_j}e^{-rWx^{-\alpha}\gamma(i_j)}\right)\right\rangle. \tag{3.2}$$

In particular,

$$\lim_{n\to\infty} \mathbf{P}\left(M_n^{(1)} \le c^{1/\alpha}\mu^{n/\alpha}x\right) = \phi(rx^{-\alpha}),\ \ x > 0,$$

where ϕ is the unique (up to a scale-change) completely monotone function on $\mathbb{R}^+$ satisfying $\phi(z) = f(\phi(z/\mu))$ with f being the probability generating function of the branching random variable Z_1.

- **Property 2** (Joint distribution of kth and $(k+1)^{th}$ upper order statistics) For all (u, v) such that $0 < u < v$,

$$\lim_{n\to\infty} \mathbf{P}(M_n^{(k+1)} \le c^{1/\alpha}\mu^{n/\alpha}u, M_n^{(k)} \le c^{1/\alpha}\mu^{n/\alpha}v) = \langle \xi_{0,(u,\infty]}(W)\rangle$$
$$+\sum_{j=1}^{k} \langle \xi_{0,(v,\infty]}(W)\, \xi_{j,(u,v]}(W)\rangle + \sum_{l=1}^{k-1}\sum_{j=0}^{k-l} \langle \xi_{l,(v,\infty]}(W)\, \xi_{j,(u,v]}(W)\rangle, \tag{3.3}$$

where for all $l \ge 0$ and for all $A \subset [-\infty,\infty] \setminus \{0\}$ such that $\nu_\alpha(A) < \infty$,

$$\xi_{l,A}(W) := \begin{cases} e^{-rW\nu_\alpha(A)} & \text{if } l = 0, \\ \displaystyle\sum_{\pi\in\Pi_l}\prod_{j=1}^{|\pi|}(rW\nu_\alpha(A)\gamma(i_j))^{y_j}\frac{1}{y_j!}e^{-rW\nu_\alpha(A)\gamma(i_j)} & \text{if } l \ge 1. \end{cases}$$

- **Property 3** (k^{th} gap statistic) Let $L : \mathbb{R}^+ \times \mathbb{R}^+ \to \mathbb{R}^+$ be the map $L(u, v) = v - u$. Then $\mathbf{P}(b_n^{-1}G_n^{(k)} \in \cdot) \to \zeta_k \circ L^{-1}$ where ζ_k is a probability measure on $\mathbb{R}^+ \times \mathbb{R}^+$ with joint cumulative distribution function (3.3).

The second term in (3.2) and the last term in (3.3) are both interpreted as zero when $k = 1$.

4. Final Remarks

We have been able to compute the asymptotic distribution of maxima and the random point configuration explicitly. Such explicit representations of the limiting point configuration is missing for light-tailed displacements. Of course, the heavy-tailed nature of the step sizes help us in obtaining such nice and useful results. This is because the extremes of the displacement random variables essentially come from one large value among the edge random variables. This intuition was already available in [17] for the rightmost particle coming from the nth generation. We had to implement this idea at the level of random point configurations and this is, by far, much more nontrivial than the result for the rightmost particle. The proof of Theorem 3.1 is based on a twofold truncation technique using multivariate extreme value theory. For details of the proof, see [7].

We have shown that the predictions in [15] remain valid in this set up with appropriate modifications. We also predict that even under moderate dependence among the heavy-tailed edge random variables coming out of the same vertex, the limiting random point configuration will satisfy superposability and can be represented as a randomly scaled scale-decorated Poisson point process.

Acknowledgement

We thank the anonymous referees for their suggestions.

References

[1] Addario-Berry L and Reed B 2009 *Ann. Prob.* **37** 1044–1079
[2] Aïdékon E, Berestycki J, Brunet É, and Shi Z 2013 *Probab. Theory Related Fields* **157** 405–451
[3] Aïdékon E 2013 *Ann. Prob.* **41** 1362–1426
[4] Arguin L.-P, Bovier A, and Kistler N 2012 *Ann. Appl. Probab.* **22** 1693–1711
[5] Arguin L.-P, Bovier A, and Kistler N 2013 *Probab. Theory Related Fields* **157** 535–574
[6] Arguin L.-P, Bovier A and Kistler N 2011 *Comm. Pure Appl. Math.* **64** 1647–1676
[7] Bhattacharya A, Hazra R. S, and Roy P November 2014 *arXiv:1411.5646*
[8] Biggins JD 1976 *Adv. Appl. Probab.* **8** 446–459
[9] Biskup M and Louidor O 2013 *arXiv:1306.2602*
[10] Biskup M and Louidor O 2014 *arXiv:1410.4676*
[11] Bramson M, Ding J and Zeitouni O 2013 *arXiv:1301.6669*
[12] Bramson M and Zeitouni O 2012 *Comm. Pure Appl. Math.* **65** 1–20
[13] Bramson M 1983 *Mem. Amer. Math. Soc.* **285** (Providence: Amer. Math. Soc.)
[14] Bramson M.D 1978 *Comm. Pure Appl. Math.* **31** 531–581
[15] Brunet É and Derrida B 2011 *J. Stat. Phys.* **143** 420–446
[16] Davydov Y, Molchanov I, and Zuyev S 2008 *Electron. J. Probab.* 1**3** 259–321
[17] Durrett R 1983 *Z. Wahrsch. Verw. Gebiete* **62** 165–170
[18] Hu Y and Shi Z 2009 *Ann. Prob.* **37** 742–789
[19] Hammersley JM 1974 *Ann. Prob.* 652–680
[20] Kingman J. F. C 1975 *Ann. Prob.* 790–801
[21] Lalley S and Sellke T 1987 *Ann. Prob.* 1052–1061
[22] Madaule T 2011 *arXiv:1107.2543*

Journal of Physics: Conference Series **638** (2015) 012007 doi:10.1088/1742-6596/638/1/012007

Meniscus arrest dominated imbibition front roughening in porous media with elongated pores

Heiko Rieger, Christian Thome and Zeinab Sadjadi

Theoretical Physics, Saarland University, D-66123 Saarbrücken, Germany

E-mail: h.rieger@physik.uni-saarland.de

Abstract. During spontaneous imbibition, a wetting liquid is drawn into a porous medium by capillary forces. Recently, anomalous scaling properties of front broadening during spontaneous imbibition of water in Vycor glass, a nanoporous medium, were reported. The mean height and the width of the propagating front increase with time t both proportional to $t^{1/2}$. We argue that this anomalously large roughening exponent of $\beta = 1/2$ is due to long-lasting meniscus arrests, when at pore junctions the meniscus propagation in one or more branches comes to a halt when the Laplace pressure of the meniscus exceeds the hydrostatic pressure within the junction. From this hypothesis we derive the scaling relations for the emerging arrest time distribution in random pore networks and show that the average front width is proportional to the height yielding a roughness exponent of exactly $\beta = 1/2$ as measured in the Vycor glass imbibition experiments. Extensive simulations of a random pore network model confirm these predictions. Finally, using a microfluidic setup as well as molecular dynamics simulations on the nanoscale, the basic hypothesis of the scaling theory is confirmed by demonstrating the existence of arrest events in Y-shaped junctions, analyzing them quantitatively and comparing them with the theoretical predictions.

1. Introduction

Fluid flow through porous media has been a topic of considerable interest due to its scientific importance and practical applications [1, 2, 3]. Imbibition, which is the displacement of one fluid by another immiscible fluid in a porous matrix, is observable in many everyday processes, for instance when dipping a sugar cube into coffee or dropping ink on a paper. It is important in nature (e.g., for water to flow through soil or to reach all parts of plants) and plays a fundamental role in various industrial areas, ranging from paper and textile treatment to oil recovery and groundwater hydrology [1, 3, 4]. Various physical aspects are involved in the process which makes it an interesting and complex phenomenon, e.g. capillarity, viscous drag, surface tension, hydrostatic pressure, and the quenched disorder of the matrix resulting in random permeability and random capillary pressure.

During imbibition the liquid-gas interface advances and broadens. The time evolution of the invading front follows simple scaling laws, which are independent of the micro- structure and the details of the fluid [5, 6, 7, 8, 9], reminiscent of the universality of critical phenomena. The often complex topology of the porous matrix induces local fluctuations in capillary pressures at the interface as well as hydraulic permeabilities in the bulk. Despite these complexities, the average position of the front $\langle h(t)\rangle$ during a purely spontaneous imbibition evolves as $\langle h(t)\rangle \propto t^{1/2}$, known as Lucas-Washburn law [11, 12]. This scaling behavior is valid down to nanoscopic pore scales [13, 14, 15].

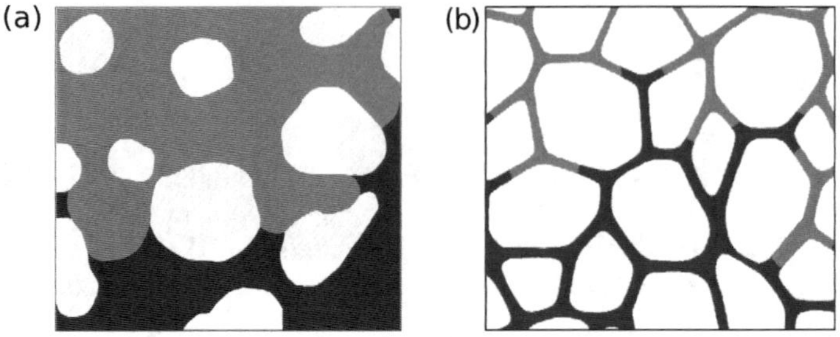

Figure 1. Schematic shape of the imbibition front in a porous structure consisting of (a) small and (b) large aspect ratio pores leading to a continuous liquid gas interface in (a) and isolated mensici in (b).

In most porous materials such as paper and sand, the propagating imbibition front forms a single connected interface between the propagating and the displaced liquid [16], as sketched in Fig. 1a. In these systems, pore space is laterally highly interconnected, resulting in a continuous liquid-gas interface, whose advancement is spatially correlated due to an effective surface tension [17]. Consequently, interface advancement beyond the average front position is slowed down while parts of the interface lagging behind are drawn forward. Hence, the roughening of the interface is reduced and gives rise to dynamical roughening behavior described by universal scaling laws [5, 6, 7, 8, 9].

For pore networks, in which the porous space forms a three-dimensional random network of interconnected pipes with a sufficiently large aspect ratio (ratio between pipe length and radius), the interface between the propagating and the displaced liquid is disconnected and consists of many isolated menisci [2, 18], see Fig.1b for a sketch. The characteristics of the imbibition dynamics of the ensemble of menisci in such pore networks can be expected to be very different from the propagation of continuous interfaces in random media.

Indeed, recent experiments on nanoporous Vycor glass (NVG) showed an anomalously fast front broadening and the absence of a growing correlation length [20]. In this paper we review the experimental situation and our current understanding of the dynamical process in these pore networks with elongated pores. The paper is organized as follows: In section 2 we summarize the experimental results on NVG, in section 3 we discuss the scaling theory for dynamical front broadening based on meniscus arrests at asymmetric pore junctions, section 4 presents the results of computer simulations of a microscopic pore network model involving the hypothesized meniscus arrests, and section 5 discusses experimental evidence for the meniscus arrests at asymmetric pore junctions.

2. Spontaneous imbibition in nano-porous Vycor glass NVG

In [20] the spontaneous imbibition of water into nanoporous Vycor glass (NVG), which is a silica substrate with an interconnected network of nanometer-sized, elongated pores [18] as sketched in Fig. 2a, was investigated. The narrow pores lead to capillary pressures of several hundred times atmospheric pressure. This means that gravity would only halt capillary rise after several kilometers and several billion years [21]. Hence, with this system, one is able to observe pure spontaneous imbibition over large length (centimeter) and long time (hours) scales. The observation of the advancing front is difficult since it is deeply buried inside the matrix, but neutron radiography allowed to image the liquid inside the porous material.

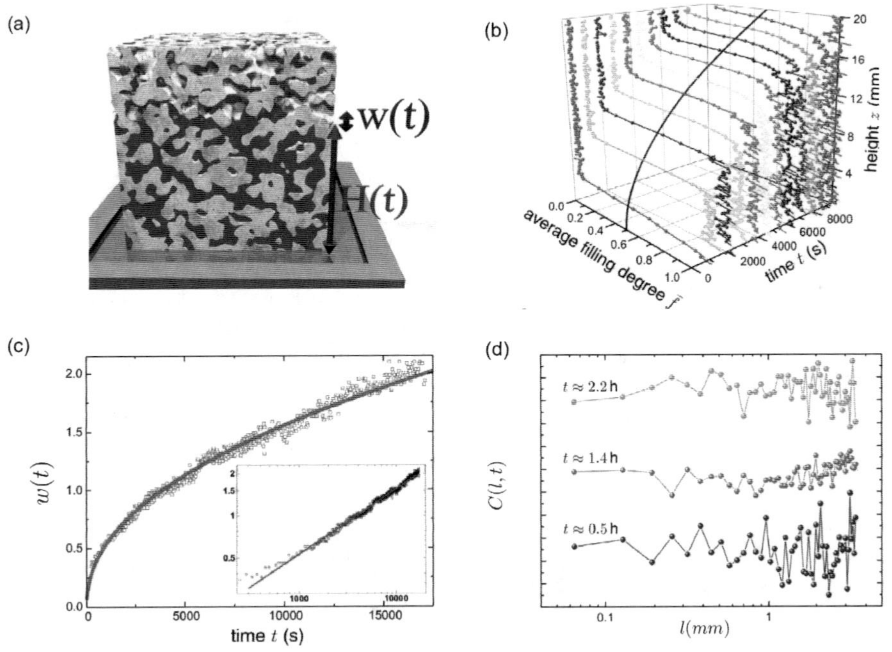

Figure 2. Spontaneous imbibition in NVG – experimental results (from [20]): a) Schematic representation of spontaneous imbibition of a fluid into a porous matrix like NVG. The arrows indicate the average rise level H(t) and the invasion front width w(t). b) Laterally averaged filling degree $\overline{f}(z,t)$ as a function of height z and time t. The Lucas-Washburn law $z \propto t^{1/2}$ is shown as solid line. c) Evolution of the front width w(t) along with a fit of $w \propto t^{\beta}$ (solid line). The inset shows the same data in a log-log representation. d) Height-height correlation function $C(l,t)$ of the invasion front at three different times, the data are shifted for clarity.

From neutron images [20] the spatial and temporal evolution of the local filling degree $0 \leq f(x,z,t) \leq 1$ are determined. Due to the projection in the y-direction, this is the average amount of filled pore space at lateral position x, height z and time t. Its lateral average, that is the vertical concentration profile $\bar{f}(z,t) \equiv \langle f(x,z,t) \rangle_{\mathrm{x}}$, is shown in Fig. 2b. The time-dependence of the front height, quantified by the mean median rise level $H(t) \equiv \langle z(f{=}0.5,x,t) \rangle_{\mathrm{x}}$, follows the Lucas-Washburn $\sqrt{t}$-law (Fig. 2b, solid lines). Fits of Gauss error functions to the profiles yield the time-dependence of the width $w(t)$ (Fig. 2c). The fit of $w(t) \propto t^{\beta}$ results in a growth exponent of the width or roughness, $\beta = 0.46 \pm 0.01$ (Fig. 2c, solid line). The value $\beta = 0.46$ significantly exceeds previous theoretical predictions, in particular those from phase-field models which are based on quenched, random fields. Such models predict slower roughening dynamics with $\beta \approx 0.19$ and a strong spatial correlation of the height fluctuations within the moving interface [10]. Note that β cannot be larger than 1/2 in NVG in particular and during spontaneous imbibition in general since otherwise the width would grow faster than than the height, which follows the Lucas-Wasburn law with exponent 1/2.

To investigate height fluctuations within the front [20] calculated the height-height correlation function:

$$C(\ell,t) = \langle (h(x,t) - h(x{+}\ell,t))^2 \rangle_{\mathrm{x}}^{1/2} \,. \quad (1)$$

The data shown in Fig. 2d exhibit neither scaling of $C(\ell,t)$ with ℓ nor any indication of spatial correlations in the experimentally accessible range 75 μm $\leq \ell \leq$ 4000 μm. Although the correlations are reduced due to the projection in y-direction, the absence of any detectable correlation is in contrast to all previously reported experiments and theories on imbibition front roughening.

3. Scaling theory

The existing theories for spontaneous imbibition [3] are unable to reconcile the large roughness exponent and in particular the absence of lateral correlations. The latter could lead us to the conclusion that the porous space of Vycor glass could be represented by an ensemble of independent pores of random but constant radius as sketched in Fig. 3a. Such an ensemble of capillaries exhibits a roughening exponent 1/2 during spontaneous imbibition since the meniscus heights evolve independently from one another ($h_i(t) = a_i\sqrt{t}$ with random pre-factors a_i). However, this model is inappropriate for Vycor glass since pores have corrugated walls and therefore the pore radii vary strongly along individual pores, as sketched in Fig. 3b. Spontaneous imbibition in an ensemble of independent pores with radii which vary randomly along their length is described by a random deposition model, for which the roughness scales as $\sqrt{\langle h(t)\rangle}$, i.e with a roughening exponent 1/4.

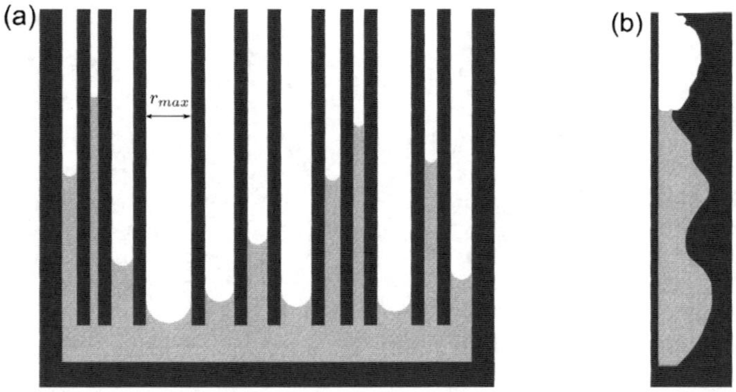

Figure 3. Sketch of independent pore models: (a) An ensemble of independent pores, each pore has a constant radius, radii of different pores vary randomly. During spontaneous imbibition the front width increases with $t^{1/2}$ (b) A pore with a corrugated wall, i.e. with a radius that varies along the pore axis. During spontaneous imbibition the front width increases with $t^{1/4}$.

For this reason independent pore models are inappropriate for NVG. Recently a lattice model for spontaneous imbibition in a silica aerogel predicted a roughening exponent $\beta \approx 1/2$. Silica aerogels have an extremely large porosity of 87–95 % which gives rise to a continuous liquid-gas interface. Consequently one expects here an effective surface tension to be present, inducing height-height correlations in the advancing imbibition front. The model details are thus also not appropriate for NVG, which must involve pore and network aspects.

With this background we proposed in [22] a scaling theory for spontaneous imbibition in porous media consisting of a network of interconnected elongated pores (Fig. 4). It is based on the observation that at pore junctions the meniscus propagation in the branch with the larger radius can come to a halt when the Laplace pressure of the meniscus exceeds the hydrostatic pressure within the junction. This leads to the emergence of voids behind the invasion front and concomitantly to anomalously fast front broadening as observed experimentally in NVG [20].

We consider a network of elongated pores with a length-to-width ratio of the order of 10, i.e. elongated, cylinder-like pores with random radii interconnected at pore junctions as sketched in Fig. 4. The bottom pores are connected to a liquid reservoir with pressure $P = 0$. We

Journal of Physics: Conference Series **638** (2015) 012007 doi:10.1088/1742-6596/638/1/012007

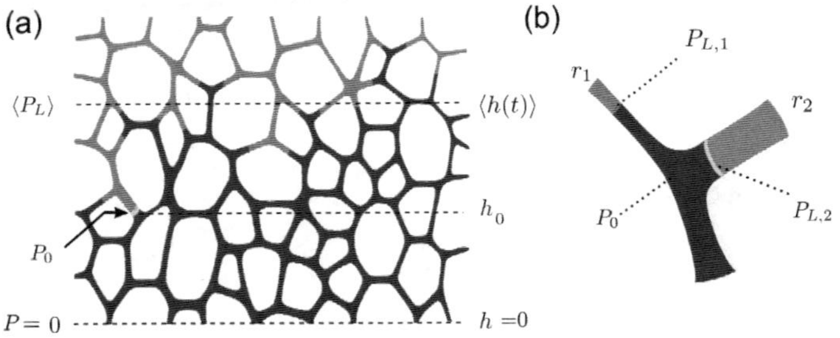

Figure 4. Sketch of a pore network with elongated pores (a) and an individual junction (b): r_i and $P_{L,i}$ denote the radius and Laplace pressure, respectively, in pore i, and p_0 denotes the hydrostatic pressure in the junction. In (b) $\langle h(t)\rangle$ and $\langle P_L\rangle$ denote the average height at time t and the average Laplace pressure.

assume that in each pore a liquid-gas interface forms, denoted as meniscus, that gives rise to a Laplace-pressure $P_L{=}-2\sigma/r$, where σ is the surface tension of the liquid and r the pore radius. If the pore radii vary between $r_{\rm min}$ and $r_{\rm max}$, the average radius is denoted by $\langle r\rangle$. Then, on large scales, the average height is expected to vary as $d/dt\,\langle h(t)\rangle = -\langle P_L\rangle/\langle h(t)\rangle$, which implies the Lucas-Washburn law $\langle h(t)\rangle \propto t^{1/2}$.

Consider now a junction at height h_0, where a pore branches into two (see Fig.4a). One branch has radius r_1, the other $r_2 > r_1$, yielding the Laplace pressures $P_{L,i}{=}-2\sigma/r_i$. Let P_0 be the hydrostatic pressure within the junction. As long as $P_{L,2} > P_0$ the meniscus in branch 2 is arrested. In the following we will answer the question how long the meniscus in branch 2 will be arrested and we will implicitly assume that it does not get annihilated by the filling of the pore from its other end. This means that we assume the radius r_2 also to be larger than the radius of the other branch of the junction of the other end. This reduces only the probability of this event by a r_2-dependent factor.

$P_0 = P_0(t)$ is a function of time and depends on how far the front has propagated and can be estimated as follows: Let the average front height be $\langle h(t)\rangle$. On average one expects the bulk pressure to decrease linearly from bottom to top:

$$P(\langle h(t)\rangle)/P_0 = \langle h(t)\rangle/h_0 \tag{2}$$

Therefore, with $P(\langle h(t)\rangle) = \langle P_L\rangle = -2\sigma\langle 1/r\rangle$ the average Laplace pressure, one obtains $P_0 = -2\sigma\langle 1/r\rangle \cdot h_0/\langle h(t)\rangle$ and the condition $P_0 = P_{L,2}$ for the arrested meniscus to resume propagation (at time $t_{\rm resume}$) reads

$$\langle h(t_{\rm resume})\rangle = h_0 r_2 \langle 1/r\rangle\ . \tag{3}$$

This equation has far reaching consequences:
1) The larger r_2 the longer the meniscus is arrested, and the average height that the front has to reach before the meniscus resumes propagation is proportional to the height where it stopped with a proportionality constant larger than one.
2) The time τ for which the meniscus is arrested is proportional to the time $t_{\rm stop}$, when it stopped

$$\tau \propto t_{\rm stop}\ . \tag{4}$$

To see this we note that with (3) one has $\langle h(t_{\rm stop}+\tau)\rangle = h(t_{\rm stop}) r_2\langle 1/r\rangle$. With Lucas-Washburn $\langle h(t_{\rm stop}+\tau)\rangle \propto (t_{\rm stop}+\tau)^{1/2}$ and assuming that $h(t_{\rm stop}) \propto t_{\rm stop}^{1/2}$, too, for the relation between the height and the time when the considered meniscus stopped, one obtains (4).
3) Consequently from (4)

$$\tau \propto h^2(t_{\rm stop}) = h_0^2;\,, \tag{5}$$

which implies that the probability distribution of arrest times for menisci arrested at height h will scale as

$$p_h(\tau) = h^{-2}\,\tilde{p}(\tau/h^2)\;. \tag{6}$$

4) The height difference $w_0(t_{\text{resume}}) = \langle h(t_{\text{resume}})\rangle - h_0$ is a measure for the local width of the propagation front (at the lateral coordinates of the position of the arrested meniscus) at time t_{resume}. The ratio of this local width and the average height is $\frac{w_0(t_{\text{resume}})}{\langle h(t_{\text{resume}})\rangle} = 1-(r_2\langle 1/r\rangle)^{-1}$, which is independent of the time t_{resume}. Thus all arrested menisci will contribute a time independent amount to the ratio of the average width $w(t)$ and average height. Since the width cannot grow faster than $h(t)$ this implies

$$w(t)/\langle h(t)\rangle = \text{const.}\;, \tag{7}$$

implying $w(t) \propto t^{1/2}$, i.e. a roughening exponent $\beta = 1/2$. Note that the invasion front dynamics is now expected to be be completely determined by the meniscus arrests, which in turn depend exclusively on the pore radii distribution and the height dependent hydrostatic pressure. Consequently one expects no lateral correlations in the meniscus heights to emerge, as observed in [20].

4. Microscopic model - computer simulations

The scaling theory presented in the last section neglects all geometric and topological details of a pore network. To test its predictions, we analyzed in [20, 22] the following microscopic model for spontaneous imbibition in a pore network with elongated pores [24, 23]: A two-dimensional square lattice of cylindrical capillaries inclined at 45° is considered, which consists of N_x and N_z nodes in horizontal and vertical directions, respectively. Capillaries, interconnected at nodes, have the same length L and random radii uniformly distributed over $[r_{\text{av}} - \delta, r_{\text{av}} + \delta]$. The average aspect ratio $2r_{\text{av}}/L$ is set to 5. The pressure at the bottom nodes attached to the liquid reservoir is set to zero, the pressure at a moving meniscus is the Laplace pressure. Here we neglect gravity, which is justified as long as capillary forces are much larger than gravitational forces $2\sigma/r \gg \rho N_z L$, where ρ the specific weight of the liquid. This is the case for instance in experiments with NVG [21].

The dynamical evolution of the meniscus positions is driven by the hydrostatic pressures at the network nodes, which themselves depend on the meniscus positions. The node pressures P_i are determined by the boundary conditions plus the conservation of volume flux at each node: $\sum_j Q_{ij} = 0$, with Q_{ij} the volume flux from node i into the capillary j attached to it. According to Hagen-Poiseuille's Law $Q_{ij} = -c_{ij}\,\Delta P_{ij}/h_{ij}$, with $c_{ij}{=}\pi r_{ij}^4/8\eta$ and $\Delta P_{ij}{=}P_i{-}P_{L,ij}$. Here, r_{ij}, h_{ij} and $P_{L,ij}$ are the radius, the length and the Laplace pressure of the meniscus in capillary j attached to node i, respectively, and η is the viscosity of the liquid. The numerically calculated values for P_i and Q_{ij} are then inserted into the equation of motion for the heights given by $Q_{ij}{=}\pi r_{ij}^2\,dh_{ij}/dt$, which is then numerically integrated with an implicit Euler scheme with variable time step.

When a meniscus reaches the end of a capillary it immediately moves an infinitesimal distance $\delta{\simeq}0.01L$ into the adjacent capillaries, creating new menisci, as shown in Fig. 5a. When two menisci meet, they vanish, thus the capillary is entirely filled. If, due to a negative pressure difference, a meniscus retracts, it proceeds backward as long as its distance from the back node is larger than δ [see Fig. 5b]. When it reaches δ, the meniscus is stuck there until the driving pressure difference is again positive.

Fig. 5c shows a series of snapshots of the dynamical evolution of the meniscus positions and the filled pore area obtained from for a specific realization of random pore radii. The average front height $h(t) = \langle h_i(t)\rangle$ and average front width $w(t) = \langle (h_i(t) - h(t))^2\rangle^{1/2}$ are obtained by averaging over all meniscus heights h_i at time t and over different disorder (radii) realizations ($N = 100$). The result is shown in Fig.5d-f: We find that the average front height obeys the

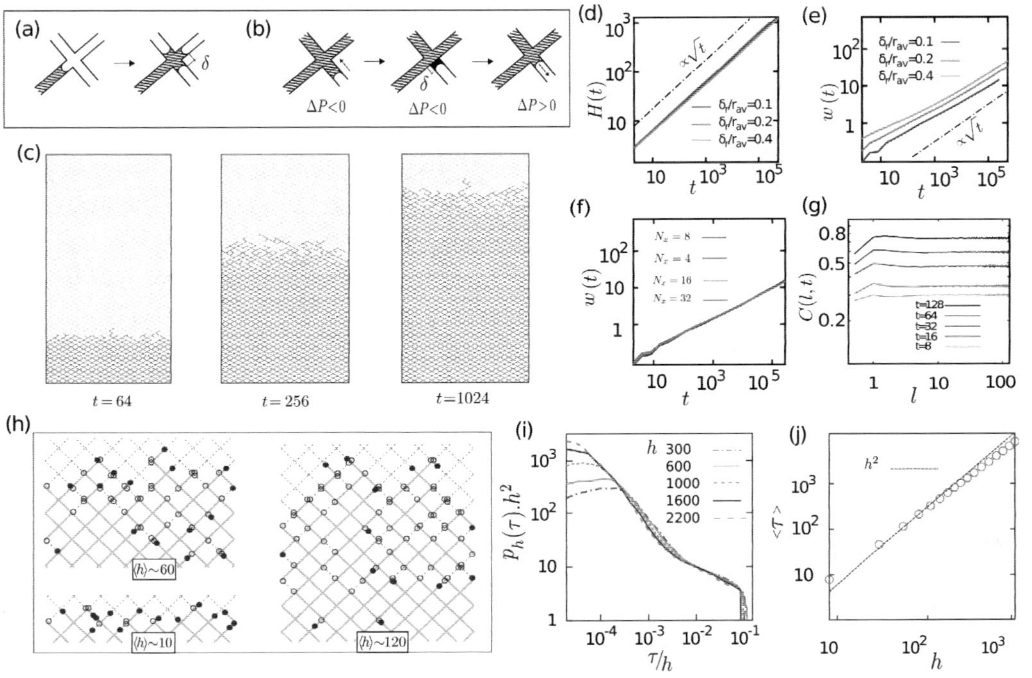

Figure 5. Micsocopic pore network model: a) After reaching an empty node, the liquid immediately fills the connected capillaries for a distance δ. b) After retracting up to δ toward a filled node, the meniscus is arrested until the pressure difference driving the liquid is again positive. c) Snapshots of meniscus configurations in a system with radius disorder strength $\delta r/\langle r\rangle = 0.3$ and lattice size 16×1000 at three different times (as indicated in units of nanoseconds). d) Evolution of the average front height $H(t)$ in a log-log plot for different radius disorder strengths $\delta r/\langle r\rangle$. The broken line indicates the Lucas-Wasburn law $H(t) \propto t^{1/2}$. e) $w(t)$ for different radius disorder strengths $\delta r/\langle r\rangle$ for a pore network with $N_x = 16$ in a log-log plot. The broken line indicates the time dependence $w(t) \propto t^{1/2}$ predicted by the scaling theory described in section 3. f) Evolution of the front width $w(t)$ for different lateral system sizes N_x with $\delta r/\langle r\rangle = 0.1$ in a log-log plot. g) Height-height correlation function $C(l,t)$ as a function of the distance l at different times t (as indicated) for a sample with $\delta r/\langle r\rangle = 0.1$ and lattice size 128×32. h) Snapshots of the arrested (empty circles) and advancing (filled circles) menisci in the invasion front at three different times in a pore-network. Broken (full) lines represent empty (full) pores, $\langle h\rangle$ is the average height at the corresponding time. The filled circles represent moving menisci, the open circles indicate arrested menisci. i) Scaling plot of the meniscus arrest time distribution $P_h(\tau)$ according to Eq. 6. $\delta r/\langle r\rangle = 0.1$ and $N_x = 8$. j) Average meniscus arrest time $\langle\tau\rangle$ at a given height h for $\delta r/\langle r\rangle = 0.1$ and $N_x = 8$ in a log-log plot. The dashed line is proportional to h^2.

Lucas-Washburn $\sqrt{t}$-behavior (Fig. 5d) and that the width increases rapidly as $w(t) \propto t^\beta$ with β close to 1/2 (Fig. 5e), consistent with the scaling prediction eq.(7).

Remarkably the time evolution of the average width does not display any dependence on the lateral system size, as seen in Fig. 5f. The absence of finite size effects is an indicator of a small or vanishing correlation length scale for height-height fluctuations. Within the framework of the scaling theory of roughening [25, 26], the interface width in a finite system of lateral size N_x is expected to grow algebraically $w(t) \sim t^\beta$ as long as the system size is larger than the correlation length, $\xi(t)/L \ll N_x$, and only saturates to a constant $w(t) \approx$ const. as soon as the correlation length exceeds the system $\xi(t)/L \gg N_x$. Since the data in Fig. 5f display an algebraic growth even for the smallest system size $N_x = 4$ the correlation length must be smaller than $4L$. This is confirmed by the height-height correlation function $C(\ell, t)$ (eq. 1, Fig. 5g), which saturates quickly (around $\ell \approx L$). Scaling theory [25, 26] predicts saturation of $C(\ell, t)$ for $\ell \gg \xi(t)$, which implies $\xi(t)/L = \mathcal{O}(1)$ independent of time t. This finding is also in agreement with the experimental result for $C(\ell, t)$ in NVG (Fig. 2d).

Our scaling theory presented in the last section explains the origin of the large roughness exponent $\beta = 1/2$ and the absence of lateral height-height correlations, namely long lasting meniscus arrests with arrest times that are of the same order as the the time when the arrest started. Fig. 5h shows snapshots of arrested and moving menisci in our model and one sees that with progressing time most menisci are actually arrested and only a few move. We analyzed the probability distribution of the arrest times $P(\tau)$ in our model [22] and found that indeed the intermediate and large time regime of the distribution, including the cut-off, scale nicely with h^2 as predicted by Eq.(6), as is shown in Fig. 5i. Consequently the average arrest time $\langle\tau\rangle$ grows with time as $h^2(t)$ as shown in Fig.5j.

The invasion front thus involves a finite fraction of the occupied volume and comprises connected clusters of empty pores whose size distribution gets broader with increasing time. Based on the conditions for meniscus arrests presented above one can derive a scaling form for the distribution of (void) cluster sizes [22].

5. Meniscus arrest in asymmetric pore junctions

The basic assumption of the scaling theory for spontaneous imbibition in porous media with elongated pores described in section 3 as well as the microscopic pore model presented in section 4 are long-lasting meniscus arrests in asymmetric pore junctions. Since such meniscus arrests have not been studied quantitatively before, we investigated in [27] the capillary rise of a fluid in a Y-shaped junction of channels with different radii experimentally and theoretically. We consider here a channel junction geometry as sketched in Fig. 6a, with cylindrical channels of radius r_f for the feeding channel that is connected to the liquid reservoir, and radius r_n and r_w for the narrow and wide branch, respectively. A simplified model for the capillary rise in this asymmetric Y-junction predicts that once the meniscus reaches the junction and splits into two menisci, the meniscus in the wide branch proceeds to propagate whereas the meniscus in the narrow channel is arrested until the meniscus in the wide channel reaches the distance l_{resume} from the junction, which is proportional to l_0, the length of the feeding channel and given by [27]:

$$l_{\text{resume}} = l_0 \cdot \alpha \quad \text{with} \quad \alpha = \frac{c_n}{c_f}\left(\frac{P_{L,n}}{P_{L,w}} - 1\right). \tag{8}$$

Here $P_{L,i} = 2\sigma\cos\theta/r_i$ is the Laplace pressure (the index $i \in \{f, n, w\}$ refers to the feeding, narrow and wide channel, respectively) and σ the surface tension, $c_i = \pi r_i^4/8\eta$ with η the viscosity of the liquid. For channels with rectangular cross section the expressions for $P_{r,i}$ and c_i have to be replaced by expressions involving the widths and the heights of the channels.

The arrest time is the time that the meniscus in the wide channel needs to reach the distance l_{resume} from the junction and is is proportional to l_0^2 and given by [27]:

$$\tau = K \cdot l_0^2, \quad \text{with} \quad K = \frac{A}{|P_{L,n}|}\frac{c_n}{c_f^2} \cdot \left\{\left(\frac{P_{L,n}}{P_{L,w}} - 1\right) + \frac{1}{2}\left(\frac{P_{L,n}}{P_{L,w}} - 1\right)^2 - \frac{c_f}{c_n} - \frac{c_f^2}{2c_n^2}\right\}. \tag{9}$$

Here A is the cross sectional area of the channel $A = \pi r^2$.

The equation (8) is equivalent to the eq.(3) occurring in the scaling theory presented in section 3. In [27] we used an experimental, microfluidic setup to demonstrate the validity of eq.(8), and concomitantly a basic ingredient of the scaling theory, eq.(3). Experimental results for capillary rise in an asymmteric Y-junction are shown in Fig. 6b, which shows a series of snapshots of the rising liquid and arrested meniscus in the microfluidic device, and Fig. 6c,

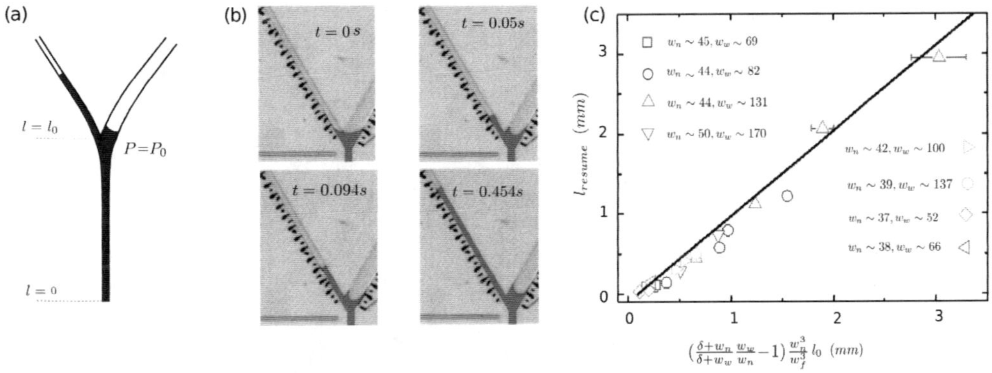

Figure 6. Capillary rise in asymmetric pore junctions – microfluidic experiments: (a) Schematic shape of a Y-junction. (b) Time series of fluorinated oil FC-70 invading a microfluidic Y-junction with a long feeding channel demonstrating the arrest of the meniscus in the wide channel during the initial propagation of the meniscus in the narrow channel. The dimensions of the device are $w_n \sim 45\mu m$, $w_w \sim 131\mu m$ and $l_0 = 8800 \pm 120\mu m$. (c) Experimental result for l_{resume}, as a function of the rescaled length of the feeding channel l_0, for different junction geometries (from [27]). The values of w_n and w_w of each device are indicated in μm. The line displays the theoretical prediction. Error bars in x- and y-direction are omitted if there are smaller than the symbol size.

which shows the experimentally measured arrest time τ as a function of the scaled variable $l_{\text{resume}} = l_0 \cdot \left(\frac{\delta+w_n}{\delta+w_w} \cdot \frac{w_w}{w_n} - 1\right) \cdot \frac{w_n^3}{w_f^3}$. Indeed the data points from different experiments (different feeding channel lengths, different channel widths) lie on a straight line as predicted by eq.(3).

Thus we have experimental evidence for the meniscus arrests and its duration in asymmetric pore junction as they occur in random pore networks, which supports the basis of the scaling theory of section 3. The simplified model for capillary rise as sketched above is based on macroscopic laws (Laplace pressure, Hagen-Poisseuille), which are indeed expected to hold for the experimental microfluidic device used to validate the model predictions. The question remains, whether meniscus arrests together with the scaling relations for arrest time etc. can also be observed on the nanometer scale, which is the appropriate length scale for the experiments on NVG described in section 2.

Experimentally it is not yet feasible to explore meniscus arrest on the nano-scale, but molecular dynamics simulations are very well suited for studying physical processes on this length scale. Therefore we considered a capillary junction formed by silica walls attached to a reservoir of argon molecules and performed molecular dynamics simulations of the capillary rise [28]. Snapshots of the particle configurations at different times are shown in Fig. 7. As it can be seen, the meniscus in the wide channel is indeed arrested for a time that is of the same order of magnitude as the time the meniscus in the feeding channel needed to reach the junction. A quantitative analysis necessitates many of these runs to obtain good statistics and is under way.

6. Discussion

We have shown that the dynamics of spontaneous imbibition in nanoporous media with elongated pores (low porosity), like NVG, is dominated by long-lasting mensicus arrests in the wide channels at pore junctions of channels with different radii. This establishes a new universality class for imbibition front roughening, which is actually predicted to be independent of the dimensionality of the pore network and mainly dependent on the pore radius distribution. The independence of dimensionality and topology of the pore network is expected due to a lack of

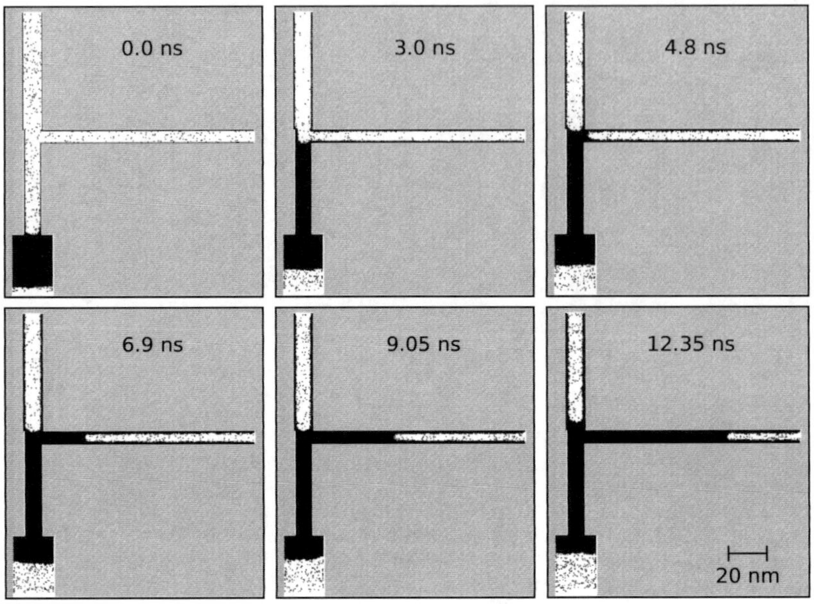

Figure 7. Capillary rise in asymmetric pore junctions – molecular dynamics simulation: Six snapshots from the molecular dynamics simulations of the spontaneous imbibition of liquid argon in silica nano-pores are taken at consecutive timesteps. The nano-pore consists of 98,540 silica atoms in three channels. The feeding channel has a diameter of 4.0 nm, the upper channel 4.8 nm and the right channel 3.2 nm. The interactions between the 191,012 argon atoms are given by a Lennard-Jones potential with the parameters $\epsilon = 0.2384 \frac{\text{kcal}}{\text{mole}}$ and $\sigma = 3.4\text{Å}$. The potential is cut off at a distance of 10.0Å. Temperature is kept at a constant value of 94 K with the help of the DPD-thermostat. The parameters of the Lennard-Jones interactions between the wall and the argon atoms are $\epsilon = 0.1521 \frac{\text{kcal}}{\text{mole}}$ and $\sigma = 3.85\text{Å}$. After filling the feeding channel the liquid penetrates the narrow channel and the rise of the meniscus is stopped for about 4.46 ns in the wide channel.

spatial correlation between the individual meniscus arrests. This meniscus arrest universality class should also be valid for forced imbibition in pore networks with elongated pores as long as the reservoir pressure is much smaller than the modulus of the Laplace pressures in the network. For applications it is interesting to note that the dependence of the imbibition front broadening on the pore radius distribution might be utilized to infer from the meniscus arrest time distribution or characteristics of the front broadening process dependent upon it back onto the radius distribution.

For decreasing aspect ratio (ratio of pore radius over pore length) one expects at some point a continuous liquid-gas interface to form and thus a crossover from the meniscus arrest universality class to a different universality class, where the liquid-gas-interface surface tension becomes relevant. In [29] we studied a model which was devised to capture some of the physics relevant for this crossover. It predicts that for non-vanishing effective surface tension the propagation of the imbibition front comes to a halt at some point.

Nanoporous materials become more and more important in fundamental research and nanotechnology [30]. Recently [31] it was shown that for aqueous electrolyte imbibition in nanoporous gold the fluid flow can be reversibly switched on and off through electric potential control of the solidliquid interfacial tension, which implies that one could accelerate the imbibition front, stop it, and have it proceed at will. In this respect it would be rewarding to generalize our scaling theory for the arrest time distribution and front broadening to field or time dependent Laplace pressures and to have a closer look with our microscopic pore network model into the imbibition dynamics in this new class of nanoporous materials.

Acknowledgments
We thank P. Huber, D.-S. Lee, M. Jung, and R. Seemann for fruitful collaborations and stimulating discussions. We acknowledge financial support from the German Research Foundation under Grant No GRK 1276.

References

[1] Bear J 1972 *Dynamics of Fluids in Porous Media* (New York: Elsevier)
[2] Sahimi M 1993 *Rev. Mod. Phys.* **65** 1393
[3] Alava M, Dube M and Rost M 2004 *Adv. Phys.* **53** 83
[4] Todd D K and Mays L W 2005 *Groundwater Hydrology* (John Wiley)
[5] Geromichalos D, Mugele F and Herminghaus S 2002 *Phys. Rev. Lett.* **89** 104503
[6] Buldyrev S V, Barabási A L, Caserta F, Havlin S, Stanley H E and Vicsek T 1992 *Phys. Rev.* A **45** R8313
[7] Horváth V K and Stanley H E 1995 *Phys. Rev.* E **52** 5166
[8] Hernandez-Machado A, Soriano J, Lacasta A M, Rodríguez M A, Ramírez-Piscina L and Ortín J 2001 *Europhys. Lett.* **55** 194
[9] Planet R, Pradas M, Hernandez-Machado A and Ortin J 2007 *Phys. Rev.* E **76** 056312
[10] Dubé M, Daneault C, Vuorinen V, Alava M and Rost M 2007 *Eur. Phys. J.* B **56** 15
[11] Lucas R 1918 *Kolloid* Z **23** 15
[12] Washburn E W 1921 *Phys. Rev.* **17** 273
[13] Dimitrov D I, Milchev A and Binder K 2007 *Phys. Rev. Lett.* **99** 054501
[14] Gruener S, Hofmann T, Wallacher D, Kityk A V and Huber P 2009 *Phys. Rev.* E **79** 067301
[15] Reyssat M, Courbin L, Reyssat E and Stone H A 2008 *J. Fluid Mech.* **615** 335
[16] Dullien F A L 1992 *Porous Media, Fluid Transport and Pore Structure* (San Diego: Academic Press)
[17] Dube M, Rost M and Alava M 2000 *Eur. Phys. J* B **15** 691
[18] Gelb L D, Gubbins K E 1998 *Langmuir 1998* **14** 2097
[19] Song Y Q, Ryu S, Sen P N 2000 *Nature* **406** 178
[20] Gruener S, Sadjadi Z, Hermes H E, Kityk A V, Knorr K, Egelhaaf S U, Rieger H, and Huber P 2012 *Proc. Nat. Acad. Sci. USA* **109** 10245
[21] Caupin F, Cole M W, Balibar S and Treiner J 2008 *Europhys. Lett.* **82** 56004
[22] Sadjadi Z and Rieger H 2013 *Phys. Rev. Lett.* **110** 144502
[23] Aker E, Maloy K J, Hansen A, and Batrouni G G 1998 *Transp. Porous. Med.* **32** 163
[24] Lam C H and Horvath V K 2000 *Phys. Rev. Lett.* **85** 1238
[25] Barabasi A L and Stanley H E 1995 *Fractal Concepts in Surface Growth* (New York: Cambridge University Press)
[26] Krug J 1997 *Adv. Phys.* **46** 139
[27] Sadjadi Z, Jung M, Seemann R, and Rieger H 2015 *Langmuir* **31** 2600
[28] Plimpton S 1995 *J. Comp. Phys* **117** 1, http://lammps.sandia.gov
[29] Lee D K, Sadjadi Z, and Rieger H 2014 *Phys. Rev. E* **90** 013016
[30] Huber P 2015 *J. Phys. Condens. Matter* **27** 103102
[31] Xue Y, Markmann J, Duan H, Weissmüller J and Huber P 2014 *Nature Comm.* **5** 4237

Journal of Physics: Conference Series **638** (2015) 012008 doi:10.1088/1742-6596/638/1/012008

Patterns, dynamics and phase transitions in Ising ferromagnet driven by propagating magnetic field wave

Muktish Acharyya
Department of Physics, Presidency University, 86/1 College street, Calcutta-700073, INDIA
E-mail: muktish.physics@presiuniv.ac.in

Abstract. The nonequilibrium behaviours of kinetic Ising ferromagnet driven by a propagating magnetic field wave have been studied by Monte Carlo simulation. Two types of propagating magnetic field waves are used here. Namely, the plane wave and the spherical wave. For plane propagating wave passing through the Ising ferromagnet, system undergoes a phase transition from a pinned phase to a propagating phase, as the temperature increases. The transition temperature is found to depend on the amplitude of the propagating magnetic field. A phase boundary is drawn in the plane described by the temperature of the system and amplitude of the propagating field. On the other hand, the nonequilibrium behaviours shown by the Ising ferromagnet driven by spherical magnetic field wave, are different. Here, the system exists in three different dynamical phases. The low temperature pinned phase, the intermediate temperature centrally localised breathing phase and the high temperature extended spreading phase. Here also, the transition temperatures are observed to depend upon the amplitude of the propagating magnetic field wave. The phase boundaries are drawn in the plane represented by temperature of the system and the amplitude of the propagating magnetic field wave. The two boundaries merge at the Onsager value of equilibrium critical temperature in the limit of vanishingly small amplitude of the propagating magnetic field. *This article is mainly a review of earlier works and is based on the invited lecture delivered in the conference STATPHYSKOLKATAVIII, held at SNBNCBS, Kolkata, India in December 1-5, 2014.* **This article is dedicated to Prof. H. Nishimori on the occasion of his 60th birthday.**

1. Introduction

The dynamical phase transition [1, 2] in Ising ferromagnet swept by oscillating magnetic field is an interesting field of research in nonequilibrium statistical mechanics. It was first observed [3] in kinetic Ising ferromagnet driven by oscillating magnetic field from the numerical solution of dynamical meanfield equation. Though this study helps to understand the phenomenon by the motion of the system in a double well free energy in the presence of oscillating field, it has a serious problem. Due to the absence of fluctuation, the dynamic transition in the meanfield study[3] would lead to the existence of the phase boundary even in the static limit. However, truely dynamic transition means such a transition where the transition should disappear in the static limit. This is only possible if the fluctuation is present, where the noise may push the system from one well to another even for inadequate amount of external magnetic field in the static limit, resulting the disappearance of dynamic phase transition. One possible way to study the effect of fluctuation, is the Monte Carlo simulation of the kinetic Ising model driven by

Journal of Physics: Conference Series **638** (2015) 012008 doi:10.1088/1742-6596/638/1/012008

oscillating magnetic field. Being motivated by this idea the Monte Carlo study was done and the disappearance of dynamic phase transition was indeed found in the static limit.

In the Monte Carlo simulation of kinetic Ising model driven by oscillating magnetic field the dynamical phase transition was found and a phase boundary was drawn in the plane described by the temperature of the system and the amplitude of oscillating magnetic field. The exponential relaxation of the dynamic order parameter was studied and critical slowing down was observed[4]. This relaxation of dynamic order parameter and the critical slowing down was verified experimentally[5]. The dynamic specific heat was also observed[4] to diverge at the dynamic transition point. The fluctuation of the dynamic order parameter was found [6] to diverge at the critical point. The dynamic transition is closely related to the hysteresis[7]. The dynamic transition by randomly varying field was also studied[8] in details. The existence of tricritical point on the dynamic phase boundary was observed and it is found to be related to the stochastic resonance[9]. The critical growth of dynamic correlation was also found[10]. Very recently, the surface and bulk dynamic critical behaviour was studied in kinetic Ising ferromagnet driven by oscillating magnetic field and the different class of universality was found[11]. The dynamical transition is associated with dynamical symmetry breaking. This is experimentally studied in highly anisotropic Co film on Cu by magneto optic Kerr effect[12].

All the studies mentioned above are of particular type. In all the cases the external magnetic field was oscillating but uniform over the space at any given instant of time. What will happen if the external field has a spatio-temporal variation ? This is precisely the subject of this article. The dynamical behaviours of planar Ising ferromagnet driven by propagating magnetic field wave, were studied recently. Here, two types of propagating magnetic waves are considered, plane[13, 14] and spherical[15].

The article is arranged in the following format: the model and the simulation scheme is described in section 2, the numerical results are described in section 3 and this ends with a summary in section 4.

2. Model and Simulation

The Hamiltonian of a two dimensional Ising ferromagnet (having uniform interaction strength among the nearest neighbours only) in presence of a propagating magnetic field wave (having spatio-temporal variation) can be represented as follows:

$$H(t) = -J\Sigma s(x,y,t)s(x',y',t) - \Sigma h(x,y,t)s(x,y,t) \quad (1)$$

The Ising spin variable, $s(x,y,t)$ assumes value ± 1 at lattice site (x,y) at time t on a square lattice of linear size L. Here, (x',y') represent the coordinates of nearest neighbors of the site (x,y). The parameter $J(>0)$ is the uniform ferromagnetic nearest neighbour interaction strength. The first term represents the Ising spin-spin ferromagnetic interaction. The spin-field interaction is represented by the second term. The $h(x,y,t)$ is the value of the magnetic field (at point (x,y) and at any time t) of the propagating magnetic field wave. The boundary condition, chosen here, is periodic in all directions.

2.1. Plane propagating wave

The plane propagating magnetic field wave is represented as:

$$h(x,y,t) = h_0\cos(2\pi ft - 2\pi y/\lambda). \quad (2)$$

The phase of the field propagates along the y direction with amplitude h_0, frequency f and wavelength λ. In the case of plane propagating field, the initial ($t = 0$) configuration, a high temperature random configuration where 50 percent spins are chosen (randomly) as

Journal of Physics: Conference Series **638** (2015) 012008 doi:10.1088/1742-6596/638/1/012008

$s(x, y, t) = -1$ The spins are updated randomly (a site (x, y) is chosen at random) and spin flip occurs (at temperature T) according to the Metropolis probability of single spin flip (W)

$$W(s \to -s) = \text{Min}[\exp(-\Delta E/kT), 1], \tag{3}$$

where ΔE is the change in energy due to spin flip and k is Boltzmann constant. Here, L^2 such random updates of spins defines the unit time step here and is called Monte Carlo Step per spin (MCSS). Here, the magnetic field and the temperature are measured in the units of J and J/k respectively. The dynamical steady state is reached by cooling the system slowly (in the presence of the propagating field) in small step ($\delta T = 0.02$ here) of temperature. It may be mentioned here that the same dynamical steady state was observed to be achieved by heating the system from a low temperature ordered configuration. The frequency and wavelength of the propagating magnetic field were kept fixed ($f = 0.01$ and $\lambda = 25.0$) throught the study. The frequency is measured in the unit of MCSS^{-1} and the wavelength is measured in the unit of lattice spacing. The total length of simulation is 2×10^5 MCS and first 10^5 MCS transient data were discarded to achieve the stable dynamical steady state. Since the frequency of the propagating field is $f = 0.01$, the complete cycle of the field requires 100 MCS. So, in 10^5 MCS, 10^3 numbers of cycles of the propagating field are present. The time averaged data over the full cycle (100 MCSS) of the propagating field are further averaged over 1000 cycles.

2.2. Spherically propagating wave

The form of spherically propagating magnetic field wave, is

$$h(x, y, t) = h_0 \frac{e^{i(2\pi f t - 2\pi r/\lambda)}}{r} \tag{4}$$

Where, $r = \sqrt{(x - x_0)^2 + (y - y_0)^2}$. The spherical wave originates at (x_0, y_0) and propagates radially outward.

The h_0, $f = \frac{\omega}{2\pi}$ and λ represent the amplitude, frequency and the wavelength respectively of the propagating magnetic field wave. In the present simulation, a $L \times L$ square lattice is considered. The boundary condition, used here, is periodic in both (x and y) directions. In this present study, $L = 100$ is considered.

In the case of spherically propagating magnetic field wave, the initial ($t = 0$) spin configuration is taken here as all the Ising spins are up ($s(x, y, t = 0) = +1$, $\forall$ x,y). Here, the dynamical steady state is obtained by heating the system slowly (δT=0.05) from fully ordered state. Here also one may achieve the same nonequilibrium steady state by cooling the system slowly from high temperature random state. The random updating of single spin flip with Metropolis rate is used here. The frequency ($f = 0.01$) and the wavelength (λ=15) were kept constant throughout the study.

Here also, the total length of simulation is 2×10^5 MCSS, where initial 10^5 MCSS is discarded and the averaging were calculated over 10^5 MCSS.

3. Results

3.1. For plane propagating wave

The fundamental physical quantity, which shows the different dynamical behaviours in this case, is the instantaneous magnetisation $m(t)$. In the case of plane propagating wave, the nonequilibrium steady states, depend on the values of the temperatre (T) of the system and the amplitude (h_0) of the propagating wave. For high temperature, the propagation of spin clusters (strip like) is observed. A typical such propagation is shown in a video (http://youtu.be/41TZPM-uxuc). However, for lower temperature no such coherent propagation of strip-like spin clusters was observed. For example, with $T = 1.5$ and $h_0 = 0.6$ the coherent

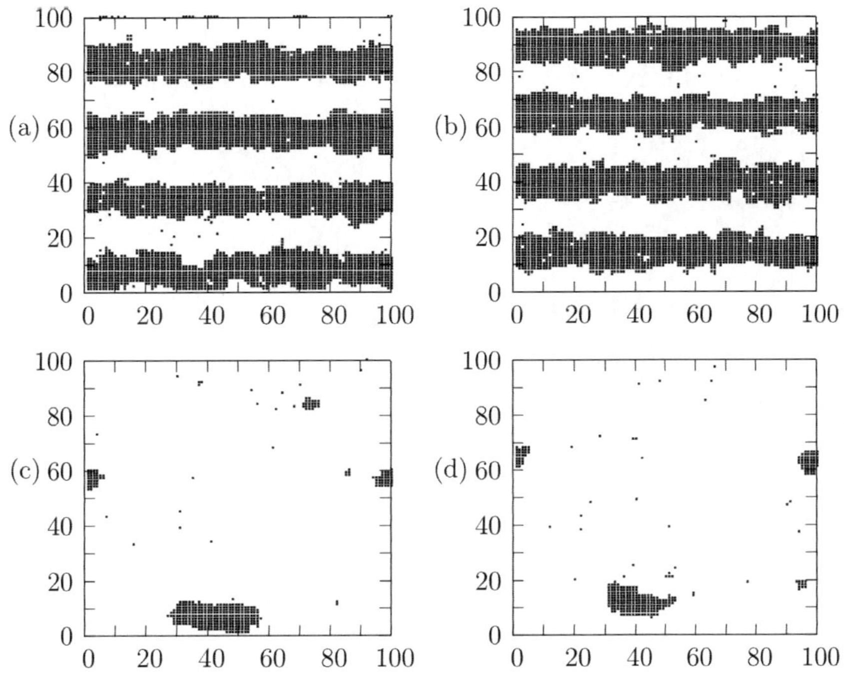

Figure 1. The motion of spin-clusters of down spins (shown by black dots), swept by propagating magnetic field wave, for different values of (a) Time = 100100 MCSS, T=1.5 and $h_0 = 0.6$ (b) Time = 100125 MCSS, $T = 1.5$ and $h_0 = 0.6$ (c) Time = 100100 MCSS, $T = 1.26$ and $h_0 = 0.6$ (d) Time = 100125 MCSS, $T = 1.26$ and $h_0 = 0.6$. See, Acharyya M. 2014 *Acta Physica Polonica B* **45** 1027.

propagation of strip-like spin clusters were found in the direction of the propagating plane magnetic field wave. No such propagation of spin clusters was detected for lower values of temperature ($T = 1.26$). Figure-1 shows some snapshots of such behaviours.

The system transits from a nonequilibrium propagating phase to a nonpropagating phase as the temperature decreases for a fixed value of amplitude of plane propagating magnetic field wave. It may be noted here that the instantaneous magnetisation $m(t) = \frac{1}{L^2}\sum s(x,y,t)$. For a quantitative analysis of this nonequilibrium phase transition, the order parameter has to be defined. Here, the time average magnetisation over the full cycle of the propagating magnetic field, is assumed to serve as the order parameter, $Q = f \oint m(t)dt$. The fluctuation in order parameter is $< (\delta Q)^2 >=< Q^2 > - < Q >^2$. The dynamic energy is $E = f \oint H(t)dt$ and the specific heat is defined as $C = \frac{dE}{dT}$. All these quantities are studied as functions of temperature to observe the nonequilibrium phase transition.

Figure-2 shows such plots for two different values of the amplitudes, i.e., $h_0 = 0.6$ and $h_0 = 0.3$. The order parameter Q is zero in the propagating phase and becomes nonzero at the transition temperature (T_d), as the system is cooled (fig-2(a)). The derivative $\frac{dQ}{dT}$, shows sharp dip (indication of scaling like $Q \sim (T_d - T)^\beta$ where $\beta < 1$) which is believed to be divergences eventually in the $L \to \infty$ limit. This is shown in fig-2(b). The transition is associated with the growth of fluctuations near the critical point and is shown in the plot $< (\delta Q)^2 >$ versus temperature in fig-2(c). The $< (\delta Q^2) >$ shows sharp peak (believed to be divergences in $L \to \infty$ limit). The dynamic specific heat C also shows sharp peak near the transition point and plotted in fig-2(d). From all these plots it is clear that the dynamic transition occurs at lower temperaure for higher values of the amplitude of the plane propagating magnetic field wave.

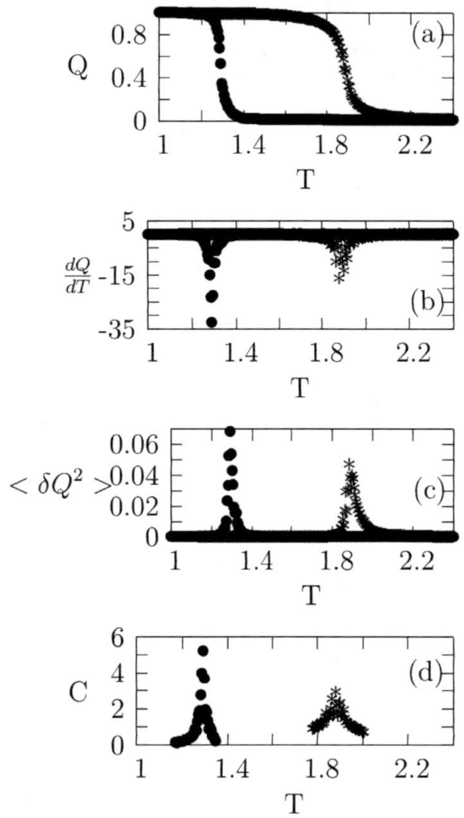

Figure 2. The temperature (T) dependences of the (a) Q, (b) $\frac{dQ}{dT}$, (c) $< (\delta Q)^2 >$ and (d) C, for two different values of h_0 for *propagating* magnetic field wave having $f = 0.01$ and $\lambda = 25$. In each figure, $h_0 = 0.3(*)$ and $h_0 = 0.6(\bullet)$. See, Acharyya M. 2014 *Acta Physica Polonica B* **45** 1027.

The transition temperatures are found to be function of the amplitude of the propagating magnetic field (for given f and λ). A comprehensive phase boundary is obtained and plotted in figure-3.

3.2. For spherically propagating wave

The dynamical responses of a two dimensional Ising ferromagnet driven by spherical magnetic field wave is also studied. The system shows three main dynamical states. For small value of field amplitude of propagating spherical wave it remains pinned. For some intermediate value of the amplitude it shows dynamically breathing phase. Figure-4 shows a typical snapshots of

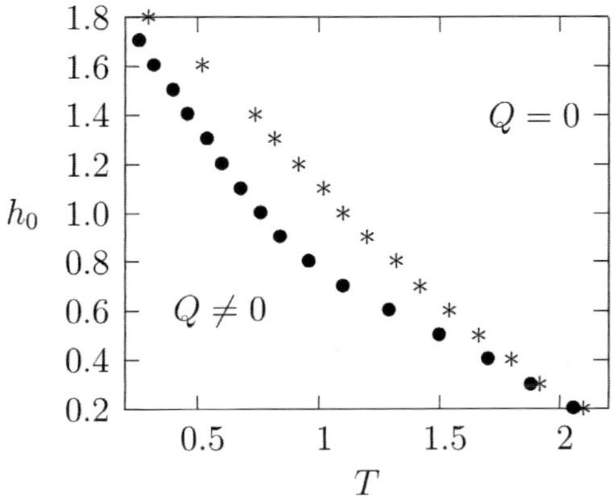

Figure 3. The phase diagram for dynamic phase transition by propagating magnetic field wave for two different values of wavelengths, $\lambda = 25(\bullet)$ and $\lambda = 50(*)$. Here, $f = 0.01$. See, Acharyya M. 2014 *Acta Physica Polonica B* **45** 1027.

these types (pinned and breathing) of dynamical phases. A video of spreading phase may be found in http://youtu.be/S_rGUQNjcig.

A centrally localised dynamical mode is found. For high value of field amplitude a spreading dynamical mode was observed. In this spreading phase the ring shaped spin clusters were found to spread over the whole lattice. Figure-5 shows the dynamically spreading phase.

The dynamical responses of a two-dimensional Ising ferromagnet driven by spherical magnetic field wave are characterised as follows: The density of instantaneous local magnetisation in the circle of radius $\lambda/2$ is $m_b(t) = \sum s(x, y, t)/(N_b)$, where the sum is carried over the number of sites (N_b) lying within the circle of radius $\lambda/2$ centered at the centre (x_0, y_0) of lattice. In the present study, the lattice size L is taken equal to 100. So, the coordinates of the centre is (x_0=50.5,y_0=50.5). and N_b is the total number of lattice sites within this circle. The order parameter of the dynamic breathing transition is defined as $Q_b = \frac{\omega}{2\pi} \oint m_b(t)dt$. The statistical fluctuations of the dynamic order (for breathing) parameter is $< \delta Q_b{}^2 >=< Q_b{}^2 > - < Q_b >^2$. The average value of Q_b and the fluctuation are calculated over $n_c = 1000$ number of random samples. It is also checked that this number of samples is sufficient to have the nonequilibrium steady state as well as to have the data within the specified accuracy.

To study the dynamical spreading transition the releavant quantities are: The instantaneous magnetisation density within a ring of width $\lambda/2$, which thouches the boundary of the lattice, is $m_s(t) = \sum s(x, y, t)/N_s$, where the summation is carried over the sites residing within the ring of width $\lambda/2$ touching the boundary, and N_s is the total number of lattice sites within that ring. It is a region bounded between two concentric ($x_0 = 50.5, y_0 = 50.5$) circles having outer radius $r_{out} = L/2$ and the inner radius $r_{in} = L/2 - \lambda/2$. The dynamic order parameter of spreading transition $Q_s = \frac{\omega}{2\pi} \oint m_s(t)dt$. The fluctuation in dynamic order parameter (for spreading) is $< \delta Q_s{}^2 >=< Q_s{}^2 > - < Q_s >^2$.

The dynamic breathing and spreading order parameters are plotted against temperature and shown in figure-6. It is observed that, for a fixed set of values of h_0, f and λ of the field wave, the breathing transition occurs at lower temperature than that for spreading transintion. As the amplitude of the magnetic field wave increases both the transitions occur at lower temperatures.

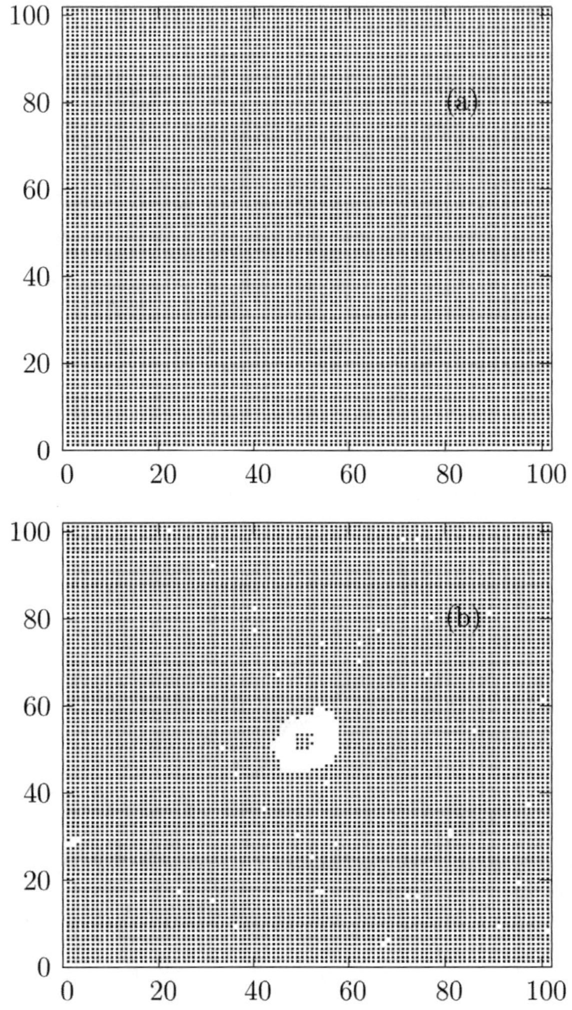

Figure 4. The pinned (a) and breathing (b) modes on the lattice. Dots represents up spins only. (a) $h_0 = 2.5$, $T = 0.30$ and (b) $h_0 = 2.5$ and $T = 1.45$. Here, $\lambda = 15.0$, $f_0 = \frac{\omega_0}{2\pi} = 0.01$ and $t = 4000$ MCSS. See, Acharyya M. 2014 *J. Magn. Magn. Mater.* **354** 349.

The fluctuations in dynamic breathing and spreading order parameters are shown in figure-7. The peaks of the fluctuations are indicating the breathing and spreading transition temperatures. The transition temperatures are observed to decreases as the amplitude of the magnetic wave increases. Hence, one can obtain the transition temperatures as functions of the amplitude of the magnetic wave and a comprehensive phase boundary can be drawn in the plane described by the temperature of the system and the amplitude of the magnetic field wave.

The comprehensive phase boundary (for fixed f and $\lambda = 15$) are shown in figure-8. Here the regions of three different nonequilibrium phases are shown in the plane described by the

Journal of Physics: Conference Series **638** (2015) 012008
doi:10.1088/1742-6596/638/1/012008

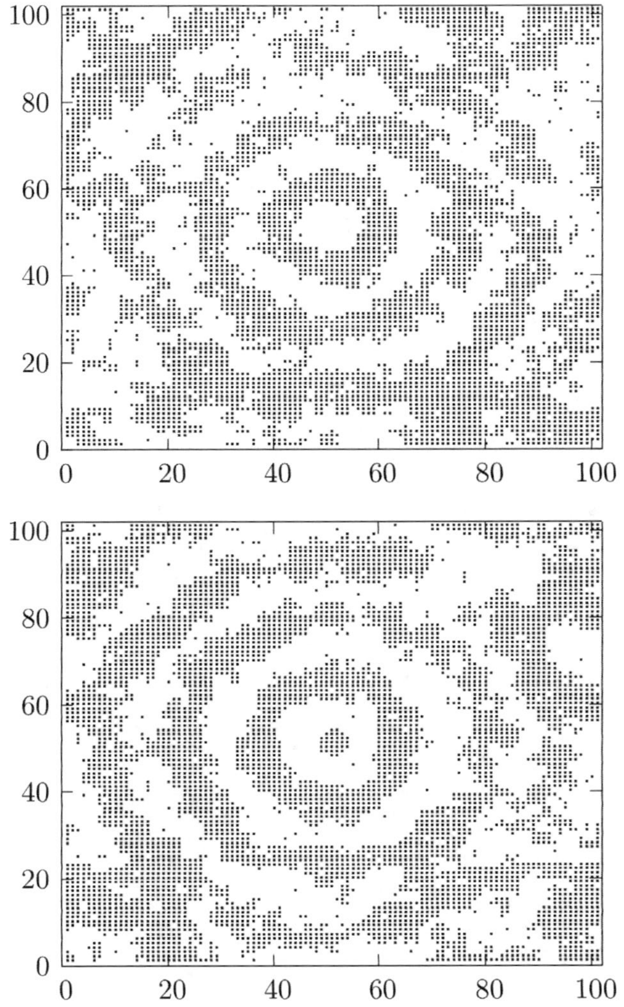

Figure 5. A high field ($h_0 = 10.0$)-high temperature ($T = 2.30$) spin wave propagation mode. The up spins ($S_i = +1$) are shown by black dot. The bottom one shows a spin configuration at 3970 MCSS and the top one shows that at 4000 MCSS. Here, $f_0 = 0.01$ and $\lambda = 15.0$. See, Acharyya M. 2014 *J. Magn. Magn. Mater.* **354** 349.

temperature of the system and the amplitude of spherically propagating magnetic field wave. It was observed that the two phase boundaries merge to a value (the equilibrium ferro-para transition temperature $T_{eq} = 2.269...$, calculated exactly by Onsager) in the limit of vanishingly small amplitude of magnetic wave.

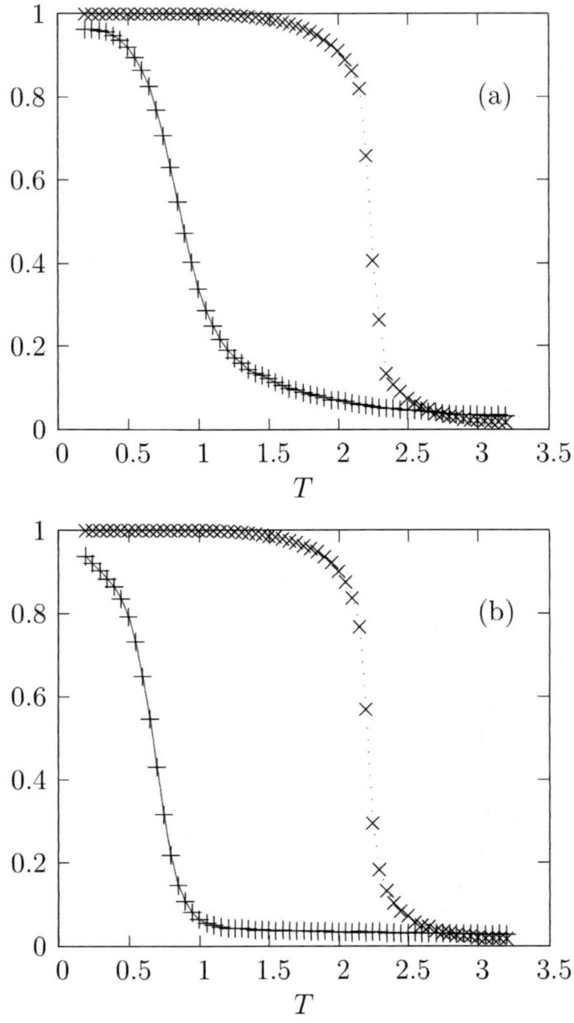

Figure 6. The plots of $Q_b(+)$ and $Q_s(\times)$ as a function of temperature T. (a) $h_0 = 2.5$, $f = 0.01$ and $\lambda = 15.0$ and (b) $h_0 = 3.5$, $f = 0.01$ and $\lambda = 15.0$. See, Acharyya M. 2014 *J. Magn. Magn. Mater.* **354** 349.

4. Summary

In this article, the nonequilibrium responses of planar Ising ferromagnet driven by propagating (plane wave and spherical wave) magnetic field wave are discussed. For plane propagating magnetic field wave, Ising ferromagnet shows two nonequilibrium phases, namely pinned and propagating phase, depending on the values of temperature of the system and the amplitude of the propagating magnetic field wave. The dynamic transitions are marked by studying the dynamic order parameter, its derivative, its fluctuations and the dynamic specific heat as functions of temperatures. A comprehensive phase boundary is plotted in the plane described

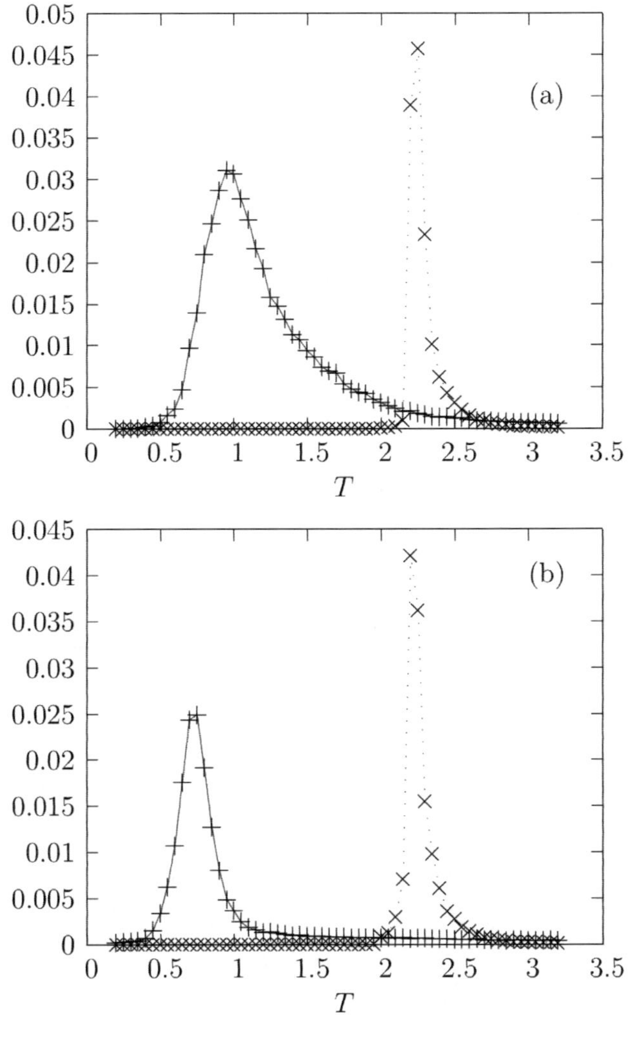

Figure 7. Fluctuations of $Q_b(+)$ and $Q_s(\times)$ plotted against temperature (T) for two different values of field amplitude (h_0) of propagating field. (a) $h_0 = 2.5$ and (b) $h_0 = 3.5$. Here, $\lambda = 15.0$ and $f = 0.01$. See, Acharyya M. 2014 *J. Magn. Magn. Mater.* **354** 349.

by the temperature and the amplitude of the plane propagating magnetic field wave.

Now, if the system is irradiated by a spherically propagating magnetic field wave whose origin is at the centre of the lattice, the nonequilibrium responses show variety of interesting phenomena. Here three different nonequilibrium phases are observed. For very small values of temperature and the amplitude of the magnetic field wave, the system remains in a pinned phase. For intermediate values of these two parameters, the system shows an interesting centrally localised dynamical phase (breathing phase) and for higher values of the temperature and field amplitude, it shows a dynamically extended spreading phase. The transition temperatures

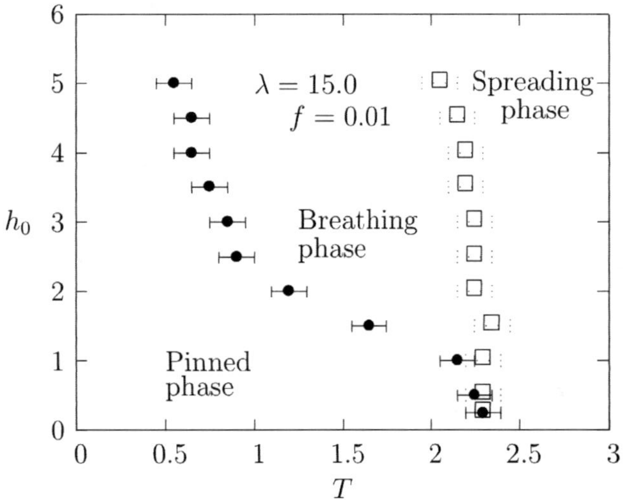

Figure 8. The phase diagram for symmetry breaking dynamic breathing and spreading transitions. Here, $f_0 = 0.01$ and $\lambda = 15.0$. See, Acharyya M. 2014 *J. Magn. Magn. Mater.* **354** 349.

for a given value of field amplitude are estimated from the peak positions of the fluctuations of dynamic breathing and spreading order parameters. The comprehensive phase boundaries are plotted in the plane described by the temperature of the system and the amplitude of the magnetic field wave. The two boundaries are observed to merge at equilibrium ferro-para transition temperature in the limit of vanishing amplitude of the magnetic field wave.

It may be mentioned here that recent experiments on permalloy (50 nm thick) excited by ultrashort (150 fs) laser pulse gives rise to the propagation of circular spin waves[16].

5. Acknowledgements

Author would like to thank S. S. Manna for helpful discussions.

6. References

[1] Chakrabarti B K and Acharyya M 1999 *Rev. Mod. Phys.* **71** 847
[2] Acharyya M 2005 *Int. J. Mod. Phys.* C **16** 1631
[3] Tome T and Oliveira M J 1990 *Phys. Rev.* A **41** 4251
[4] Acharyya M 1997 *Phys. Rev. E* **56** 2407
[5] Berger A, Idigorus O and Vavassori P 2013 *Phys. Rev. Lett.* **111** 190602
[6] Acharyya M 1997 *Phys. Rev. E* **56** 1234
[7] Acharyya M 1998 *Phys. Rev. E* **58** 179
[8] Acharyya M 1998 *Phys. Rev. E* **58** 174
[9] Acharyya M 1999 *Phys. Rev. E* **59** 218
[10] Sides S W, Rikvold P A, Novotny M A 1998 *Phys. Rev. Lett.* **81** 843
[11] Park H and Pleimling M 2012 *Phys. Rev. Lett.* **109** 175703
[12] Jiang Q, Yang H Y and Wang G C 1995 *Phys. Rev.* B **52** 14911; see also Suen J S and Erskine J L 1997 *Phys. Rev. Lett.* **78** 3567
[13] Acharyya M 2011 *Physica Scripta* **84** 035009

[14] Acharyya M 2014 *Acta Physica Polonica B* **45** 1027
[15] Acharyya M 2014 *J. Magn. Magn. Mater.* **354** 349
[16] Au Y, Dvornik M, Davison T, Ahmed E, Keatley P S, Vansteenkitse A, Van Waeyenberg B, Kruglyak V V, 2013 *Phys. Rev. Lett.* **110** 097201

Journal of Physics: Conference Series **638** (2015) 012009 doi:10.1088/1742-6596/638/1/012009

Commensurate - Incommensurate vortex phase in a nanopatterned superconductor

Gorky Shaw[1, †], S S Banerjee[1,*], T Tamegai[2] and Hermann Suderow[3]

[1] Department of Physics, Indian Institute of Technology, Kanpur-208016, India
[2] Department of Applied Physics, The University of Tokyo, Hongo, Bunkyo-ku, Tokyo 113-8656, Japan
[3] Laboratorio de Bajas Temperaturas, Unidad Asociada UAM, CSIC, Instituto N. Cabrera, Condensed Matter Physics Center, Facultad de Ciencias Universidad Autónoma de Madrid, E-28049 Madrid, Spain

*Email: satyajit@iitk.ac.in

[†] Present address: Université Grenoble Alpes, Institut Néel, F-38042 Grenoble, France CNRS, Institut Néel, F-38042 Grenoble, France.

Abstract. Magneto-optical imaging studies on a high-quality $Bi_2Sr_2CaCu_2O_8$ single crystal partially patterned with a triangular array of holes reveal enhanced flux shielding in the patterned region of the sample. By mapping local magnetic field and shielding current density distributions at different applied magnetic fields and temperatures we determine the regime where pinning from the patterned holes dominates over the intrinsic pinning in the sample. In this regime, the flux density near the center of the patterned region is observed to increase when the applied field is varied from below the matching field to just above it, while significant magnetic field gradients are sustained in the patterned region. Our measurements indicate heterogeneous pinning properties of the vortex population, exhibiting signatures of both weak and strong pinning, in the nanopatterned region of the superconductor.

1. Introduction

In pinning-free superconductors vortices are arranged in a well-ordered triangular Abrikosov vortex lattice configuration with a lattice period (inter-vortex separation) of $a_0 \sim \sqrt{\phi_0/B}$, where ϕ_0 is the magnetic flux quantum and B is the magnetic field. The competition and interplay between vortex - vortex interactions and vortex - pin interactions leads to metastable glassy vortex phases, like the well-known Bragg glass and vortex glass [1,2,3] phases. In a superconductor dissipation arises due to normal cored vortices drifting under the influence of a Lorentz force acting on them whenever current is sent through the superconductor. From applications point of view enhancing pinning strength is important as it helps to strongly localize vortices. Strong localization of vortices enhances the threshold current value of a superconductor (viz., the critical current density) beyond which dissipation appears in the superconductor due to delocalized vortices. Recent studies on superconductors with array of holes with diameter of the size of the superconducting coherence length suggest a different interesting mechanism of vortex localization. Popularly studied artificial pinning centers are antidots, which are cylindrical holes open at both ends patterned in a superconducting film. In the antidot array two different mechanisms of vortex pinning have been identified. One type is the pinning of interstitial vortices present in between the antidots when each antidot in the array is saturated with vortices. The strong localization of these interstitial vortices [4] is brought about by the nucleated interstitial vortex itself inducing a modulation of the superconducting order parameter within the interstitial mesoscopic superconducting patch. Recent studies on superconductors patterned with an array of blind holes (cylindrical holes open only on one end) have exhibited properties associated with significant enhancement in pinning [5,6,7,8]. The presence of periodicity in the artificially generated periodic pinning array induces another interesting effect. Superconductors with artificially generated periodic array of antidots show periodic enhancements in the pinning force experienced by the vortices as a function of applied magnetic field (H) [9,10,11,12,13,14,15,16,17]. The periodic enhancements are related to the inter-vortex separation (a_0) becoming commensurate with the period (d) of the lattice of pins at integral multiples of the first matching field (B_ϕ), where $a_0 \sim d$, $B_\phi(a_0 = d) \sim \frac{\phi_0}{d^2}$. As a function of the H a variety of different commensurate and incommensurate phases appear.

In superconductors with a natural random distribution of pinning a well-known critical state exists viz., the Bean critical state [18]. The Bean critical state is an example of organization setting in a system out of equilibrium. An example of self-organized critical state is well known for sand-piles, wherein there is a critical slope of the sand-pile. The addition of sand to enhance the slope beyond this critical value leads to destabilizing the pile and the generation of an avalanche until the slope returns back to its critical value. The vortex configuration attained by increasing the field is not a system in equilibrium. The presence of pinning centers which localize vortices keeps the vortex configuration from achieving its ground state configuration. In superconductors with random distribution of pinning centers, when the applied field is increased from zero to a finite field value, the penetrating vortices organize themselves into a configuration with a gradient in density. The vortex density profile from the sample edge to the center has a critical slope akin to the sand-pile. The vortex density decreases from the edge to the center of the superconductor. The vortex density gradient from all the sample edges results in a nonzero curl of the local field $\vec{B} \propto$ current density (J). According to the Bean critical state model the value of this shielding current density J is equal to the critical current density (J_c) of the superconductor (which is a measure of the pinning strength in the superconductor). The diamagnetic shielding currents lead to a uniform gradient in the magnetic field distribution across the superconductor. Thus external magnetic field sets up diamagnetic shielding currents which determine the magnetization response of the superconductor. The diamagnetic shielding currents set up in the superconductor are in turn related to field distribution in the superconductor which we investigate in this paper.

In superconductors with periodic arrangement of pins, the Bean like state is expected to be modified as a result of a balance [19,20] between two opposing tendencies. One tendency is associated with forming an ordered arrangement of vortices commensurate with the periodic lattice of pins and thereby generating a uniform (gradientless) field distribution. The competing tendency favours a uniform constant gradient in the field distribution or a constant bulk critical current density in the superconductor. The competition between these tendencies leads to domains of vortices which are ordered around the periodic pinning sites. Unlike the antidots, blind holes are relatively weaker pins, where the vortex - blind hole interaction does not overwhelm the inter-vortex interaction inside the superconductor. In superconductors patterned with relatively weak pinning blind holes the inter-vortex and vortex – weak pin interactions are comparable in magnitudes. Thus such a system is potentially favourable to study a variety of different vortex configurations arising out of competition and interplay between the sample's intrinsic pinning and the pinning from patterned blind holes.

In this paper we investigate a triangular array of blind holes patterned on a superconducting single crystal instead of thin films to ensure intrinsic pinning in the superconductor doesn't dominate the vortex pin interaction. Furthermore we restrict our study to low $B < B_\phi$, viz., the dilute vortex density case to keep the inter-vortex interaction low. For this situation at first glance it may be expected that due to comparable inter-vortex and vortex - blind hole pin interaction, vortices will be uniformly distributed across the blind hole array with a uniform filling fraction of ~ B/B_ϕ. In such a vortex configuration on an average only B/B_ϕ of the total number of blind holes will be occupied with single vortices. In this work using Magneto-Optical Imaging (MOI) technique we image the nature of field distribution in the blind hole array patterned on a $Bi_2Sr_2CaCu_2O_8$ (BSCCO) single crystal. MO imaging of the magnetic flux distribution over the patterned region reveals enhanced flux shielding in the patterned region. We have mapped the local magnetic field and shielding current density distributions at different applied magnetic fields and temperatures. Using these we determine the regime where blind hole pinning dominates over the intrinsic pinning. In this regime, the effective local magnetic flux penetration field is found to be higher in the patterned region. It is observed that in this scenario, when the applied field is varied from below the matching field to just above it, the flux density near the center of the patterned region increases while significant magnetic field gradients are sustained in the patterned region. Our measurements indicate heterogeneous pinning properties of the vortex population in the nanopatterned region of the superconductor, with features suggestive of both weak pinning and strong pinning inside the patterned region. We suggest a simple model to explain features in our data, as well as compare our results with predictions from more rigorous recent theories, which are partially related to our work.

2. Experimental details

2.1. Sample preparation

For our experiments we use a high-quality single crystal of $Bi_2Sr_2CaCu_2O_8$ (BSCCO) [5,21] of dimensions ($0.8 \times 0.5 \times 0.03$ mm^3) and $T_c = 90$ K (figure 1(a)). The sample surface (*'ab'* plane of the single crystal) was milled using Focused Ion Beam (FIB) machine (dual beam FEI make Nova 600 NanoLab) with a focused Ga ion beam (diameter ~ 7 nm) to create a triangular array of circular blind holes covering an area of ~ 40 μm × 40 μm. Figure 1(b) shows a magnified Scanning Electron Microscopic (SEM) image of the blind hole array. Mean diameter of the holes is 170 nm, mean center-to-center spacing between the holes (d) is 350 nm (BSCCO penetration depth λ_{ab} ~ 200 nm for *'ab'* crystal orientation) and depth of the holes is ~ 500 nm (thickness of single crystal = 30 μm). For this blind hole array the matching field is $B_\phi \approx 195$ G.

2.2. Magneto-Optical Imaging

The sample was studied using a high-sensitivity magneto-optical imaging (MOI) technique [5,21,22]. In the MOI setup a sensitive CCD camera captures the Faraday light intensity distribution, $I(x, y)$, across a superconductor placed in a constant applied magnetic field H_z (applied parallel to the c-axis of the single crystal). Linearly polarized light reflected off a Faraday active garnet film with a high Verdet constant (V) placed in close contact with the sample surface helps produce a distribution in the Faraday rotated light intensity reflected off different regions on the sample surface where there is a distribution in the local magnetic field $B_z(x, y)$. Recall the Faraday rotation angle $\theta(x, y) \propto VB_z(x, y)$. The reflected Faraday rotated light intensity contains information of $B_z(x, y)$ across the sample surface (Faraday rotated light intensity $I(x,y) \propto (B_z(x,y))^2$, $\hat{z}$ is an orientation perpendicular to the crystal surface which is the ab-plane of the single crystal). In our measurements we also perform differential MOI (DMOI) [23,24] wherein, differential Faraday rotated light intensity, $\delta I(x, y)$, is captured in response to step increase in the H, $\delta H = 1$ Oe, (viz., $\delta I(x, y) = \langle I(x, y)$ at $H + \delta H\rangle - \langle I(x, y)$ at $H\rangle$, where <..> represents averaging over n (= 20) images at each H). By measuring $\delta I(x, y)$ in the DMOI images produced in response to a calibrated change in the H far away from the sample being investigated, we determine the local magnetic field changes $\delta B_z(x, y)$ across the superconductor produced in response to δH = 1 Oe. Here we emphasize that due to irreversible magnetization response within the nanopatterned region we do not repeatedly modulate δH about H, , which is unlike conventional DMOI technique [23,24].

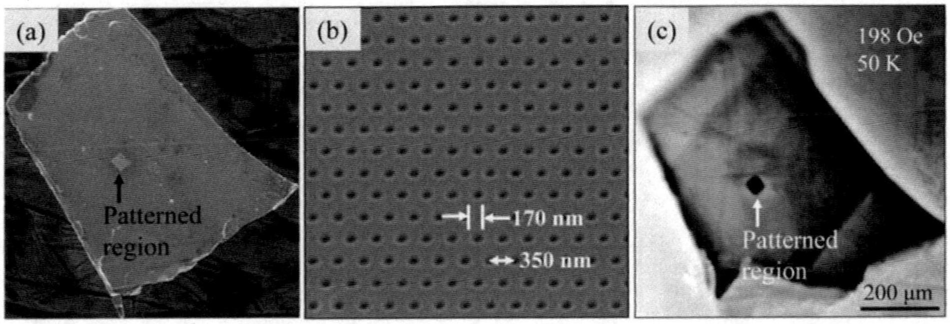

Figure 1. (a) Scanning electron microscope (SEM) image of the sample. The patterned region is the bright rectangular area (~ 40 μm × 40 μm) indicated by the arrow. (b) Magnified portion of the patterned region, showing the triangular array of blind holes of diameter 170 nm and with center to center hole spacing of 350 nm. Depth of the holes is ~ 500 nm. The triangular array of holes correspond to a matching field B_ϕ = 195 G. (c) Conventional MO image of the zero field cooled sample (ZFC) at H = 198 Oe and T = 50 K. The patterned region is the dark rectangular area indicated by the arrow.

Figure 1(c) shows a conventional MO image showing the full sample to identify different features in such an image. This image was obtained at H = 198 Oe and T = 50 K after zero field cooling (ZFC) the sample. The contrast variations in the image represent changes in $B_z(x, y)$ across the sample. In this MO image the nanopatterned region is readily identified by the square shaped region with dark contrast compared to the surrounding brighter contrast in the unpatterned regions of the sample. The darker contrast inside the patterned region represents a lower average $\langle B_z(x, y)\rangle$ inside the patterned region compared to the brighter contrast in the unpatterned regions.

3. Results and discussion

3.1. Local isothermal magnetization response

We study the local isothermal magnetization ($M(H)$) response in different regions of the sample using conventional MOI (viz., we use conventional MOI to obtain the $B_z(x, y)$ distribution in the sample without using δH modulation). The $M(H)$ behaviour is deduced using $4\pi M(H) = (\int_A [B_z(x,y) - H]dxdy)/A$, where A is the area over which averaging is performed and the local field $B_z(x, y)$ is determined from conventional MO images. Typically, A = 25 μm^2. Figure 2 shows $M(H)$ curves for the nanopatterned region (red, squares) and the whole sample (bulk, using A = entire sample area (blue, circles)), at different T. Clearly, the local $M(H)$ behaviour in the patterned region is significantly different from the bulk response at lower T, and at higher T the two magnetization responses merge with each other.

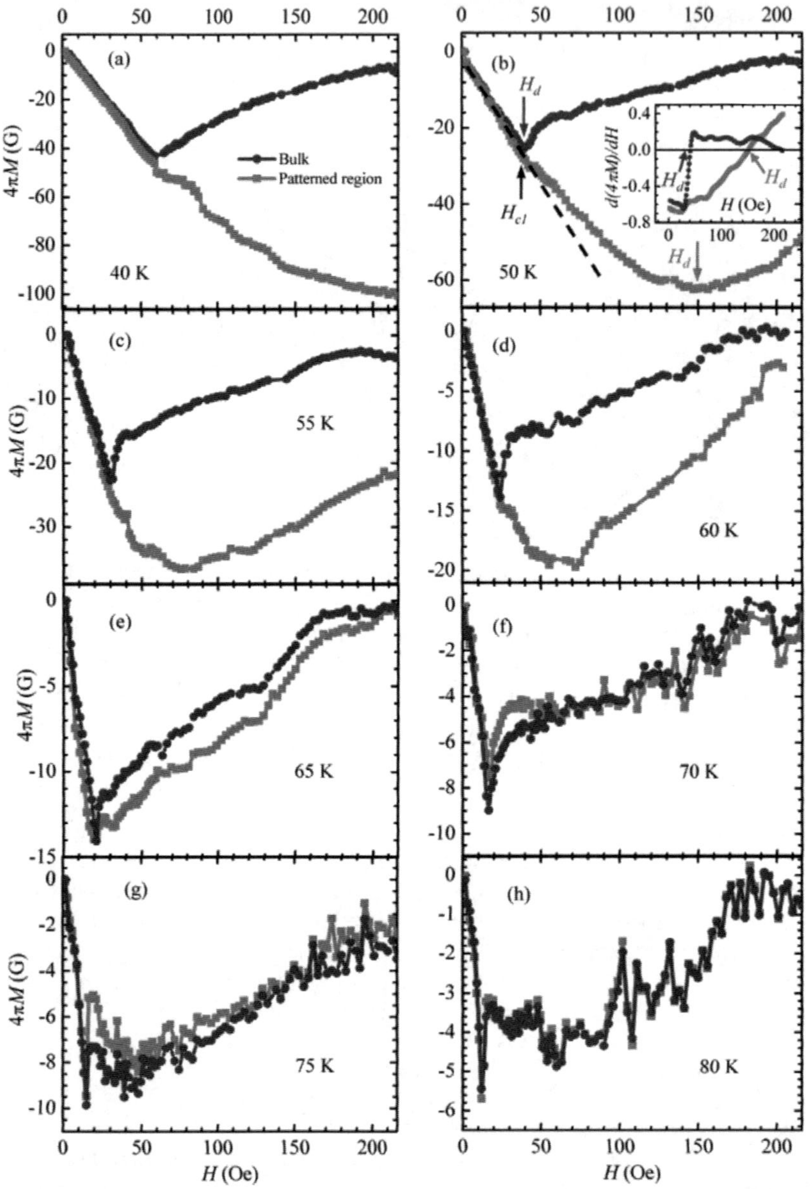

Figure 2. (a)-(h) *M*(*H*) curves obtained from MO images at different *T* as indicated in each panel. The blue curves (circles) represent *M*(*H*) response for the entire sample area. The red curves (squares) represent local *M*(*H*) response in the patterned region. Inset of (b) shows *dM/dH* vs. *H* plots at 50 K obtained from the *M*(*H*) curves shown in the main panel. The mean penetration fields for the patterned region and the bulk at 50 K are indicated in (b) by the dashed and dotted arrows, respectively.

We further analyse the *M*(*H*) responses at 50 K. The dashed line in figure 2(b) identifies the location in the vicinity of which the slope of the bulk *M*(*H*) curve changes from linear behaviour. The field (*H*) up to which the linear *M*(*H*) behaviour is maintained identifies the lower critical field (H_{c1}) of the superconductor. At $H < H_{c1}$ the local field $B_z = 0$ G inside the superconductor (Meissner state), due to which $4\pi M = - H$ (up to demagnetization factor), viz., a linear behaviour in $M(H)$.

At $H > H_{c1}$ the $M(H)$ begins to deviate from the linear behaviour and the negative slope (dM/dH) which was at its maximum value below H_{c1} begins to decrease. At a higher H the slope changes from negative to positive value as shown in dM/dH vs. H plots in inset of figure 2(b). At $H > H_{c1}$ the vortices begin penetrating into the superconductor, however they do not achieve a uniform distribution. There are gradients in the local vortex distribution and the center of the sample (or patterned region in our case) is free of vortices up to a field H_d. The regime where the dM/dH changes curvature helps us identify the location of the mean penetration field H_d ~ 40 Oe at 50 K where vortices have penetrated into the bulk of the unpatterned regions of the sample. By noting the deviation from linearity in $M(H)$ the bulk H_{c1} is estimated to be ~ 40 Oe (at 50 K), and it is found to be almost the same for the patterned region also. The local $M(H)$ curve of the patterned region is also found to be linear up to the local H_{c1} ~ 40 Oe (indicated by the black arrow in figure 2(b)). However the local magnetization curve for the patterned region shows enhanced diamagnetic response due to enhanced pinning compared to the unpatterned regions of the sample. This in turn indicates that the patterned region is associated with enhanced screening currents [18], which are stronger than those present in the unpatterned regions. However, even beyond 40 Oe, the deviation is from linearity is gradual and the $M(H)$ response is strongly diamagnetic. The gradual deviation from linearity indicates that the flux has not penetrated fully up to the center of the patterned region. Above H_d, the diamagnetic magnetization response monotonically decreases with increasing H as the flux has reached the center of the superconducting region under consideration. The dotted arrow in figure 2(b) identifies the penetration field H_d ~ 150 Oe (determined from the dM/dH vs. H plot in inset of figure 2(b)) at 50 K for the 25 μm^2 area inside the patterned region as compared to a value ~ 40 Oe in the unpatterned regions of the sample. As $H_d \propto J_c$ [18], the ratio of H_d at 50 K shows J_c of the patterned region is at least three to four times larger than that in the unpatterned regions.

Note from figure 2 that with increasing T the mean penetration field H_d for the patterned region decreases from 150 Oe at 50 K to 72 Oe at 60 K. H_d for both curves are identical at higher T as the local and bulk magnetization responses merge with each other. From these observations, we can assert that the effective pinning strength of the patterned region is ~ 70 K. For $T < 70$ K, pinning from the patterned blind holes dominates over the intrinsic pinning of the sample (as observed in the enhanced flux shielding inside the patterned region in this regime). We note that at 60 K, local $M(H)$ response of the patterned region indicates higher flux shielding inside the patterned region compared to the bulk of the sample and approaches the bulk $M(H)$ curve at higher H. Hence, for further analysis, we choose 60 K as the regime of interest. At 60 K the pinning properties of the patterned blind holes are prominent, while not being overwhelming, and the intrinsic pinning in the unpatterned regions of the sample are weakened. We further investigate local shielding response inside the patterned using DMOI.

3.2. Measurement of the local shielding response using DMOI

We now investigate the nature of the field gradients inside the patterned region and how susceptible the gradients are to field changes. Shown in figures 3(c) and 3(d) are DMOI images of the patterned BSCCO sample. To aid in understanding the DMOI images in figures 3(c) and 3(d), we show two schematics in figures 3(a) and 3(b). The images at the bottom of the figures 3(a) and 3(b) show schematic DMOI images showing two regions with two different gray shades which correspond to a larger change (light gray) and a smaller change (dark gray) in B_z in response to field change $\delta H = 1$ Oe, respectively. The main panels in figures 3(a) and 3(b) show two possible Bean like critical $B_z(r)$ profiles across the white dashed line drawn in the insets of figures 3(a) and 3(b). In these figures, the black and red curves represent the B_z profile at $H = 10$ and $H + \delta H = 11$ (arb. units), respectively. In figure 3(a) main panel the difference between the $B_z(r)$ profiles across the region marked unpatterned is larger than that in the region marked patterned. Therefore in a differential image where the contrast in the image is proportional to the difference in B_z at H and $H + \delta H$, one obtains a comparatively darker contrast within the patterned region (cf. figure 2(a) inset). The situation is opposite in figure 3(b).

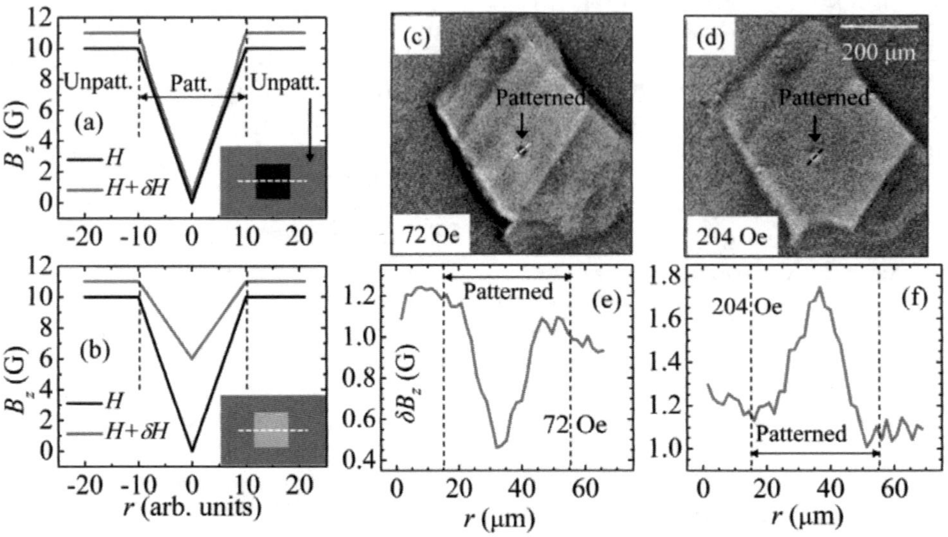

Figure 3. (a) and (b) main panels are schematics representing possible behaviour of $B_z(r)$ profiles across the patterned region at applied fields H and $H + \delta H$ (black and red curves, respectively) across a region corresponding to the white dashed line drawn in the respective insets. The insets of (a) and (b) show the schematic differential magneto-optical image (DMOI) where the contrast difference in the image is obtained by taking the difference between the $B_z(r)$ profiles at H and $H + \delta H$ in the (a) and (b) main panels, respectively. The different shades of gray correspond to a larger difference (light gray) and a smaller difference (dark gray) in $B_z(r)$ in response to $\delta H = 1$ Oe, respectively. (c) and (d) are actual DMOI images measured at 60 K with $H = 72$ and 204 Oe, respectively. (e) and (f) are the $\delta B_z(r)$ behaviour measured across the patterned region at 60 K at $H =$ 72 and 204 Oe, corresponding to (c) and (d), respectively. The patterned region is identified by the dashed vertical lines in (e) and (f).

Figures 3(c) and 3(d) shows actual experimental DMOI images of the patterned region of the sample captured at 60 K and $H = 72$ and 204 Oe, respectively. Figures 3(e) and 3(f) show the $\delta B_z(r)$ measured along the dashed lines in figures 3(c) and 3(d), respectively. We can see that by comparing figures 3(e) and 3(a), that the $B_z(r)$ profile (viz., the vortex distribution) inside the patterned region doesn't get significantly altered due to δH of 1 Oe when compared to the unpatterned regions of the sample. From figures 3(e) and 3(f) we observe that outside the patterned region the $\delta B_z(r) \approx \delta H = 1$ Oe, indicating nearly reversible response in these unpatterned regions of the sample. In comparison to figure 2(c), the situation changes quite dramatically at higher fields, e.g. at 204 Oe (cf. figures 3(d) and 3(f)). A comparison of figures 3(f) and 3(d) at 204 Oe shows significantly larger changes present in vortex density distribution inside the patterned region, leading to a bright contrast inside the patterned region in response to the δH modulation. From the conventional Bean model [18], as compared to field distribution at low H, at higher H the average B_z inside the superconductor is higher. Also, the gradient in the field distribution decreases w.r.t. gradient at lower H since the vortex distribution approaches closer to a uniform distribution with increasing inter-vortex interaction. Hence the brightening observed in figure 3(d) implies that at higher fields the vortex state inside the patterned region is approaching the expected behaviour in a weak pinning superconductor. Note that in the unpatterned

Journal of Physics: Conference Series **638** (2015) 012009 doi:10.1088/1742-6596/638/1/012009

region of the superconductor as the sample is reversible with $B_z \sim H$ hence the maximum δB_z change expected in the local field will be of the order of δH. Note the appearance of bright contrast inside the patterned region in DMOI images begins only above 132 Oe. It appears that at low fields when the inter-vortex interaction is low and comparable with the vortex - blind hole pinning the dilute vortex configuration is quite stable.

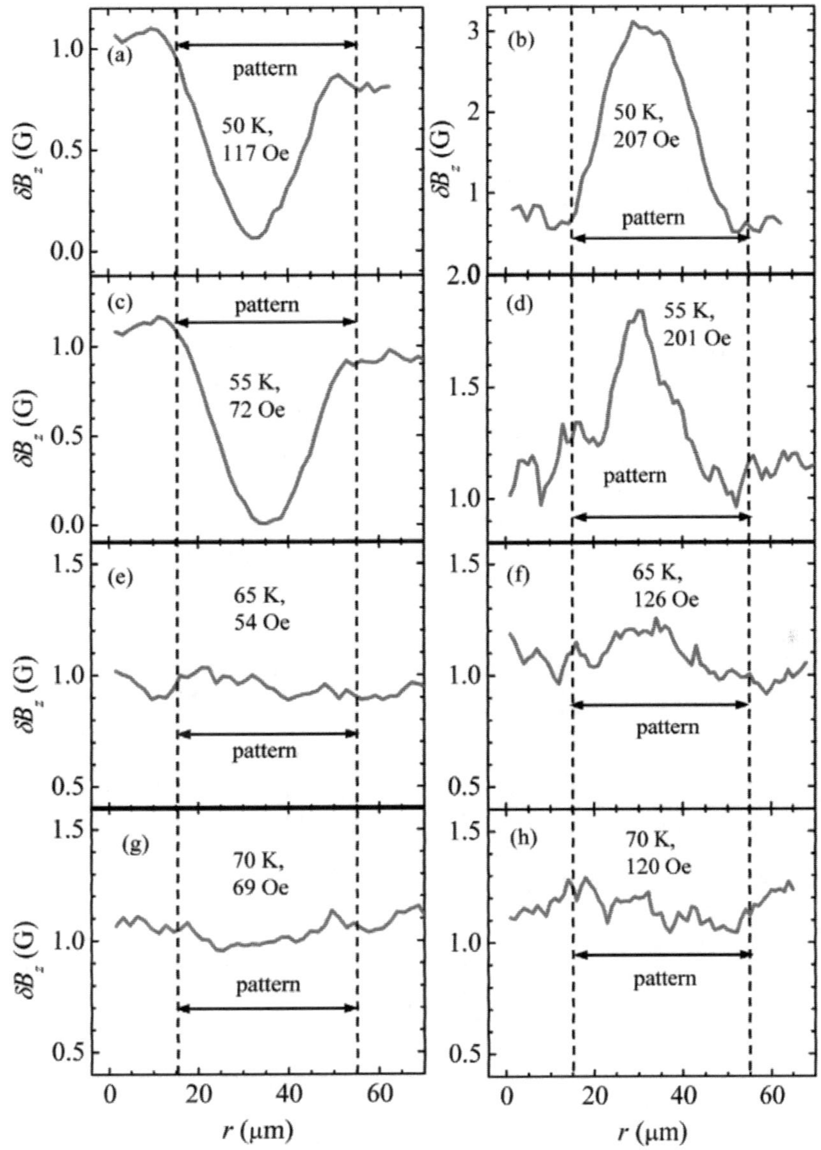

Figure 4. $\delta B_z(r)$ behaviour across the patterned region at various T above and below 60 K. (a)-(b) 55 K, (c)-(d) 60 K, (e)-(f) 65 K and (g)-(h) 70 K, at different H as indicated in each panel. The patterned region is identified by the dashed vertical lines in (a)-(h).

Figure 4 shows the $\delta B_z(r)$ measured across the patterned region in similar fashion as in figure 3 at different T above and below 60 K. As expected from the observations in figure 2, the $\delta B_z(r)$ profiles at

50 K and 55 K (< 60 K) (cf. figures 4(a)-(d)) follow similar trends of lower δB_z inside the patterned region transforming to higher δB_z as H is increased, as observed at 60 K (cf. figure 3). And at 65 K and 70 K (> 60 K) (cf. figures 4(e)-(h)), the δB_z inside the patterned region is seen to be almost indistinguishable from that outside it. By comparing figures 2, 3 and 4 we see that at each T near the local penetration field H_d for the patterned region the DMOI response (δB_z) changes from small to larger values inside the patterned region in comparison changes in rest of the sample, i.e., δB_z inside the patterned region changes from ~ 0 G to value greater than δH (the field modulation = 1 Oe).

3.3. Gradients in vortex density in the vicinity of the patterned region

To further analyse the magnetization response, we obtain the variation of local field $B_z(x, y)$ profile with H across the nanopatterned region determined from conventional MOI at different T. Figures 5(a)-5(d) show maps of the $B_z(x, y)$ (as deduced from MO measurements) in and around the patterned region shown with a resolution of 0.1 G at T = 60 K and H = 51, 96, 144, and 216 Oe, respectively. The patterned region is indicated by the dashed square in figure 5(d) for clarity. The scale bar beside each image represents the variation in $B_z(x, y)$ in the corresponding image. Note that the intensities (contrast) in a conventional MO image represent variations in the distribution of B_z, unlike a DMOI image in which the intensities represent variations in δB_z produced in response to a H modulation (cf. discussion on MOI technique in section 2.2).We note that in each image, the lowest B_z value (darkest) is observed near the center of the patterned region, and highest B_z value (brightest) is observed outside its periphery. Figure 5(e) shows $B_z(r)$ profiles across the patterned region (along the dashed line shown in figure 5(a)) at 60 K, at different H as indicated along the profiles. The vertical dashed lines in figure 5(e) indicate the boundaries of the patterned region. It is observed that as H is increased; significant gradients (slope) in the B_z profile start developing across the patterned region.

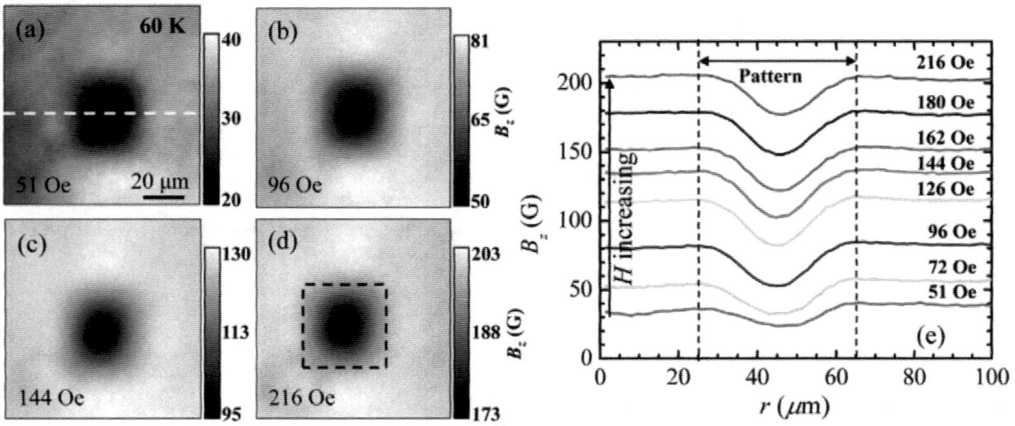

Figure 5. (a)-(d) Contour maps of the $B_z(x, y)$ distribution in a ~ 80 μm × 80 μm region around the patterned region at 60 K and H = 51, 96, 144 and 216 Oe, respectively. The patterned region is indicated by the dashed square in (d) for clarity. (e) $B_z(r)$ profiles across the patterned region along the dashed line shown in (a), at 60 K at different H as indicated on the figure. The patterned region is identified by the dashed vertical lines in (e).

The presence of significant large gradients around the patterned region suggests that any change in B_z inside the patterned region in response to a change in H is shielded by the patterned region. This is consistent with what has been discussed previously in context of figures 2 and 3. However, from the scale bars, it is noted that at H < and close to B_ϕ, the local B_z inside the patterned region is non-zero and it increases with H. This indicates that the patterned region does not completely shield magnetic field changes even at lower H. Note that the $B_z(r)$ data figure 5(e) corresponds to the particular line along which the profile has been obtained. However the same might not be the case along all the directions of the patterned area, as clearly from the images in figures 5(a)-5(d) B_z distribution across the patterned region is not uniform (cf. the contrast variation along the boundary of the patterned region indicated by the dashed square in figure 5(d)). At 60 K, the gradients in the B_z profiles remain almost unchanged for higher H (> B_ϕ) and are sustained up to the maximum H applied in our experiments. From the above observations, we conclude that the nanopatterned region is characterized by an unusual heterogeneous pinning response. On one hand it exhibits characteristics of strong pinning, with significant gradients in B_z being sustained near the periphery (cf. figure 5), while local B_z increases with relative ease with increase in H near the center (cf. figures 3 and 5), which is characteristic of weak pinning.

3.4. Mapping the distribution of shielding currents

By numerically inverting the field distribution [25,26] measured across the patterned region using MOI we determine the absolute value of the screening current distribution $|j(r)|$ inside the patterned region at 60 K. Figures 6(a)-6(d) show maps of the $|j(r)|$ distribution in and around the patterned region.

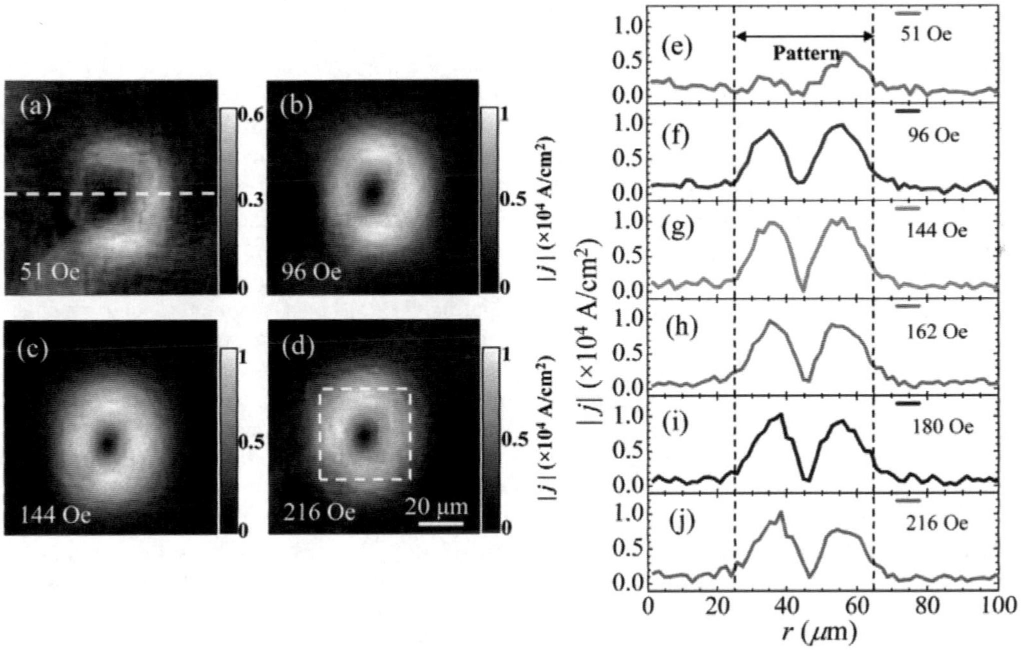

Figure 6. (a)-(d) Contour maps of the $|j(r)|$ distribution, obtained from the $B_z(x, y)$ distribution, in a ~ 80 μm × 80 μm region around the patterned region at 60 K and H = 51, 96, 144 and 216 Oe, respectively. The patterned region is indicated by the dashed square in (d) for clarity. (e)-(j) $|j(r)|$ profiles across the patterned region along the dashed line shown in (a), at 60 K at different H as indicated on the figure. The patterned region is identified by the dashed vertical lines in (e)-(j).

The patterned region is indicated by the dashed square in figure 6(d) for clarity. The graphs in figures 6(e)-(j) show the $|j(r)|$ profiles along the dashed line (cf. figure 6(a)) at different H. The scale bar along each image in (a) to (d) indicates the range of $|j|$ in the corresponding image (in units of 10^4

A/cm^2). At H = 51 Oe, (cf. figure 6(a)) large $|j(r)|$ around the periphery of the patterned region (indicated by the bright contrast in these regions) is observed. This corresponds to these regions possessing relatively high gradients in $B_z(r)$. Far outside the patterned region $|j(r)|$ is almost zero (indicated by the dark contrast in these regions), indicating that $B_z(r)$ distribution in the unpatterned regions of the sample is almost constant. As H is increased, $|j(r)|$ at the boundary of the patterned region increases, as indicated by the appearance of bright contrast over larger regions, and also higher value corresponding to the bright contrast in figures 6(b)-6(d). Also note that the $|j(r)|$ distribution is not uniform across the patterned region. At 51 Oe, larger $|j(r)|$ is observed along the lower right edge of the patterned region, whereas at higher H, larger $|j(r)|$is observed along the upper right edge. These features are also seen in the $|j(r)|$ profiles in figures 6(e)-(j). At 51 Oe (cf. figure 6(e)), $|j(r)|$ near the right edge of the patterned region is higher than the left edge. Subsequently at higher H (cf. figures 6(f)-(j)), the overall $|j(r)|$ values near the periphery of the patterned region are enhanced and the large $|j(r)|$ values are sustained at higher H. These observations are consistent with those in figure 5 above as gradients in $B_z(r)$ are proportional to $j(r)$ [18] (Note that $|j(r)|$ represents the absolute value only, and not the sign of j).

3.5. Heterogeneous pinning in the patterned region

As discussed above, we observed in figures 3-6 that with increasing H there is an increase in B_z near the center of the patterned region while large gradients near the periphery of the patterned region are sustained. $B_z(r)$ and $|j(r)|$ distributions inside the nanopatterned region are also seen to vary non-uniformly across the patterned region (cf. figures 5 and 6) at low H. These results suggest that the vortex population inside the nanopatterned region of the sample exhibits signatures associated either with weak or strong pinning viz., a heterogeneous pinning response at low H below B_ϕ. We believe that the heterogeneous pinning response indicates an unusual distribution of vortices inside the patterned region, contrary to the expectation that at $H < B_\phi$ vortices are mostly pinned on blind holes.

In the absence of a general model which allows a direct comparison with the peculiar nature of field gradients and current distribution determined in our nanopatterned pinning array study (specifically in the specific H-T regime we investigate), we present a speculative one dimensional microscopic model. The proposed microscopic model is a simplified one dimensional model which is similar to a 1-d model proposed to understand general features commensurate incommensurate states in condensed matter systems [27]. We adapt this model [27] to describe a one dimensional distribution of vortices over a one dimensional periodic array of pins (cf. figure 7(a)). In its present form the model helps make qualitative connections with the data. The model presently lacks rigor for direct quantitative comparisons with maps of field distributions over a pinning array in effectively two dimensions. We hope future work will help evolve the model allowing for more quantitative comparisons with actual data. Infact a recent theoretical study [20] has considered the peculiarities of the vortex distribution in mesoscopic superconductors with correlated pins. While this study [20] doesn't directly allow for a direct quantitative comparison with the gradients we measure however there are interesting inferences one can draw from this work which we will discuss subsequently.

Returning back to the model we propose: In this periodic configuration of pins the vortices are expected to separate into a commensurate - incommensurate (CI) vortex state as shown in figure 7(a). The vortices are represented as red circles and the periodic pinning potential created by periodic array of pins (or patterned holes as in our experiment) as a solid wavy line. If the periodic pinning potential has a periodicity b, then the vortex location in the CI phase at the nth pinning site is described in terms of the inter-pin distance b by a phase term [27] φ_n, $x_n = nb + \frac{2\pi}{b}\varphi_n$. In a continuum limit by considering x as a continuous variable one can write a differential equation for $\varphi(x)$ [27]. The one dimensional differential equation has a Solitonic solution for $\varphi(x)$. We describe the generation of an incommensurate vortex state separating the commensurate vortex phase as a domain wall (DW) in

Journal of Physics: Conference Series **638** (2015) 012009 doi:10.1088/1742-6596/638/1/012009

terms of the Solitonic solution [27] with the phase $\varphi(x)$ behaving as $\varphi(x) \propto \tan^{-1}[\exp(x)]$. The phase, $\varphi(x)$ changes from 0 to 2π from left to right across the domain wall as shown in figure 7(b) as per $\varphi(x) \propto \tan^{-1}[\exp(x)]$ (note, $\varphi(-x)$ is an equally possible solution, with the $\varphi(x)$ changing from 0 to 2π from right to left). Thus the incommensurate region in the vortex state is identified with the nucleation of a Domain wall in the system as shown in figures 7(a) and 7(b).

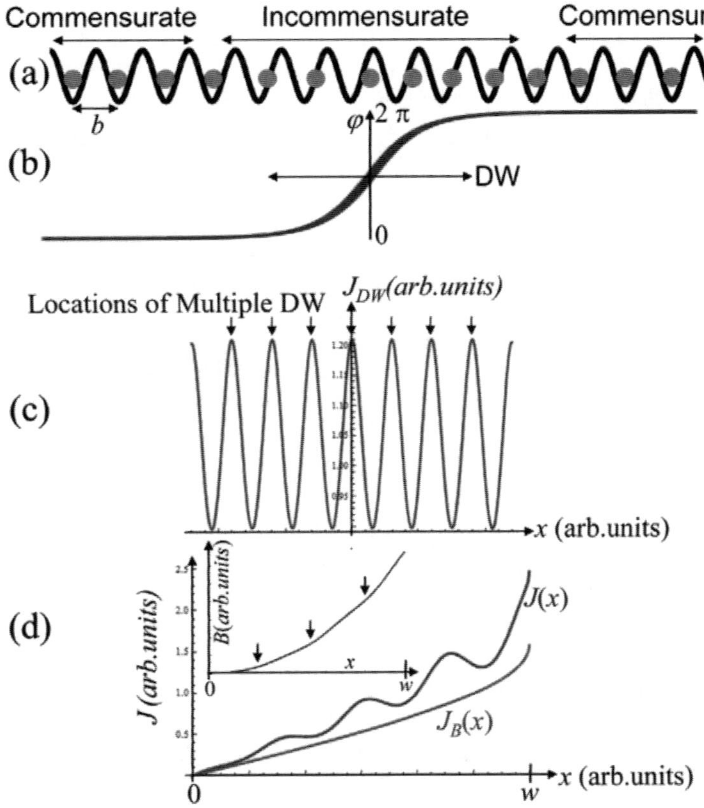

Figure 7. (a) A schematic representation of commensurate – incommensurate (CI) phase of vortices (represented as red circles) on a periodic pinning force background (black solid lines). (b) The behaviour of the phase $\varphi(x)$ describing the domain wall (DW) a CI phase, (c) Current distribution profile associated with multiple domains walls (J_{DW}) in the one dimensional model. (d) Current density profile $J(x)=J_{DW}(x)J_B(x)$. For comparison we show the Bean current density profile $J_B(x)$. In the inset we show the $B(x) = \int_0^x J(x)dx$ behaviour calculated numerically from the $J(x)$. The location of breaks in slope of $B(x)$ is indicated by arrows. In the graphs shown, w represents the width of the patterned region from its center.

From figure 7(a) it is clear that across the DW, viz., in the incommensurate region, there will be a change in the density of vortices and hence a magnetic field gradient will develop in this region compared to the region where there is a commensurate phase (the local field distribution here will be is uniform). We propose the behaviour of the vortex density, as one moves from the commensurate to the incommensurate region in figure 7(a), can be modelled as the vortex number density behaving as

Journal of Physics: Conference Series **638** (2015) 012009 doi:10.1088/1742-6596/638/1/012009

$\rho[x] = \delta(x-nb) + c_{incom}(1-Cos[\phi(x)])$ where the $\delta(x\text{-}nb)$ represents a delta function for the number density at each pin site in the commensurate region and c_{incom} is a positive constant and the term gives the extent of change in number density in the incommensurate DW region. In the incommensurate region the factor (1-Cos[$\varphi(x)$]) is zero near the edges of the DW or the incommensurate region where $\varphi(x) = 0$ or 2π and is maximum near the center where $\varphi(x) = \pi$.

We now attempt to estimate the slope of the local vortex density at a given point inside DW: Across the DW the density of vortices changes. The incommensurate DW region is characterised by change in φ, therefore around a given value of $\varphi(x)$ we estimate the local slope of $\varphi(x)$ and hence the changes in vortex density. For an infinitesimal $\delta\varphi$ increment in φ at a point inside DW, corresponds to an incremental change in (1-$Cos[\varphi(x)]$) ~ $2\varphi(x)\delta\varphi$. This in turn leads to a change in local vortex density at a point inside the DW, by an amount $\delta\rho[x]$ ~ $c_{incom}\phi(x)\delta\phi$. Note as in general $B[x] = \phi_0\rho[x]$ is valid, therefore from this 1-D model one may determine the Bean like currents associated with local changes in gradients using $dB/dx \propto d\rho/dx$. From the above discussion one can find that the local current density at a location inside a domain wall (J_{DW}) by $J_{DW} \propto \frac{dB}{dx} \propto \frac{d\varphi(x)}{dx} \propto \frac{d}{dx}\{\tan^{-1}[\exp(x)]\}$ (here we have considered the dominant contribution to Bean critical current as coming from the DW region).

Above we have considered only one DW, next we consider multiple domain walls which can be excited in this system. We model the current density associated with equally spaced multiple domain walls as

$$J_{DW}(x) \propto \frac{d}{dx}\{\sum_k \tan^{-1}[\exp(x-k)]\} \tag{1}$$

where k is an integer index identifying the location of a domain wall. For multiple domain walls in this 1-D model the J_{DW} is shown in figure 7(c), where the different peaks located with arrows are associated with current peaks at domain walls at the location index k. If the domain walls are reasonably well separated then the modulation in J_{DW} is much larger as compared to that shown in figure 7(c). Apart from J_{DW}, inside the patterned region there is an overall Bean current density profile which is of a well-known form obtained from the modified Bean profile [28]

$$J_B(x) \propto \tan^{-1}\left[\frac{x}{\sqrt{w^2-x^2}}\right] \tag{2}$$

where w is half the width of the patterned region. Using equations (1) and (2), we get overall current density profile inside the patterned region as

$$J(x) \propto \frac{d}{dx}\left\{\sum_k \tan^{-1}[\exp(x-k)]\right\} \times \tan^{-1}\left[\frac{x}{\sqrt{w^2-x^2}}\right] \tag{3}$$

as the overall Bean current density profile modulated by the domain wall current profile (J_{DW}). The behaviour of $J(x)$ is shown in figure 7(d) along with the behaviour of $J_B(x)$ also shown on the same plot for comparison (note in figure 7(d), w represents the sample width, and we plot the behaviour for only half the sample). It is clear from figure 7(d) that the current profile inside the patterned region gets modulated by the presence of multiple domain walls and the current profile appears uneven and rides above the Bean profile. Note from figure 7(d) that the $J(x)$ appears higher than the conventional Bean like $J_B(x)$. The reason for this is that we have considered very closely spaced domain walls due to which there is a significant overall contribution of the currents from the domain walls in the $J(x)$ expression and therefore the average current density increases. By considering sufficiently well spaced out DW's the above feature does not appear. From the above model, it appears at the location of a domain wall there is a peak in the current density $J(x)$.

Using equation (3), we calculate the behaviour of the magnetic field profile $B(x)$ in the one dimensional system, by numerically integrating, $B(x) \propto \int J(x)dx$. The behaviour of $B(x)$ is shown in figure 7(d) inset. As expected the peaks in the $J(x)$ naturally produce breaks in the slope of the magnetic field distribution $B(x)$ (cf. location of arrows in figure 7(d) inset). From figure 7(d) it can be summarized that while the breaks in slope of $B(x)$ due to DW is less easy to discern, their evidence in $J(x)$ behaviour is seen more strongly. Due to the presence of domain walls inside the patterned region there is a distribution of current densities inside the patterned region and the Bean like profile gets modified. The above is characteristic of heterogeneous pinning properties which can be found due to CI transition in the vortex state in the presence of nanopatterned pinning array.

Based on the above calculations, the heterogeneous pinning inside the patterned region may arise from domain wall boundaries excited in the vortex configuration inside the nanopatterned region of the sample. A detailed recent theoretical study [20] also suggests a nested domain like structure for the vortex configuration within a mesoscopic superconductor with correlated pinning centers. The study also showed a domain like structure with a hierarchical organization of vortices over pinning centers. Each domain is associated with different filling fractions. Such a multi-domain like vortex configuration described in the theoretical study [20] may also exhibit a heterogeneous pinning character. In one of our recent detailed studies, [29] detailed maps of the shielding current distribution inside the patterned region upon reducing the applied field from a high field value suggest regions with different critical currents co-habiting within the patterned region, which supports the recent theoretical scenario of vortex phase separation [20] inside the patterned region. The above scenarios pave the way for future investigations into the peculiarities of the vortex configurations possible inside the nanopatterned regions and also ways to enhance pinning.

4. Conclusion

In conclusion in a BSCCO single crystal patterned with a periodically spaced blind hole array we have identified signatures of enhanced flux shielding and heterogeneous pinning response of the vortex configuration inside the patterned region at low H below B_ϕ. We have identified the regime where the pinning from the blind hole array dominates over the intrinsic pinning in the sample and shown that in this regime, large field gradients are sustained near the periphery of the patterned region while at the same time flux density near the center of the patterned region increases with increasing H. The distribution of local field gradients and shielding current density across the patterned region is also found to be non-uniform in the dilute vortex density regime of our measurements. We believe our work paves the way for future explorations into the nature of vortex configurations possible in nanopatterned systems.

Acknowledgments

SSB acknowledges FIB related staff and help from Dr. Amit Banerjee and funding support from JSPS, Japan- DST India and Indo – Spain joint research projects.

References

[1] Blatter G, Feigel'man M V, Geshkenbein V B, Larkin A I and Vinokur V M 1994 *Rev. Mod. Phys.* **66** 1125
[2] Giamarchi T and Le Doussal P 1994 *Phys. Rev. Lett.* **72** 1530; 1995 *Phys. Rev. B* **52** 1242
[3] Fisher M P A 1989 *Phys. Rev. Lett.* **62** 1415
[4] Córdoba R, Baturina T I, Sesé J, Mironov A Yu, De Teresa J M, Ibarra M R, Nasimov D A,

Gutakovskii A K, Latyshev A V, Guillamón I, Suderow H, Vieira S, Baklanov M R, Palacios J J and Vinokur V M 2013 *Nature Communications* **4** 1437
[5] Shaw Gorky, Bag Biplab, Banerjee S S, Suderow Hermann and Tamegai T 2012 *Supercond. Sci. Technol.* **25** 095016
[6] Shaw Gorky, Sinha Jaivardhan, Mohan Shyam and Banerjee S S 2010 *Supercond. Sci. Technol.* **23** 075002
[7] Shaw Gorky, Mandal Pabitra, Bag Biplab, Banerjee S S, Tamegai T and Suderow Hermann 2012 *Appl. Surf. Sci.* **258** 4199
[8] Banerjee S S, Shaw Gorky, Sinha Jaivardhan, Mohan Shyam and Mandal Pabitra 2010 *Physica C* **470** S817
[9] Hebard A F, Fiory A T and Somekh S 1977 *IEEE Trans. Magn.* **1** 589
[10] Baert M, Metlushko V V, Jonckheere R, Moshchalkov V V and Bruynseraede Y 1995 *Phys. Rev. Lett.* **74** 3269
[11] Moshchalkov V V, Baert M, Metlushko V V, Rosseel E, van Bael M J, Temst K, Jonckheere R and Bruynserade Y 1996 *Phys. Rev. B* **54** 7385
[12] Moshchalkov V V, Baert M, Metlushko V V, Rosseel E, van Bael M J, Temst K, Bruynserade Y and Jonckheere R 1998 *Phys. Rev. B* **57** 3615
[13] Thakur A D, Ooi Shuuichi, Chockalingam Subbaiah P, Jesudasan John, Raychaudhuri Pratap and Hirata Kazuto 2009 *Appl. Phys. Lett.* **94** 262501
[14] Tamegai T, Tominaga T, Naito D, Nakajima Y, Ooi S, Mochiku T and Hirata K 2010 *Physica C* **470** S784
[15] Silhanek A V, Gutierrez J, Kramer R B G, Ataklti G W, Van de Vondel J and Moshchalkov V V 2011 *Phys. Rev. B* **83** 024509
[16] Dai H, Liu J and Lieber C M 1994 *Phys. Rev. Lett.* **72** 748
[17] Fasano Y and Menghini M 2008 *Supercond. Sci. Technol.* **21** 023001
[18] Bean C P 1964 *Rev. Mod. Phys.* **36** 31
[19] Cooley L D and Grishin A M 1995 *Phys. Rev. Lett.* **74** 2788
[20] Laroshenko O, Rybalko V, Vinokur V M and Berlyand L 2013 *Scientific Report (Nature)* **3** 1758
[21] Shaw Gorky, Mandal Pabitra, Banerjee S S and Tamegai T 2012 *New J. Phys.* **14** 083042
[22] Mandal Pabitra, Chowdhury Debanjan, Banerjee S S and Tamegai T 2012 *Review of Sci. Instr.* **83** 123906
[23] Soibel A, Zeldov E, Rappaport M, Myasoedov Y, Tamegai T, Ooi S, Konczykowski M and Geshkenbeink V B 2000 *Nature (London)* **406** 282
[24] Banerjee S S, Soibel A, Myasoedov Y, Rappaport M, Zeldov E, Menghini M, Fasano Y, de la Cruz F, van der Beek C J, Konczykowski M and Tamegai T 2003 *Phys. Rev. Lett.* **90** 087004
[25] Wijngaarden R J, Heeck K, Spoelder H J W, Surdeanu R and Griessen R 1998 *Physica C* **295** 177
[26] Banerjee S S, Goldberg S, Soibel A, Myasoedov Y, Rappaport M, Zeldov E, de la Cruz F, van der Beek C J, Konczykowski M, Tamegai T and Vinokur V M 2004 *Phys. Rev. Lett.* **93** 097002
[27] Bak P 1982 *Rep. Prog. Phys.* **45** 587
[28] Brandt E H and Indenbom M 1993 *Phys. Rev. B* **48** 12893
[29] Shaw Gorky, Banerjee S S, Suderow Hermann and Tamegai T 2015 (*submitted*)

Journal of Physics: Conference Series **638** (2015) 012010 doi:10.1088/1742-6596/638/1/012010

Complex ordering in spin networks: Critical role of adaptation rate for dynamically evolving interactions

Anand Pathak and Sitabhra Sinha

The Institute of Mathematical Sciences, CIT Campus, Taramani, Chennai 600113, India

E-mail: anandp@imsc.res.in, sitabhra@imsc.res.in

Abstract. Many complex systems can be represented as networks of dynamical elements whose states evolve in response to interactions with neighboring elements, noise and external stimuli. The collective behavior of such systems can exhibit remarkable ordering phenomena such as *chimera order* corresponding to coexistence of ordered and disordered regions. Often, the interactions in such systems can also evolve over time responding to changes in the dynamical states of the elements. Link adaptation inspired by Hebbian learning, the dominant paradigm for neuronal plasticity, has been earlier shown to result in structural balance by removing any initial frustration in a system that arises through conflicting interactions. Here we show that the rate of the adaptive dynamics for the interactions is crucial in deciding the emergence of different ordering behavior (including chimera) and frustration in networks of Ising spins. In particular, we observe that small changes in the link adaptation rate about a critical value result in the system exhibiting radically different energy landscapes, viz., smooth landscape corresponding to balanced systems seen for fast learning, and rugged landscapes corresponding to frustrated systems seen for slow learning.

1. Introduction

Many natural and technological complex systems can be described as networks connecting a large number of elements whose states evolve in time [1, 2]. The collective behavior of a system resulting from interactions between its components can exhibit non-trivial features, including critical phenomena [3]. For instance, a system of binary-state elements (e.g., representing individuals having opposing opinions on an issue) connected through a network having modular organization can exhibit ordering dynamics at very distinct time-scales [4] and under certain circumstances, self-organize into locally aligned clusters that correspond to the communities of the network (i.e., subnetworks characterized by a significantly higher connection density compared to the overall density of the network) [5]. In many situations, the links of the network (representing the interactions) can also evolve over time as a result of the changes in the states of the components that they connect. Such connections may not only be characterized by weights (indicating the strength of interaction) but also sign (representing the nature of the interaction). For instance, in the context of a network of synaptically-connected model neurons, positive links may correspond to excitatory interactions (whereby activation of one element can result in activation of other connected elements) while negative links can give rise to inhibition (i.e., activation of an element tends to suppress subsequent activation of neighboring elements) [6].

The occurrence of negative links can lead to the emergence of *frustration* because of the existence of inconsistent relations within cycles in the network [7, 8]. A network is said to

Journal of Physics: Conference Series **638** (2015) 012010 doi:10.1088/1742-6596/638/1/012010

be structurally balanced if its positive and negative links are arranged such that frustration is absent. Such a network can always be represented as two communities, with interactions within each community being exclusively positive while those between communities can only be negative [9]. This is of interest not only for social systems in the context of which the concept of balance was first introduced [10], but also for biological systems. For instance, it has been recently observed that the resting human brain is organized into a pair of dynamically anti-correlated subnetworks [11]. This suggests that the network responsible for this behavior may be structurally balanced. This possibility is intriguing in view of the observation of non-trivial collective behavior in balanced networks of binary-state dynamical elements, such as, the coexistence of ordered and disordered regions referred to as "chimera" order, in the presence of an external field [12]. The process by which networks can evolve to a balanced configuration has been explored by considering a link adaptation process inspired by Hebb's principle, the basis behind neural plasticity [13]. According to this mechanism, weights associated with each link change in proportion to the correlation in the activity of the connected elements. Therefore, any initial frustration in the network can be removed by modifying the connection weights in accordance with the dynamical states of the elements. Systems undergoing such link adaptation in the presence of environmental noise show high variability in the convergence time required to reach the balanced state [14].

In this paper we consider how the rate of dynamical evolution of interactions in accordance with the Hebbian adaptation principle affects the collective behavior of the network. Starting from a fully connected network that is structurally balanced and introducing noise and external field so that the system exhibits chimera order, we show that different rates of adaptation can result in distinct outcomes. Fast learning rates result in persistence of the chimera regime, although the balanced network now comprises communities with asymmetric sizes, viz., the ordered sub-network containing a much larger fraction of elements of the network. Slow learning, on the other hand, results in a completely ordered state with the interactions becoming only of positive nature. On removing the external field, fast learning results in retrieving a network that is similar to the original one, i.e., structurally balanced with communities of almost equal size. However, slow learning gives rise to a fully frustrated network exhibiting disorder. We observe that there exists a critical interval for values of the learning rate, around which small changes result in the system converging to distinct final states. In the following sections we first discuss the model and adaptation dynamics, followed by description of the results of numerical simulations. We conclude with a brief summary of our results and a discussion of their implications.

2. The Model

For the purpose of investigating how collective behavior of a complex system is affected by adaptive dynamics of interactions, we use one of the well-known spin models of statistical physics which are generic systems for analyzing cooperative phenomena. In particular, we use the binary-state Ising spin, the spin orientations ("up" or "down") representing a pair of mutually exclusive choices. The simplicity of the model makes it applicable to not just magnetic materials (in the context of which it was originally proposed) but any system where the elements choose between the two competing states based on interactions with neighboring elements, noise (represented by thermal fluctuations characterized by a temperature T) and external stimulus (often represented by a magnetic field H). The interactions J_{ij} between any pair of spins (i, j) can be either positive (promoting connected spins to have the same state) or negative (promoting connected spins to have opposite states) in nature. For instance, in neuronal networks, one can view the neurons as binary-state devices that are either firing (active) or quiescent (inactive). Correspondingly, as pointed out earlier, the excitatory and inhibitory connections between neurons can be represented by positive and negative interactions, respectively. At a different scale, one can view genetic regulatory networks in a similar vein, with genes being in either of two states, viz., getting expressed (active) or not(inactive). In this setting, genes promoting the expression of other genes correspond to positive interaction, while inhibiting the expression of other genes correspond to negative interactions. In a different context, the interactions between Ising spins can also be used to represent social intercourse [15]. Here, the spin states are considered to be analogous to two competing opinions, while the interactions may represent the nature of the relation between a pair of individuals - positive corresponding to affiliative and negative corresponding to antagonistic relations.

We consider a system of $2N$ globally coupled Ising spins arranged into two sub-populations (called modules or communities) each having N spins, at a constant temperature T. For the simulations reported below we have chosen $N = 100$. The interaction between a pair of spins belonging to the same module is positive having strength J (> 0), while that between spins belonging to different modules is negative with strength $-J'$ (where $J' > 0$). In the absence of an external field and thermal fluctuations (i.e., $H = 0$, $T = 0$), the two modules will be completely ordered in opposite orientations. When subjected to an external field of strength H, the energy of the system is described by

$$E = -J\sum_{i,j,\alpha}\sigma_{i\alpha}\sigma_{j\alpha} - J'\sum_{i,j,\alpha,\beta}\sigma_{i\alpha}\sigma_{j\beta} - H\sum_{i,\alpha}\sigma_{i\alpha}, \tag{1}$$

where $\sigma_{i\alpha} = \pm 1$ is the Ising spin on the i-th node ($i, j = 1, 2, \ldots N$) in the α-th module ($\alpha, \beta = 1, 2$). As every spin interacts with every other spin, a mean-field treatment should describe the system accurately. For convenience we define $a = J(N-1)/2$ and $b = J'N$ as system parameters. The state of all spins in the system are updated stochastically at discrete time-steps using the Metropolis Monte Carlo (MC) algorithm with temperature T expressed in units of k_BT/J where k_B is the Boltzmann constant. The initial state of the system we have chosen is one exhibiting chimera order [12], to obtain which we use $a = 1$, $b = H = 10$ and $k_BT/J = 5$.

We also allow the possibility that the interaction strengths can change over time through an adaptation process inspired by the principle of Hebbian learning, a classical concept in the area of neural networks [13]. Colloquially often described as "neurons that fire together, wire together", in this mechanism the strength of synaptic connection between a pair of neurons is incremented when the two exhibit correlated activation. In Ref. [14], this idea was used to propose a link adaptation rule that applies to a system of spins coupled via positive, as well as, negative interactions as:

$$J_{ij}(t+1) = (1-\epsilon)J_{ij}(t) + \epsilon\sigma_i(t)\sigma_j(t), \tag{2}$$

which is applied after every MC step. The adaptation rate, ϵ, decides the time-scale over which the interaction strength changes relative to the spin dynamics. Starting with an initially frustrated spin system, implementing the above J_{ij} dynamics in absence of thermal fluctuations (i.e., $T = 0$) results in the system converging to a structurally balanced state [14]. This can be intuitively understood in terms of changes in the energy landscape over which the state of the spin system evolves. Frustrated systems have rugged energy landscapes comprising a large number of local minima (a fact which is exploited in neural network models of associative memory where these minima are used to embed desired patterns one wishes the network to "memorize" [16, 17]. Thus, beginning at any randomly chosen initial state of spin configurations, the spin dynamics drives the system into the nearest local energy minimum. The subsequent evolution of the J_{ij}s converts this into the global minimum of the system. However, in the presence of noise ($T > 0$), fluctuations in the spin dynamics can prevent the system from being trapped in any state for sufficiently long duration. Thus, the adaptation dynamics fails to alter the energy landscape

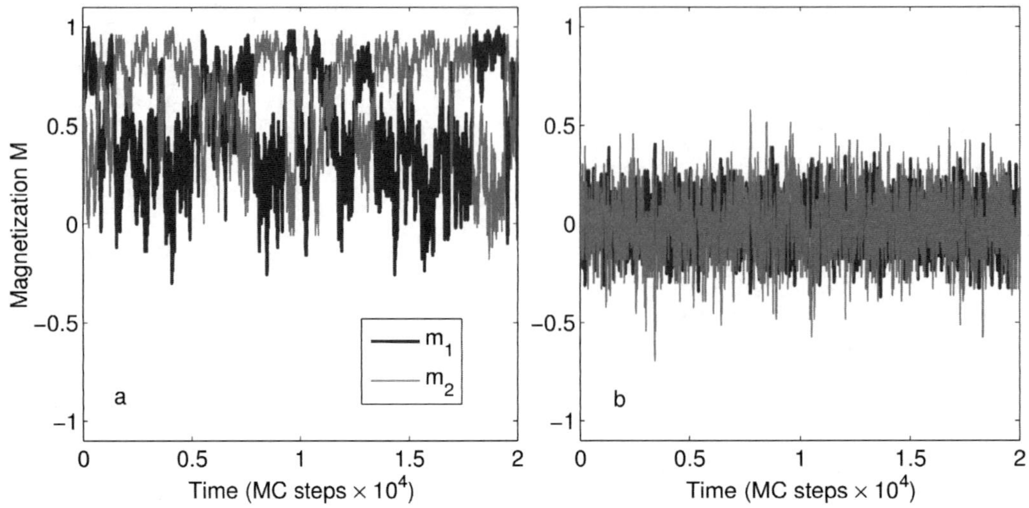

Figure 1. Typical time-evolution of the magnetizations per spin of the two modules, m_1 and m_2, in (a) chimera ordered state and (b) disordered state, shown for MC simulations with $N = 100$ at (a) $k_BT/J = 5$ and (b) $k_BT/J = 8$, respectively.

sufficiently so as to turn the configuration into the global minimum. Therefore, with increasing temperature, an extremely long time may be required to reach structural balance.

In this paper we use a modified form of the Hebb-inspired adaptation rule for interaction strengths that was introduced in Ref. [14]. This is to take into account the asymmetry in the strengths of positive and negative interactions necessary for observing chimera order (typically $J' \simeq 10J$) [12]. We change the adaptation rule such that the upper and lower bounds for the evolving interactions have the same values as that of the positive and negative interactions (respectively) which give rise to the chimera ordered state. Thus after each MC step, interaction strengths are changed according to:

$$J_{ij}(t+1) = J_{ij}(1-\epsilon) + \epsilon[(J+J')\frac{\sigma_i\sigma_j}{2} + \frac{1}{2}(J-J')]. \tag{3}$$

The state of the system at any time is characterized by two quantities. One of these is the frustration associated with the interactions, measured by calculating the fraction of frustrated triads (i.e., connected sets of 3 spins with an odd number of negative interactions) in the system:

$$F = \sum_{i \neq j \neq k} \frac{J_{ij}J_{jk}J_{ki}}{{}^NC_3}, \quad \forall J_{ij}J_{jk}J_{ki} < 0, \tag{4}$$

which varies between 0 (no frustration) and 1 (completely frustrated).

The other quantity is an order parameter $p = |m_1 - m_2|$ that is used to identify a chimera state in a spin system with two modules that have magnetizations m_1 and m_2, respectively. To numerically estimate the order parameter p, we are confronted with a few potential complications. First, as the magnetizations of the two modules are stochastically fluctuating (especially the one which is disordered, i.e., having lower magnetization), we can use their average values. However, this gives rise to an additional problem as the modules can frequently exchange their order/disorder status, especially when the modules are of the same size [Fig. 1 (a)]. Thus, a method is required to measure p that is unaffected by the modules switching their

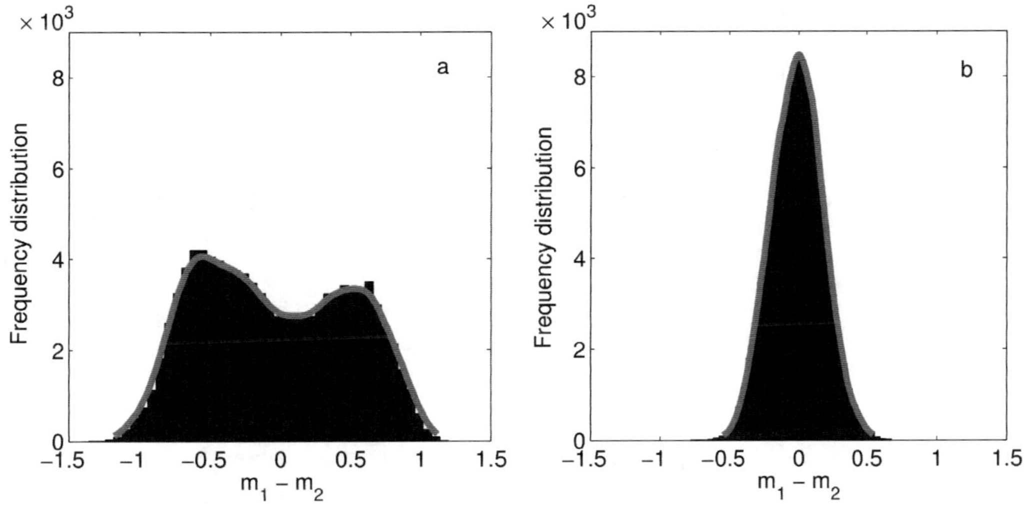

Figure 2. Frequency distribution of the quantity $m_1 - m_2$ shown when the spin system exhibits (a) chimera order and (b) disorder. The kernel smoothened distribution function is represented by a thick curve. Results shown for MC simulations with $N = 100$ at (a) $k_B T/J = 5$ and (b) $k_B T/J = 8$, respectively.

magnetizations, while being able to determine if the time-averaged magnetizations of the two modules are different (the signature of chimera order). For instance, a simple time average of the absolute value of differences between m_1 and m_2 will not give correct results as one obtains a finite value (because of stochastic fluctuations) even when the time averaged magnetizations are same for both modules [Fig. 1 (b)]. Therefore, to estimate p we have used the following algorithm. First, the frequency distribution of $m_1 - m_2$ is computed. In the absence of chimera, the distribution is unimodal [Fig. 2(a)] which is approximated as a Gaussian function, while for chimera ordering the distribution is bimodal [Fig. 2(b)] which is approximated as a superposition of two Gaussian functions. For a unimodal distribution, the value of $m_1 - m_2$ corresponding to the peak is used to calculate the order parameter p. For bimodal distributions, a weighted average of the values corresponding to the peaks is used. For example, if one of the peaks occurs at p_1 with value h_1 while the other is at p_2 with value h_2, the order parameter is calculated as $p = (h_1|p_1| + h_2|p_2|)/(h_1 + h_2)$. In order to make the automated detection of the peak locations accurate, the frequency distributions need to be smoothed of all fluctuations in the frequencies so that the local maxima at the peaks can be uniquely determined. For this purpose we have used a kernel smoothing technique [18] that gives a single peak location for unimodal distributions and locations for two peaks in the case of bimodal distributions.

3. Results

Starting from a structurally balanced network comprising two modules, we first introduce field and thermal fluctuations so as to drive the system into chimera order - i.e., one of the modules becomes ordered while the other is disordered. Once this is achieved we allow adaptation dynamics of the interactions to take place. As the evolution of the interactions are related to the degree of fluctuations in the spin states, these would occur mostly in the disordered module where the spins are subject to the competing forces of negative interactions with the spins belonging to the ordered module and the influence exerted by the uniform external magnetic field which tries to align the spins in parallel with those of the ordered module. Thus, at any instant, a few spins in the disordered module will get aligned with the spins in the other module and consequently may develop positive interactions with the latter as a result of link adaptation. This means that they will now no longer be part of the disordered module but become part of the ordered module. As a result, the size of the modules would change over time - the ordered module increasing at the expense of the disordered one. This has to be taken into account when measuring system properties, such as, magnetization per spin of the two modules. In order to have a consistent definition of module size that would be valid even when the system is no longer in structural balance, we heuristically measure it as follows. At each update of the J_{ij}s, we randomly choose a spin from the ordered module and consider all spins that have positive interactions with it to belong to the ordered module, with the remaining spins comprising the disordered module. Following this process, we can follow the time-evolution of modular membership of individual spins, as well as, that of the module sizes. However, this measure becomes less useful as the system becomes increasingly frustrated.

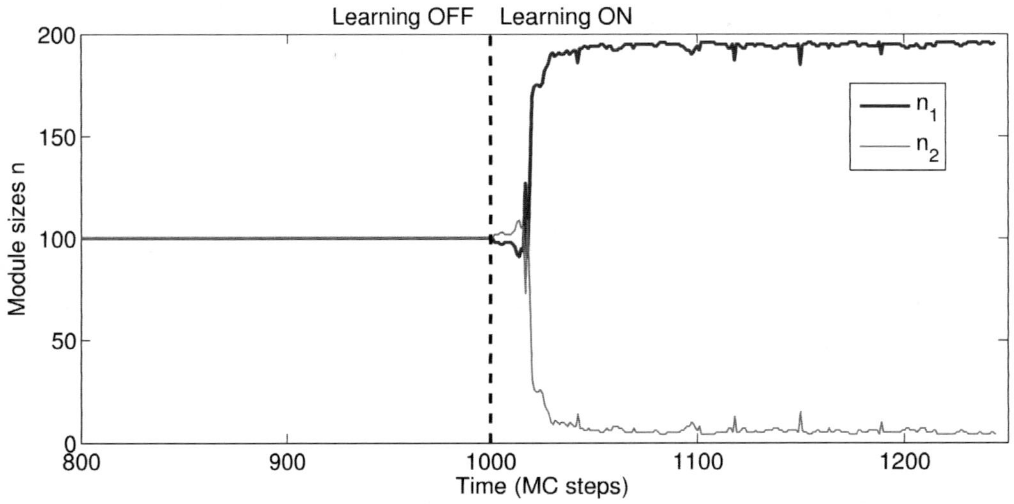

Figure 3. Time-evolution of sizes of the modules resulting from link adaptation dynamics that is introduced at time $t = 1000$ MC steps (indicated by broken line). The ordered module expands in size (n_1) while the disordered one shrinks. Learning rate $\epsilon = 0.1$

Fig. 3 illustrates the evolution of module sizes with time following the introduction of link adaptation dynamics in the system. We can now measure the magnetizations per spin, m_1 and m_2, of the two modules by taking into account the modular membership of each spin and the module sizes. Fig. 4 shows these order parameters, as well as, the frustration in the network, as a function of time before and after the link adaptation dynamics is introduced. As expected, we observe that following the start of link adaptation dynamics, the ordered module begins to grow in size, which is also observed in the time-evolution of the interaction matrix (shown as snapshots at regular intervals in Fig. 5).

The time-evolution of the initial chimera-ordered network subjected to link adaptation dynamics with various learning rates ranging from $\epsilon = 0.1$ to 10^{-4} are indicated in Figs. 6-9. For fast learning rates (e.g., $\epsilon = 0.1$), the system immediately converges to a balanced state characterized by a large ordered module (comprising about 95% of all the elements) and a small disordered one. By observing the magnetization time-series of the two modules in Fig. 6 it is easy to infer that the system still exhibits chimera order. We note that a rise in frustration is seen for a very brief duration during the initial period immediately following initiation of the link adaptation dynamics. For a slower learning rate, viz. $\epsilon = 0.01$, the process of expansion in

Journal of Physics: Conference Series **638** (2015) 012010 doi:10.1088/1742-6596/638/1/012010

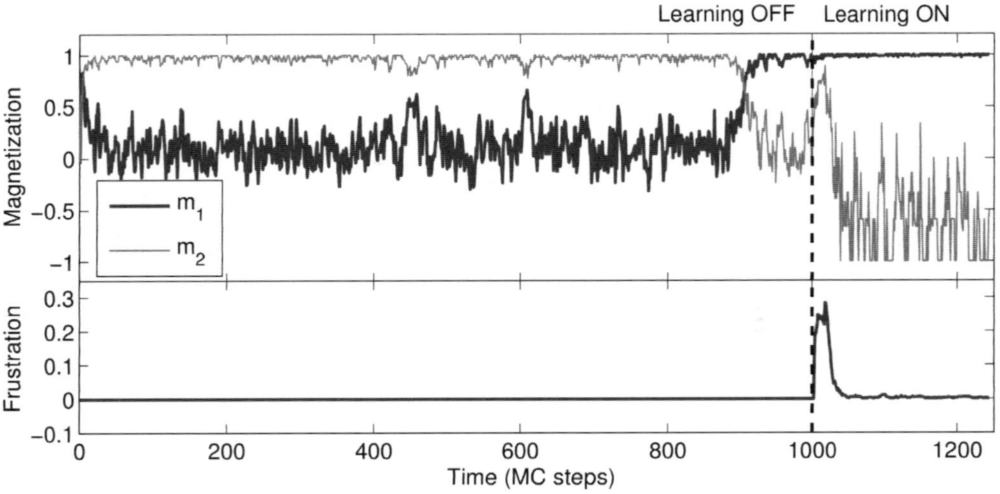

Figure 4. (top) Time-evolution of magnetizations m_1 and m_2 of the two modules in the network (having sizes n_1 and n_2, respectively) before and after link adaptation dynamics is introduced at time $t = 1000$ MC steps (indicated by broken line). In absence of adaptation, magnetizations show the characteristic signature for chimera order. Once adaptation dynamics is operational, the ordered module with magnetization m_1 is seen to increase in size. (bottom) In absence of learning, the network exhibits no frustration as it is structurally balanced. Following the introduction of adaptation dynamics, there is a transient increase in frustration, before the system once again achieves balance, but with modules of asymmetric sizes.

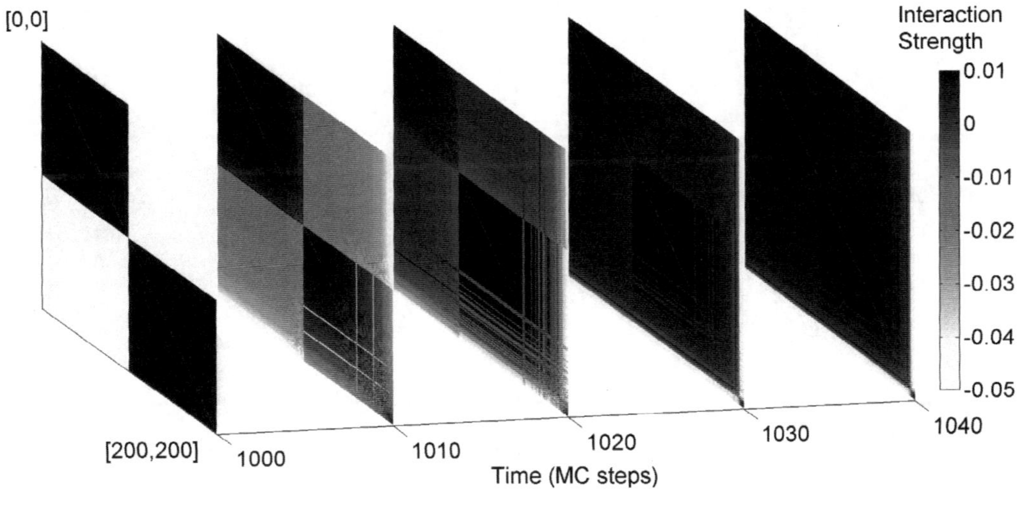

Figure 5. Snapshots of the time-evolution of the interaction matrix **J** after adaptation dynamics is initiated at $t = 1000$ MC steps. Note that the time interval shown here corresponds to the same period over which the frustration in the system initially rises and then decreases again as the system once more reaches a balanced state, as indicated in Fig. 4.

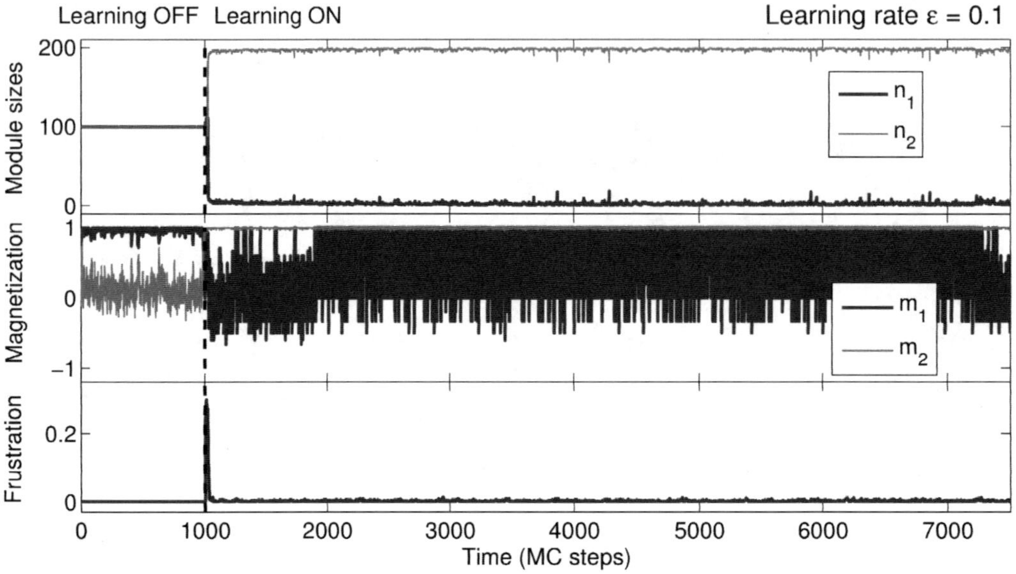

Figure 6. Time-evolution of module sizes (top panel), magnetization (middle panel) and frustration (bottom panel) for an initially chimera ordered system in the presence of an external field before and after link adaptation dynamics with rate $\epsilon = 0.1$ is introduced (the time at which the link adaptation is initiated is indicated by the broken line). Following a brief transient rise in the frustration, the system again attains a balanced state although with asymmetric module sizes.

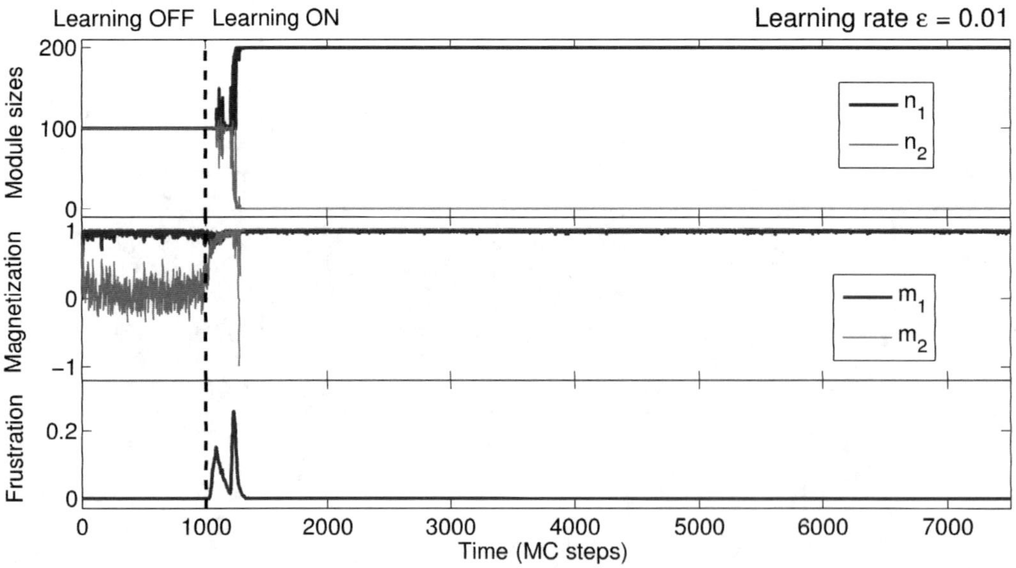

Figure 7. Time-evolution of module sizes (top panel), magnetization (middle panel) and frustration (bottom panel) for an initially chimera ordered system in the presence of an external field before and after link adaptation dynamics with rate $\epsilon = 0.01$ is introduced (the time at which the link adaptation is initiated is indicated by the broken line). The system eventually becomes completely ordered with all interactions becoming positive (resulting in merging of the two modules into a single one).

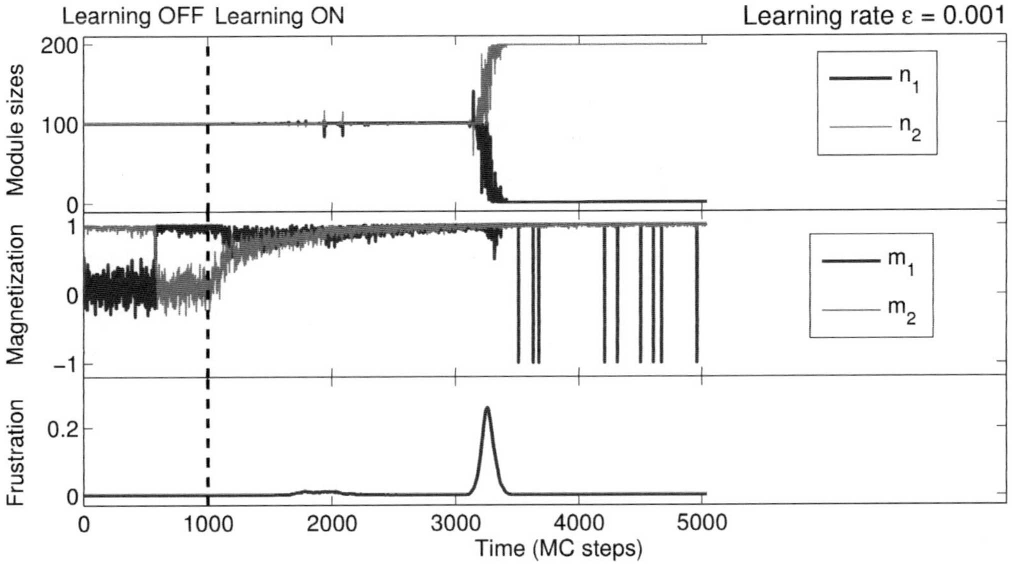

Figure 8. Time-evolution of module sizes (top panel), magnetization (middle panel) and frustration (bottom panel) for an initially chimera ordered system in the presence of an external field before and after link adaptation dynamics with rate $\epsilon = 10^3$ is introduced (the time at which the link adaptation is initiated is indicated by the broken line). Initially, the two modules both become ordered. However, eventually the two modules merge into a single one as all interactions become positive. The time of merging coincides with the transient rise in frustration.

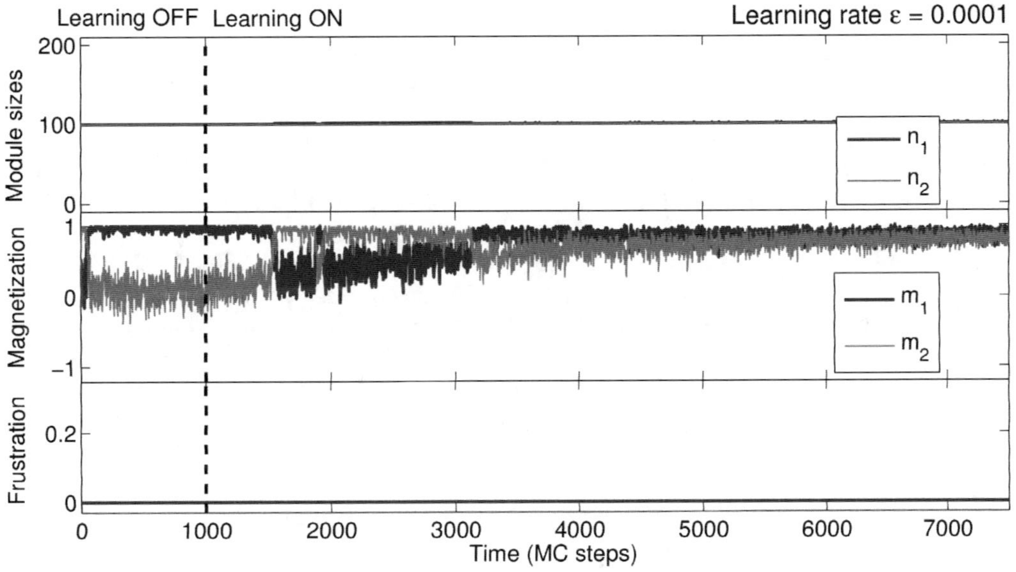

Figure 9. Time-evolution of module sizes (top panel), magnetization (middle panel) and frustration (bottom panel) for an initially chimera ordered system in the presence of an external field before and after link adaptation dynamics with rate $\epsilon = 10^4$ is introduced (the time at which the link adaptation is initiated is indicated by the broken line). We observe that while one of the modules remains ordered, the initially disordered modules is gradually also tending towards complete order. Eventually one expects the two modules to merge into a single one.

size of the ordered module is different (Fig. 7). We notice that soon after the beginning of link adaptation, the chimera ordering is lost as the disordered module becomes ordered. However, the modules still retain their different identities as they are defined based on the sign of interactions of the spins within them (which should be positive) and that with spins in the other module (which should be negative). For even slower learning rates such as $\epsilon = 10^{-3}$ or 10^{-4} (Figs. 8-9) the duration over which the two modules exist while both being ordered is seen to increase. During this period the modular membership of the spins do not change significantly (apart from small fluctuations). Chimera ordering is lost as the interaction strengths for spins in the disordered module becomes weaker as a result of frequent switchings of their orientations. As a result the effect of the external field becomes dominant, which causes the spins in the disordered module to align with it. As spins in both modules are now aligned parallel to each other, adaptation of their interactions with time would eventually make all interactions positive, thereby merging the two modules into one. The time at which this happens is indicated by a transient rise in the frustration (Figs. 7 and 8), occurring much later for slower rates of adaptation.

The different scenarios we observe at fast and slow learning rates suggest that there is a competition between two processes: if the modular membership of the spins can change rapidly following initiation of link adaptation so that the system adopts a new structurally balanced configuration, the interaction strengths of the spins in the disordered module do not decrease significantly - thereby allowing the two modules to coexist with the chimera ordering intact (although the modules now have very different sizes). Note that the modular membership of the spins will change rapidly as they continually change their interaction strengths because of the fast adaptation rate. However, the average number of spins in each module will remain fairly steady. If the learning rate is slow, the interaction strengths will not have time to change sufficiently rapidly to allow the system to reach a new structurally balanced state following the initiation of link adaptation dynamics. As a result the field dominates over the weakened interactions of the spins in the disordered module, thereby eliminating chimera order and eventually causing the two modules to merge.

The existence of chimera order even in situations where the two modules have very different sizes is a novel observation as earlier it had only been observed in a system with symmetric module sizes [12]. In order to see how the region in the field-temperature parameter space where we observe chimera ordering varies with asymmetry in the module sizes, we show in Fig. 10 the dependence of the order parameter $|m_1 - m_2|$ on the strength of the external field (H), the scaled temperature ($k_B T/J$) and the ratio of sizes of the two modules (n_1/n_2) in the absence of link adaptation dynamics. We observe that the region in parameter space where we find chimera order (higher values of $|m_1 - m_2|$) increases significantly as we go towards higher module size ratios (i.e., n_1/n_2) especially for lower values of temperature.

We now consider the situation when the external field is withdrawn while allowing the link adaptation dynamics to continue. A difference is expected with the results shown in the earlier work [14] where the effect of link adaptation on the evolution of a system subject to thermal fluctuations was investigated, as the learning rule is different on account of the asymmetry in the strengths allowed for negative and positive interactions. As the adaptation dynamics used here allows for much stronger negative interactions (as compared to positive), we would expect the system to become more frustrated as negative interactions are much more likely to contribute frustrated triads. This is indeed what is observed for slow learning rates, e.g., $\epsilon = 0.1$ and $\epsilon = 0.01$ (Fig. 11). Note that in a frustrated system, one cannot define modules in a meaningful way, and thus the distinction between modules indicated in the time-series for magnetization is an artifact of the measurement method. For both of the slow learning rates we observe that the system converges to a fully frustrated state corresponding to the maximum value of the measure of frustration in the system (= 1).

For faster link adaptation rates, however, we observe unexpected behavior in the system. For

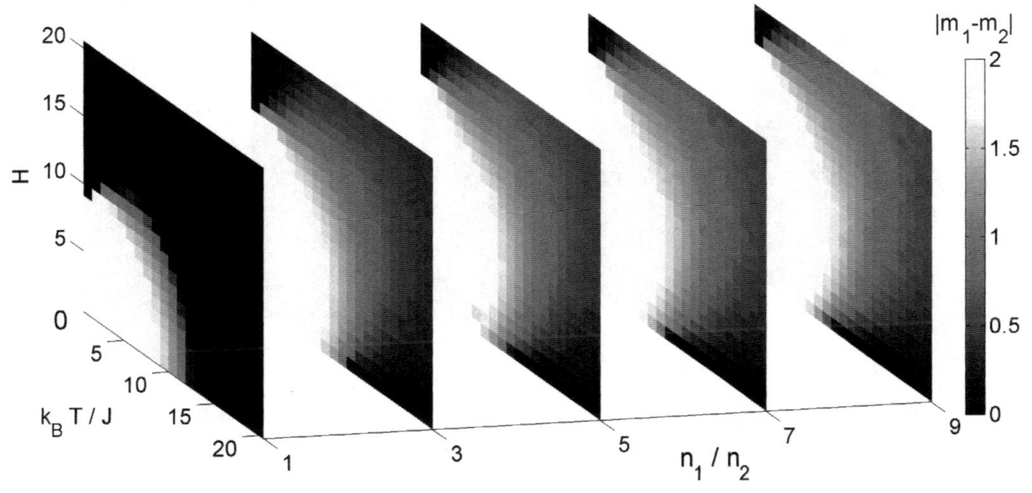

Figure 10. The region of chimera order in the (H, T) parameter space shown for different ratios of the sizes of the two modules. Chimera is indicated in terms of high values of the order parameter $|m_1 - m_2|$ (represented by different colors).

a learning rate $\epsilon = 0.13$ [Fig. 12], which is only slightly higher than $\epsilon = 0.1$ [shown in Fig. 11 (a)], the system initially becomes fully frustrated after the external field is withdrawn. Surprisingly, after a period of $\sim$ 2000 MC steps during which the system remains frustrated, it suddenly converges to a balanced state. This intervening period between the field being switched off and the system spontaneously splitting into two modules having positive interactions among all elements within them (and correspondingly, negative interactions between elements belonging to different modules) becomes even shorter with increasing learning rates (e.g., see Fig. 13 for $\epsilon = 0.15$). Thus, we can identify a critical value of around 0.13 for the link adaptation rate ϵ, below which the system remains frustrated and above which the system converges to a balanced state, on withdrawal of the external field. Indeed our simulations show that at $\epsilon = 0.13$ there is a wide variation in the time required for the system to converge to balance after the field is switched off.

In the previous study of structural balance [14], it had been observed that the time required to converge to a structurally balanced state exhibits a bimodal distribution for a range of temperatures. We observe similar features in our model also - for instance, for a link adaptation rate of $\epsilon = 0.13$. For slightly slower learning rates (e.g., $\epsilon = 0.12$), the convergence time increases significantly. In our simulations, many realizations did not converge to the balanced state within the time of observation (10^4 MC steps). For higher learning rates, the system rapidly converges to balance. However, unlike the convergence to balance seen in the previous study [14], the modules in the structurally balanced state that our model system reaches are of almost equal size (see Figs. 12 and 13) independent of the initial state of the network (including the initial distribution of interactions which was seen to affect the nature of the balanced state that the system converges to in the earlier work). This may appear surprising as starting from a balanced state, one expects that the network will remain in that state as the interactions adapt so as to make the corresponding energy minimum even deeper. However, what we observe is that, on withdrawing the field the balanced state characterized by asymmetric module sizes is destabilized and the energy minimum moves in the configuration space eventually reaching the region corresponding to balanced state with modules having equal size. This difference between our results and that of the previous study suggests that the adaptation rule used here (which

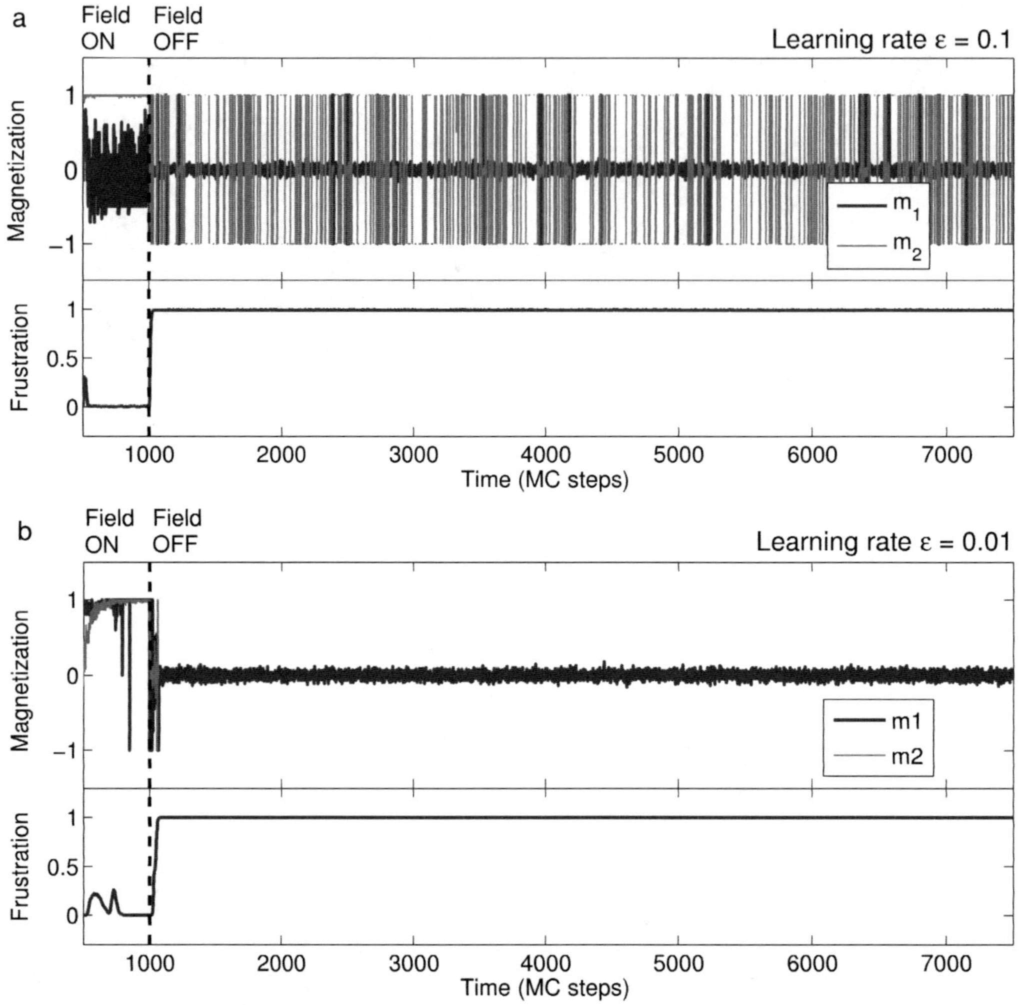

Figure 11. (a) Time-evolution of magnetization (top panel) and frustration (bottom panel) for an initially chimera ordered system with link adaptation rate $\epsilon = 0.1$, before and after an external field is withdrawn (the time at which the field is switched off is indicated by the broken line). We observe that the system becomes completely frustrated on withdrawal of the field. (b) The corresponding time-evolution shown for link adaptation rate $\epsilon = 0.01$.

is biased in favor of negative interactions) is responsible for an intriguing meta-dynamics of the energy landscape itself.

We have also observed how the structure of the energy landscape underlying our system evolves as the interactions change through the adaptation rule. In the model studied earlier [14], the frustration of the initially disordered system is around 0.5. On initiating link adaptation, this value decreases as the system tends towards balance. When frustration reaches a value around 0.49, we observe that the energy landscape becomes such that starting from any spin configuration one can reach the same minimum energy spin arrangement. This suggests that a small decrease in the frustration can accompany a a very large change in the basin of attraction of an energy minimum (which now encompasses a significant fraction of the configuration space).

Journal of Physics: Conference Series **638** (2015) 012010 doi:10.1088/1742-6596/638/1/012010

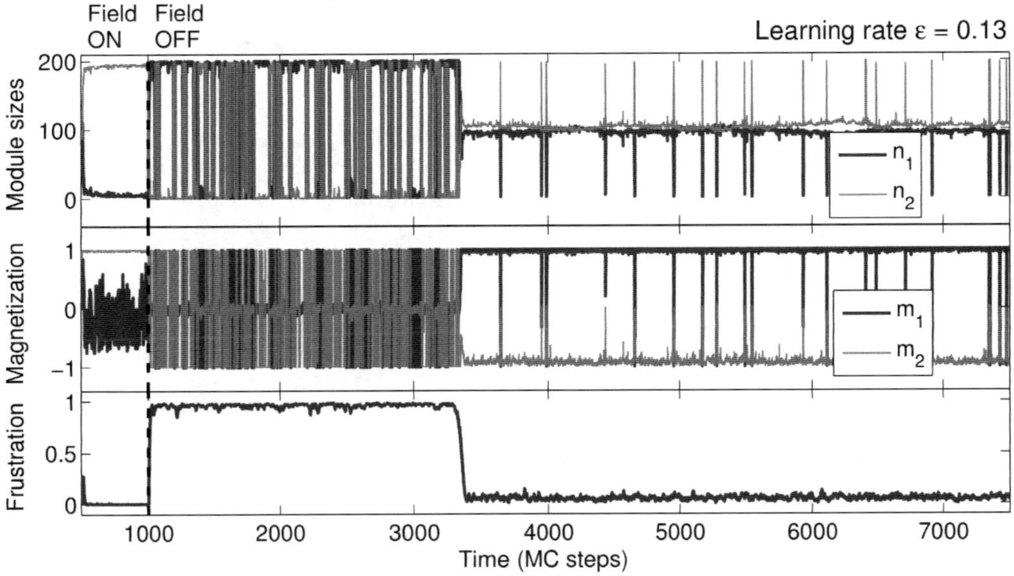

Figure 12. Time-evolution of module sizes (top panel), magnetization (middle panel) and frustration (bottom panel) for an initially chimera ordered system with link adaptation rate $\epsilon = 0.13$, before and after an external field is withdrawn (the time at which the field is switched off is indicated by the broken line). We observe that the system initially becomes completely frustrated on withdrawal of the field. However after about 2000 MC steps the system suddenly becomes balanced.

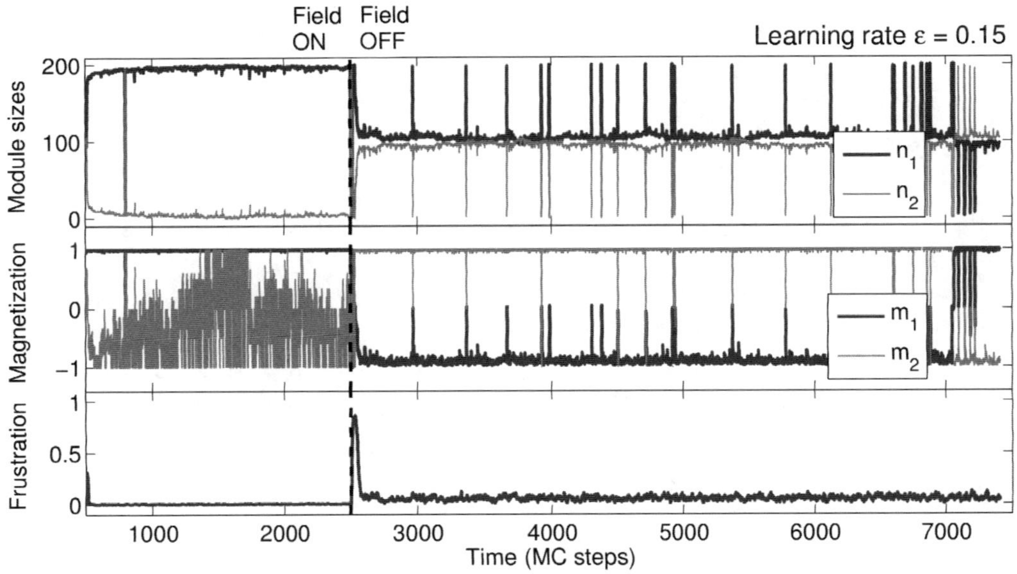

Figure 13. Time-evolution of module sizes (top panel), magnetization (middle panel) and frustration (bottom panel) for an initially chimera ordered system with link adaptation rate $\epsilon = 0.15$, before and after an external field is withdrawn (the time at which the field is switched off is indicated by the broken line). We observe that the system quickly becomes balanced after a brief transient rise in frustration following the withdrawal of the field.

We observe similar behavior in our model system where the adaptation occurs in the presence and subsequent absence of an external field. For example, for an adaptation rate of $\epsilon = 0.13$, when the field is withdrawn the system initially becomes completely frustrated with the value of frustration measured as around 0.97. The corresponding energy landscape has a very large number of minima, each having very small basins of attraction. As the system evolves towards a balanced state, we note that even when the frustration decreases by a very small amount, e.g., to 0.93, the energy landscape transforms to one having an extremely deep energy minimum with a very large basin of attraction. We believe that this radical transformation of the energy landscape of the system at the initial stages of approach to balance points to features of landscape evolution that deserve further study.

4. Conclusions

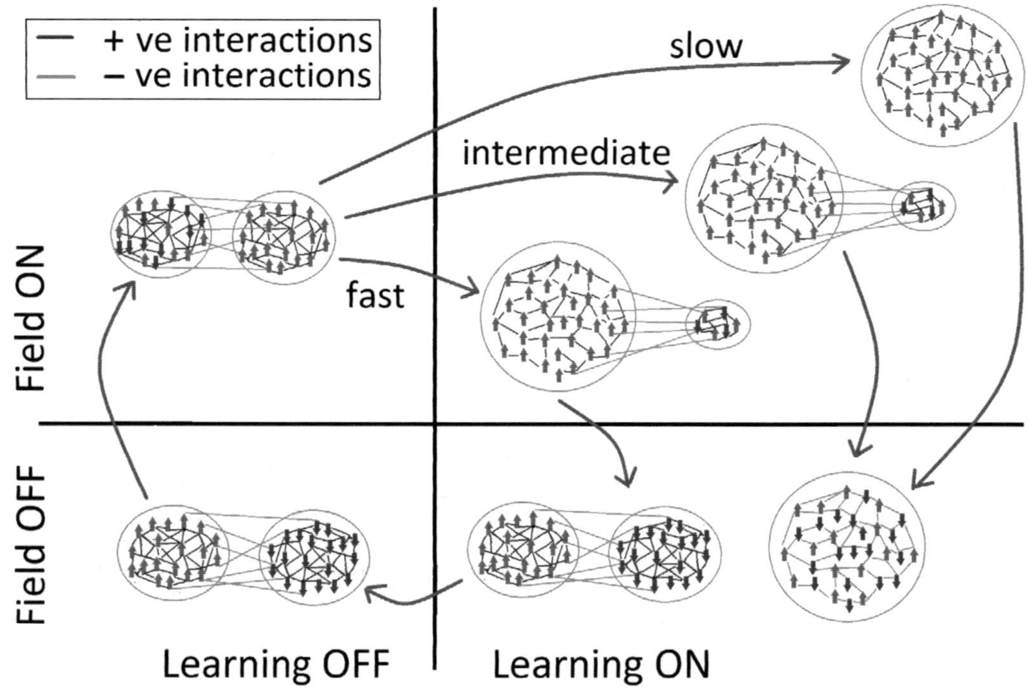

Figure 14. Schematic representation of our key results. Starting from the structurally balanced network (with the different communities showing opposite ordering) at the bottom left quadrant, we obtain chimera state on introducing an external field of requisite strength (top left quadrant). Allowing the interactions to adapt with different learning rates result in the network maintaining balance but with different outcomes for the ordering. Fast and intermediate learning rates retain chimera order, although the disordered region is much diminished, while slow learning leads to complete order (top right quadrant). On withdrawing the field, the system subject to fast learning rate again converges to a structurally balanced network with oppositely ordered communities of approximately equal size, similar to the original network. Intermediate and slow learning, however, results in a fully frustrated network with complete disorder (bottom right quadrant).

In this paper we have explored the behavior of a spin system that is subjected to an external field, while the interactions adapt in response to the system dynamics. The results obtained through our simulations that is described above is summarized in Fig. 14. From an earlier study, we had known that introducing link adaptation inspired by the Hebbian principle in a frustrated system can result in it converging to a structurally balanced state - corresponding to evolution

Journal of Physics: Conference Series **638** (2015) 012010 doi:10.1088/1742-6596/638/1/012010

of the underlying energy landscape - that depends on the initial state of the system. Applying this adaptation dynamics to a system exhibiting chimera order (as done in here) allows us to explore how the dynamics of the energy landscape is affected by an external field, as well as, asymmetry in the adaptation rule regarding negative and positive interactions.

We observe that when system is adapting through faster learning rates, the chimera order is maintained throughout its evolution, with the ordered module increasing in size at the expense of the disordered module. For slower learning rates, first the interactions of the elements belonging to the disordered module become weak which results in loss of chimera order. This is followed by the system converging to a fully ordered state where almost all the interactions are positive. The sizes of the two modules defining the system where one observes chimera order can be varied to see how the ratio of module sizes can affect the subsequent evolution of the system. Indeed, we see that the region in the field-temperature parameter space over which chimera order is seen increases as the ratio increases from 1. When the external field is withdrawn, the system returns to a structurally balanced state with modules of similar sizes if the adaptation rate is high. However, for slower learning rates, the system becomes completely frustrated.

We have also tried to explore how the structure of the energy landscape of the system evolves during learning. Starting from a chimera ordered state that is subjected to adaptation rate and external field, we observe changes in the landscape after the field is switched off. Initially the system becomes completely frustrated. However, after an interval (which decreases with increasing learning rate) the system shows a sudden large increase in the basin of attraction of an energy minimum, although the frustration of the system has decreased only negligibly. This surprisingly radical change in the landscape resulting from relatively small degree of change in the interaction structure of the network is a question that needs to be explored further.

It is of interest to consider the implications of the results reported here for adaptation in biological systems. As connections in the brain evolve according to long-term potentiation that embodies the Hebbian principle that has inspired the link adaptation dynamics used in our study, it is natural to expect that the brain may exhibit at least some of the features observed here. Indeed, as mentioned earlier, the experimental observation of two dynamically anti-correlated subnetworks in the resting human brain [11] strongly suggests that in the absence of strong external stimulation the underlying network is structurally balanced. On the other hand, when exposed to stimuli, correlated brain activity is indeed observed - although not encompassing the entire network. This is, to some extent, reminiscent of the chimera ordered state that is seen in our model system. Thus, the brain may be seen as corresponding to a system subject to relatively fast adaptation rate. However, the Hebbian principle can apply to a much broader class of biological systems, e.g., gene regulation networks where the co-expression of genes has been suggested to result in their co-regulation over evolutionary time-scales [19]. As the adaptation in this case is provided by natural selection, which is orders of magnitude slower than the learning process in the brain mentioned earlier, it is probably not unreasonable to conclude that this can be seen as one corresponding to our model system subject to slow rate of adaptation. It is easy to see that the frustrated system with a large number of energy minima can be considered to be analogous to the cellular differentiation process that allows convergence to any one of a large number of possible cell fates dependent upon initial conditions. Indeed this analogy has been used earlier by Kauffman to motivate Boolean network models for explaining differentiation during biological development [20]. Extending this analogy, one may wonder whether the system has a state corresponding to the ordered state seen on exposing it to an external stimulus. As this state is unique and will be attained by the system independent of all initial conditions, it is tempting to suggest that induced pluripotency in cells exposed to chemical stimuli [21] may be the biological analogue. Therefore, it may be of interest to study the phenomena reported here in biologically realistic models of networks adapting at different time scales.

Acknowledgments
We thank Rajeev Singh, R. Janaki and Shakti N. Menon for helpful discussions. This work was supported in part by IMSc Complex Systems Project. The numerical work was carried out in machines of the IMSc High-Performance Computing facility, including "Satpura" which is partially funded by DST (Grant No. SR/NM/NS-44/2009).

References
[1] Newman M E J 2010 *Networks: An Introduction* (Oxford: Oxford University Press)
[2] Barrat A, Barthélemy M and Vespignani A 2008 *Dynamical Processes on Complex Networks* (Cambridge: Cambridge University Press)
[3] Dorogovtsev S N, Goltsev A V and Mendes J F F 2008 *Rev. Mod. Phys.* **80** 1275
[4] Pan R K and Sinha S, *EPL* **85** 68006
[5] Dasgupta S, Pan R K and Sinha S 2009 *Phys. Rev. E* **80** 025101
[6] Hertz J A, Krogh A and Palmer R G 1991 *Introduction to the Theory of Neural Computation* (Redwood City, CA: Addison-Wesley)
[7] Mezard M, Parisi G and Virasoro M A 1987 *Spin Glass Theory and Beyond* (Singapore: World Scientific)
[8] Fischer K H and Hertz J A 1991 *Spin Glasses* (Cambridge: Cambridge University Press)
[9] Cartwright D and Harary F 1956 *Psychol. Rev.* **63** 277
[10] Heider F 1946 *J. Psychol.* **21** 107
[11] Fox M D, Snyder A Z, Vincent J L, Corbetta M, Van Essen D C and Raichle M E 2005 *Proc. Natl. Acad. Sci. USA* **102** 9673
[12] Singh R, Dasgupta S and Sinha S 2011 *EPL* **95** 10004
[13] Hebb D O 1949 *The Organization of Behavior* (New York: John Wiley)
[14] Singh R, Dasgupta S and Sinha S 2014 *EPL* **105** 10003
[15] Facchetti G, Iacono G and Altafini C 2011 *Proc. Natl. Acad. Sci. USA* **108** 20953
[16] Hopfield J J 1982 *Proc. Natl. Acad. Sci. USA* **79** 2554
[17] Pradhan N, Dasgupta S and Sinha S 2011 *EPL* **94** 38004
[18] Bowman A W and Azzalini A 1997 *Applied Smoothing Techniques for Data Analysis* (Oxford: Oxford University Press)
[19] Fernando C T, Liekens A M L, Bingle L E H, Beck C, Lenser T, Stekel D J and Rowe J E 2009 *J. R. Soc. Interface* **6** 463
[20] Kauffman S A 1991 *Scientific American* **265** (2) 78
[21] Obokata H, Wakayama T, Sasai Y, Kojima K, Vacanti M P, Niwa H, Yamato M and Vacanti C A 2014 *Nature (Lond.)* **505** 641

Journal of Physics: Conference Series **638** (2015) 012011 doi:10.1088/1742-6596/638/1/012011

Stochastic models of classical particle pumps : Density dependence of directed current

Debasish Chaudhuri

Indian Institute of Technology Hyderabad, Yeddumailaram 502205, Telengana, India

E-mail: debc@iith.ac.in

Abstract. We present and compare different versions of a simple particle pump-model that describes average directed current of repulsively interacting particles in a narrow channel, due to time-varying local potentials. We analyze the model on discrete lattice with particle exclusion, using three choices of potential-dependent hopping rates that obey microscopic reversibility. Treating the strength of the external potential as a small parameter with respect to thermal energy, we present a perturbative calculation to obtain the expression for average directed current. This depends on driving frequency, phase, and particle density. The directed current vanishes as density goes to zero or close packing. For two choices of hopping rates, it reaches maximum at intermediate densities, while for a third choice, it shows a curious current reversal with increasing density. This can be interpreted in terms of a particle-hole symmetry. Stochastic simulations of the model show good agreement with our analytic predictions.

1. Introduction

In everyday life we often use mechanical devices that utilize oscillatory force along with a valve mechanism to generate unidirectional flow, e.g., a simple hand pump. In this, the valve breaks space-inversion symmetry, as a result breaking time-reversal symmetry, leading to unidirectional flow. The same basic principle of oscillatory forcing, and time-reversal symmetry breaking leading to directed motion, is utilized in operation of various molecular motors [1, 2, 3], and ion pumps [4, 5] in biological cell, and to generate overall unidirectional flow of electrons in quantum pumps [6, 7, 8, 9]. However, in all these cases noise, stochastic or quantum, plays important role in the resultant dynamics. In molecular motors, e.g., repeated hydrolysis of ATP leads to a stochastic oscillatory energy input, and the intrinsic head-tail directionality of polymeric track on which the motors move, breaks the inversion symmetry to act like a valve [1]. Most theoretical studies of stochastic pumps have discussed properties of noninteracting system of particles, apart from a few exceptions [10, 11, 12, 13].

In this paper, we present a stochastic particle pump model in which an external time-varying potential pumps energy, and a spatially varying phase factor of the oscillation breaks time-reversal symmetry to generate a directed current. We particularly focus on the effect of inter-particle interaction on the dynamics. This interaction may be incorporated via an exclusion process in a spatially discretized version of the dynamics. Two variants of this model have already been proposed and analyzed in some detail [14, 15, 16]. Here we present a unified description of the model on discrete lattice. The model allows for several choices of hopping rates dependent on instantaneous local potential, where each choice obeys microscopic reversibility.

We show that depending on this choice, one obtains different forms of density dependence of average directed current.

2. Model

We consider particles hopping on a ring, discretizing space into $s = 1, \ldots, N$ lattice sites, such that the system size is $L = N$ in units of lattice spacing b. We assume that particles evolve under a position dependent weak oscillatory potential $\beta V_s = \lambda_s \sin(\Omega t + \phi_s)$ where $\beta = 1/k_B T$ with Boltzmann constant k_B and temperature T, and ϕ_s denote local phase factor. This potential drives the system out of equilibrium. If the driving frequency Ω is slow with respect to the diffusion time scale $1/f_0 = a^2/D$ where D denotes the diffusivity, the system of particles would come to local thermal equilibrium with the instantaneous local potential. We assume microscopic reversibility, i.e., the hopping rates are such that given the value of local potential at any instant of time the detailed balance condition, $n_s w_{s,s+1} = n_{s+1} w_{s+1,s}$, is obeyed. In this relation n_s stands for the occupation number of s-th site and $w_{s,s\pm1}$ is the time-dependent hopping rate from s-th to $s \pm 1$-th site. At each moment the system tries to reach equilibrium distribution corresponding to the instantaneous potential, but lags behind as the potential itself changes with time. This keeps the system out of equilibrium. Following three choices of particle hopping rates $w_{s,s\pm1}$ obey local detailed balance:

(A) $w_{s,s\pm1} = f_0 \exp[\beta V_s]$, a symmetric hopping rate that depends only on the on-site potential energy [14, 15];

(B) $w_{s,s\pm1} = f_0 \exp[-\beta(V_{s\pm1} - V_s)/2]$, depends on relative strength of the potential energies [16];

(C) $w_{s,s\pm1} = f_0 \exp[-\beta V_{s\pm1}]$ depends only on the potential energy at the site where the particle hops to.

In this paper we present a *unified derivation* of the time averaged DC current obtained for these three cases. While models B and C are able to pump DC current even in a non-interacting system of particles, in model A interaction is crucial in order to achieve pumping.

We present analytic results using a perturbation theory proposed in Ref. [16], and compare them with numerical simulations. We present a detailed comparative analysis of the three models of $w_{s,s\pm1}$ proposed above, two of which (models A and B) were already discussed earlier [15, 16], and the third one (model C) being the main new result of this paper.

A hard core repulsion between particles is modeled by exclusion process, in which two particles can not occupy the same lattice site. With this restriction, the local density $\rho_s = \langle n_s \rangle$ and two-point correlation functions $C_{s,p} = \langle n_s n_p \rangle$ obey the following dynamics,

$$\begin{aligned}
\frac{d\langle n_s\rangle}{dt} &= w_{s-1,s}\langle n_{s-1}(1-n_s)\rangle + w_{s+1,s}\langle n_{s+1}(1-n_s)\rangle \\
&- w_{s,s-1}\langle n_s(1-n_{s-1})\rangle - w_{s,s+1}\langle n_s(1-n_{s+1})\rangle. \qquad (1)\\
\frac{d\langle n_s n_p\rangle}{dt} &= w_{s-1,s}\langle n_{s-1}(1-n_s)n_p\rangle + w_{s+1,s}\langle n_{s+1}(1-n_s)n_p\rangle \\
&+ w_{p-1,p}\langle n_s n_{p-1}(1-n_p)\rangle + w_{p+1,p}\langle n_s n_{p+1}(1-n_p)\rangle \\
&- w_{s,s-1}\langle n_s(1-n_{s-1})n_p\rangle - w_{s,s+1}\langle n_s(1-n_{s+1})n_p\rangle \\
&- w_{p,p-1}\langle n_s n_p(1-n_{p-1})\rangle - w_{p,p+1}\langle n_s n_p(1-n_{p+1})\rangle \qquad (2)\\
\frac{d\langle n_s n_{s+1}\rangle}{dt} &= w_{s-1,s}\langle n_{s-1}(1-n_s)n_{s+1}\rangle + w_{s+2,s+1}\langle n_s n_{s+2}(1-n_{s+1})\rangle \\
&- w_{s,s-1}\langle n_s(1-n_{s-1})n_{s+1}\rangle - w_{s+1,s+2}\langle n_s n_{s+1}(1-n_{s+2})\rangle \qquad (3)
\end{aligned}$$

The last equation is for the special case of nearest neighbor correlations. Writing $\rho_s = \langle n_s \rangle$ and the multi-point correlations as $C_{s,p,\ldots} = \langle n_s n_p \ldots \rangle$ we can re-express the above relations as

$$\frac{d\rho_s}{dt} = w_{s-1,s}(\rho_{s-1} - C_{s-1,s}) + w_{s+1,s}(\rho_{s+1} - C_{s,s+1})$$

Journal of Physics: Conference Series **638** (2015) 012011 doi:10.1088/1742-6596/638/1/012011

$$
\begin{aligned}
&- \; w_{s,s-1}(\rho_s - C_{s-1,s}) - w_{s,s+1}(\rho_s - C_{s,s+1}). \qquad (4)\\
\frac{dC_{s,p}}{dt} &= w_{s-1,s}(C_{s-1,p} - C_{s-1,s,j}) + w_{s+1,s}(C_{s+1,p} - C_{s,s+1,p})\\
&+ w_{p-1,p}(C_{s,p-1} - C_{s,p-1,p}) + w_{p+1,p}(C_{s,p+1} - C_{s,p,p+1})\\
&- w_{s,s-1}(C_{s,p} - C_{s-1,s,p}) - w_{s,s+1}(C_{s,p} - C_{s,p,p+1})\\
&- w_{p,p-1}(C_{s,p} - C_{s,p-1,p}) - w_{p,p+1}(C_{s,p} - C_{s,p,p+1}) \qquad (5)\\
\frac{dC_{s,s+1}}{dt} &= w_{s-1,s}(C_{s-1,s+1} - C_{s-1,s,s+1}) + w_{s+2,s+1}(C_{s,s+2} - C_{s,s+1,s+2})\\
&- w_{s,s-1}(C_{s,s+1} - C_{s,s-1,s+1}) - w_{s+1,s+2}(C_{s,s+1} - C_{s,s+1,s+2}). \qquad (6)
\end{aligned}
$$

Thus dynamics of each order of correlation depends on correlations of higher order, following a Bogoliubov-Born-Green-Kirkwood-Yvon (BBGKY) hierarchy. Note that the evolution of local density may be represented as $d\rho_s/dt = J_{s-1,s} - J_{s,s+1}$ where the local current

$$J_{s-1,s} = (w_{s-1,s}\rho_{s-1} - w_{s,s-1}\rho_s) - (w_{s-1,s} - w_{s,s-1})C_{s-1,s} \qquad (7)$$

In the time periodic steady state, local current averaged over the period $\tau = 2\pi/\Omega$ is independent of position. Therefore the net time- and space- averaged directed current is given by

$$\bar{J} = \frac{1}{N\tau}\sum_{s=1}^{N}\int_0^{\tau} dt J_{s-1,s}. \qquad (8)$$

For potential strength $\lambda_s = 0$ at all lattice sites, the above model reduces to homogeneous symmetric exclusion process, characterized by the following local density and correlation functions

$$
\begin{aligned}
\bar{\rho} &= \frac{n}{L} = \rho,\\
C^{(2)} &= \rho\frac{n-1}{L-1}\\
C^{(3)} &= C^{(2)}\frac{n-2}{L-2} \qquad (9)
\end{aligned}
$$

etc. [17], where n is the total number of particles. As we show in the following, the BBGKY hierarchy separates by order, if one expands local quantities like ρ_s, $C_{s,p}$ etc. in perturbative expansion around $\lambda_s = 0$. This allows one to obtain exact expressions within perturbative expansion.

3. Perturbative calculation

We consider driving at all the sites with constant potential strength and frequency

$$\beta V_s = \lambda \sin(\Omega t + \phi_s) = \lambda \times u_s \qquad (10)$$

where

$$
\begin{aligned}
u_s &= 2\text{Re}\,[\eta_s e^{i\Omega t}],\\
\eta_s &= -\frac{i}{2}e^{i\phi_s}. \qquad (11)
\end{aligned}
$$

For small values of λ we can linearize the transition rates to obtain : (A) $w_{s,s\pm1} = f_0(1+\lambda u_s)$, (B) $w_{s,s\pm1} = f_0\{1 - \frac{1}{2}\lambda(u_{s\pm1} - u_s)\}$ (C) $w_{s,s\pm1} = f_0(1 - \lambda u_{s\pm1})$ and the corresponding bond currents are expressed as

(A) $J_{s-1,s} = -f_0(\rho_s - \rho_{s-1}) - \lambda f_0(u_s\rho_s - u_{s-1}\rho_{s-1}) + \lambda f_0(u_s - u_{s-1})C_{s-1,s}$,
(B) $J_{s-1,s} = -f_0(\rho_s - \rho_{s-1}) - (\lambda f_0/2)(u_s - u_{s-1})(\rho_{s-1} + \rho_s - 2C_{s-1,s})$
(C) $J_{s-1,s} = -f_0(\rho_s - \rho_{s-1}) - \lambda f_0(u_s\rho_{s-1} - u_{s-1}\rho_s) + \lambda f_0(u_s - u_{s-1})C_{s-1,s}$.

Therefore the space-time averaged directed current is

$$\begin{aligned} \bar{J}_A &= \frac{\lambda f_0}{N\tau}\sum_{s=1}^{N}\int_0^{\tau} dt\,(u_s - u_{s-1})C_{s-1,s} \\ \bar{J}_B &= -\frac{\lambda f_0}{2N\tau}\sum_{s=1}^{N}\int_0^{\tau} dt\,(u_s - u_{s-1})(\rho_{s-1} + \rho_s - 2C_{s-1,s}) \\ \bar{J}_C &= -\frac{\lambda f_0}{N\tau}\sum_{s=1}^{N}\int_0^{\tau} dt\,[(u_s\rho_{s-1} - u_{s-1}\rho_s) - (u_s - u_{s-1})C_{s-1,s}] \end{aligned} \tag{12}$$

Note that, for non-interacting particles the correlation function $C_{s-1,s} = 0$ leads to $\bar{J}_A = 0$. That means, within model-A, the pump will drive particles in an averaged unidirectional fashion only in the presence of interaction – free particles can not be pumped within this model [15]. However, for the other two models this requirement is absent. Even free particles may be pumped in a unidirectional manner.

Let us write down the density and correlation functions as perturbative expansion in potential strength $\lambda\,(\ll 1)$

$$\begin{aligned} \rho_s &= \rho + \sum_{k=1,2,\ldots} \lambda^k \rho_s^{(k)} \\ C_{s,p} &= C^{(2)} + \sum_{k=1,2,\ldots} \lambda^k C_{s,p}^{(k)}. \end{aligned} \tag{13}$$

We consider the case where the phase factor of driving potential $\phi_s = \phi s$ where $\phi = 2\pi/L$, remembering L is expressed in units of lattice parameter. Within the perturbative expansion, the above relations for directed current may be calculated exactly [15, 16]. The results for the first two cases were derived earlier, and the third one is presented in this paper. In what follows we derive all three results. However, before we start deriving them, let us enlist the expressions for current corresponding to the three variants of the model here,

$$\begin{aligned} \bar{J}_A &= -2\lambda^2 k_0 f_0^2 \frac{\Omega\sin\phi(1-\cos\phi)}{\Omega^2 + 4f_0^2(1-\cos\phi)^2} \\ \bar{J}_B &= \lambda^2(q_0 - 2k_0) f_0^2 \frac{\Omega\sin\phi(1-\cos\phi)}{\Omega^2 + 4f_0^2(1-\cos\phi)^2} \\ \bar{J}_C &= 2\lambda^2(q_0 - k_0) f_0^2 \frac{\Omega\sin\phi(1-\cos\phi)}{\Omega^2 + 4f_0^2(1-\cos\phi)^2}, \end{aligned} \tag{14}$$

with $q_0 = \rho - C^{(2)}$, $k_0 = C^{(2)} - C^{(3)}$. Note that in the limit of large n and L keeping $\rho = n/L$ constant, $q_0 \approx \rho(1-\rho)$ and $k_0 \approx \rho^2(1-\rho)$. Thus $\bar{J}_A \sim \rho^2(1-\rho)$, and $\bar{J}_B \sim \rho(1-\rho)(1-2\rho)$, and $\bar{J}_C \sim \rho(1-\rho)^2$ (See Fig.1).

3.1. Time evolution and solution

Let us first consider model C, which constitutes the main new contribution of this paper. Using the perturbative expansion of Eq.(13), time evolution of the first order perturbations can be

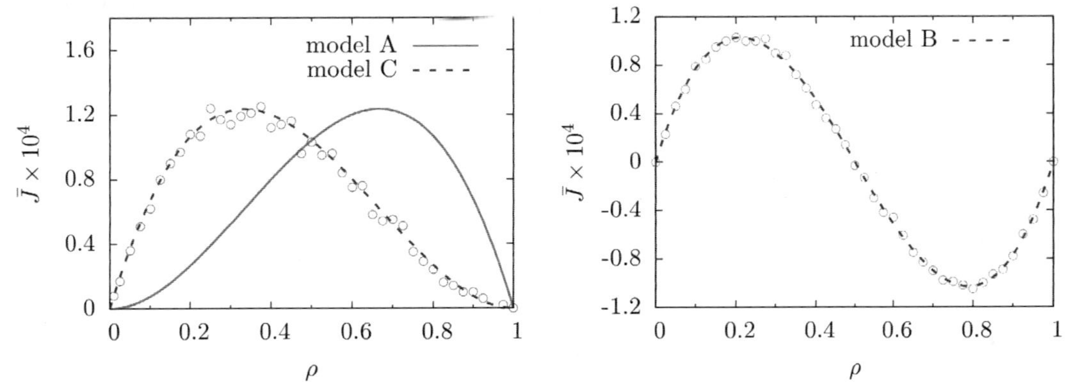

Figure 1. (Color online) Directed current $\bar{J}$ as a function of mean density $\bar{\rho}$. The points denote Monte-Carlo result and the lines are plot of the functions in Eq.14. In plotting the functions we replaced q_0 by $\rho(1-\rho)$ and k_0 by $\rho^2(1-\rho)$. Left panel: Shows plots of analytic functions for models A and C, and simulation data for model C. Right panel: Comparison of simulation results of model B with analytic prediction. The parameters used are system size $L = 16$, bare hopping rate due to free particle diffusion $f_0 = 0.34$, potential strength $\lambda = 0.5$, frequency of oscillation $\Omega = 0.2\pi$ such that the time period $\tau = 10$. We choose phase difference between consecutive lattice points $\phi = \pi/2$. The data (points) were collected over 100τ after equilibration over 100τ. All the data for model C were averaged over 10^5 initial conditions, and for model B over 10^6 initial conditions.

written as,

$$\begin{aligned}
\frac{d\rho_s^{(1)}}{dt} &= f_0\Delta_s\rho_s^{(1)} + f_0 q_0 \Delta_s u_s, && (15)\\
\frac{dC_{s,p}^{(1)}}{dt} &= f_0(\Delta_s + \Delta_p)C_{s,p}^{(1)} + f_0 k_0(\Delta_s u_s + \Delta_p u_p) \quad \text{for } p \neq s \pm 1, \\
\frac{dC_{s,s+1}^{(1)}}{dt} &= f_0(C_{s-1,s+1}^{(1)} + C_{s,s+2}^{(1)} - 2C_{s,s+1}^{(1)}) + f_0 k_0(u_{s-1} + u_{s+2} - u_s - u_{s+1}) && (16)
\end{aligned}$$

where $\Delta_s g_{s,p} = g_{s+1,p} + g_{s-1,p} - 2g_{s,p}$. Note that the above time evolution, for first order terms in perturbative expansion, remains the same for all the three variants of the model considered above. This is easy to see by comparing with Ref.s [15, 16].

These linear differential equations can be solved exactly to find long time limit of time-varying steady state [16],

$$\rho_s^{(1)}(t) = 2\mathrm{Re}[A_s^{(1)} \exp(i\Omega t)]. \tag{17}$$

Using this in Eq.(15) we find, for all s,

$$-A_{s-1}^{(1)} - A_{s+1}^{(1)} + (2 + i\Omega/f_0)A_s^{(1)} = -q_0(-\eta_{s+1} - \eta_{s-1} + 2\eta_s). \tag{18}$$

Clearly this equation can be written in the operator form,

$$\hat{Z} \mid A\rangle = -q_0\hat{\Delta} \mid \eta\rangle \tag{19}$$

with matrix elements

$$\begin{aligned} \Delta_{s,p} &= \delta_{s,p+1} + \delta_{s,p-1} - 2\delta_{s,p}. \\ Z_{s,p} &= \Delta_{s,p} - \frac{i\Omega}{f_0}\delta_{s,p} \end{aligned} \quad (20)$$

where $\hat{\Delta}$ satisfies the eigenvalue equation $\hat{\Delta}|q\rangle = \epsilon_q|q\rangle$ with $\epsilon_q = -2(1 - \cos q)$ and the eigenfunction $\psi_s(q) = \langle s|q\rangle = (1/\sqrt{L})\exp(-iqs)$ where $q = 2\pi k/L = \phi\, k$ where $k = 1, 2, \ldots, N$ such that $\psi_{s+N} = \psi_s$. Similarly, $\hat{Z}|q\rangle = (\epsilon_q - i\Omega/f_0)|q\rangle$. Thus the solution may be expressed as $|A\rangle = -q_0\hat{Z}^{-1}\hat{\Delta}|\eta\rangle$.

In the real-space representation $\langle s|A\rangle = -q_0\sum_{q,m}\langle s|\hat{Z}^{-1}|q\rangle\langle q|\hat{\Delta}|m\rangle\langle m|\eta\rangle = -q_0\sum_{q,m}(\epsilon_q - i\Omega/f_0)^{-1}\langle s \mid q\rangle\epsilon_q\langle q \mid m\rangle\eta_m = -q_0\sum_{q,m}(\epsilon_q - i\Omega/f_0)^{-1}\epsilon_q\psi_s(q)\psi^*_m(q)\eta_m$. Therefore,

$$\begin{aligned} A_s^{(1)} &= -q_0\sum_{m=1}^{N}\sum_{k=1}^{N}\frac{\epsilon_{\phi k}}{\epsilon_{\phi k} - i\Omega/f_0}\psi_s(\phi k)\psi^*_m(\phi k)\eta_m \\ &= \frac{iq_0}{2}e^{i\phi s}\frac{\epsilon_\phi}{\epsilon_\phi - i\Omega/f_0}, \end{aligned} \quad (21)$$

where $\epsilon_\phi = -2(1-\cos\phi)$. The equation for two point correlation function can also be solved [16]

$$C_{s,p}^{(1)}(t) = \frac{k_0}{q_0}[\rho_s^{(1)}(t) + \rho_p^{(1)}(t)] = 2\mathrm{Re}[A_{s,p}^{(1)}e^{i\Omega t}] \quad (22)$$

where $A_{s,p}^{(1)} = (k_0/q_0)(A_s^{(1)} + A_p^{(1)})$.

3.2. Averaged directed current

Using the above relations, one can calculate the space-time averaged directed current in the time-periodic steady state for all three variants of the model, through the relations shown in Eq.(12). For the particular case of model C, using the last relation in Eq.(12), we have

$$\begin{aligned} \bar{J}_C &= -\frac{\lambda^2 f_0}{N\tau}\sum_{s=1}^{N}\int_0^\tau dt\left[\left(1 - \frac{k_0}{q_0}\right)(u_s\rho_{s-1}^{(1)} - u_{s-1}\rho_s^{(1)}) - \frac{k_0}{q_0}(u_s\rho_s^{(1)} - u_{s-1}\rho_{s-1}^{(1)})\right] \\ &= -\frac{\lambda^2 f_0}{N}\sum_{s=1}^{N}2\,\mathrm{Re}\left[\left(1 - \frac{k_0}{q_0}\right)(\eta_s^* A_{s-1}^{(1)} - \eta_{s-1}^* A_s^{(1)}) - \frac{k_0}{q_0}(\eta_s^* A_s^{(1)} - \eta_{s-1}^* A_{s-1}^{(1)})\right] \end{aligned} \quad (23)$$

where in the last step we have used the fact that after integration over a period $\tau = 2\pi/\Omega$, only the time-independent combinations of $u_s\rho_s^{(1)}$ terms which can be expressed as $2\,\mathrm{Re}[\eta_s^* A_s^{(1)}]$ etc. remain non-zero.

Given that $\eta_s = (-i/2)e^{i\phi s}$, and as we may write $A_s^{(1)} = (iq_0/2)e^{i\phi s}a$ where $a = \epsilon_\phi/[\epsilon_\phi - i\Omega/f_0]$, the terms in Eq.(23) may be evaluated. We find that $\eta_s^* A_s^{(1)} = -q_0 a/4$, and $\eta_{s-1}^* A_{s-1}^{(1)} = -q_0 a/4$. Thus the second term in the parentheses $(\eta_s^* A_s^{(1)} - \eta_{s-1}^* A_{s-1}^{(1)}) = 0$. Similarly one can show that $(\eta_s^* A_{s-1}^{(1)} - \eta_{s-1}^* A_s^{(1)}) = (q_0/2)\sin\phi\,(ia)$. Thus the terms $2\mathrm{Re}(\eta_s^* A_{s-1}^{(1)} - \eta_{s-1}^* A_s^{(1)}) = q_0\sin\phi\,\mathrm{Im}(a)$. Therefore,

$$\begin{aligned} \bar{J}_C &= -\lambda^2 f_0\left(1 - \frac{k_0}{q_0}\right)q_0\sin\phi\,\mathrm{Im}(a) \\ &= -\lambda^2(q_0 - k_0)f_0^2\frac{\Omega\sin\phi\epsilon_\phi}{\Omega^2 + f_0^2\epsilon_\phi^2} \end{aligned}$$

$$= 2\lambda^2(q_0 - k_0)f_0^2 \frac{\Omega \sin\phi(1-\cos\phi)}{\Omega^2 + 4f_0^2(1-\cos\phi)^2}, \tag{24}$$

where in the last step we used the expression for ϵ_ϕ. In the limit of large system size $q_0 - k_0 \approx \rho(1-\rho)^2$, and thus $\bar{J}_C \sim \rho(1-\rho)^2$.

Similar arguments may be used to derive the results corresponding to models A and B. For example using relations in Eq.(12),

$$\begin{aligned} \bar{J}_A &= \frac{\lambda^2 f_0}{N\tau} \sum_{s=1}^{N} \int_0^\tau dt \frac{k_0}{q_0}(u_s - u_{s-1})(\rho_s^{(1)} + \rho_{s-1}^{(1)}) \\ &= \frac{\lambda^2 f_0}{N} \frac{k_0}{q_0} \sum_{s=1}^{N} 2\,\mathrm{Re}\left[(\eta_s^* - \eta_{s-1}^*)(A_s^{(1)} + A_{s-1}^{(1)})\right]. \end{aligned} \tag{25}$$

One can show that $(\eta_s^* - \eta_{s-1}^*)(A_s^{(1)} + A_{s-1}^{(1)}) = (q_0/2)\sin\phi\,(ia)$, and thus $2\,\mathrm{Re}\left[(\eta_s^* - \eta_{s-1}^*)(A_s^{(1)} + A_{s-1}^{(1)})\right] = q_0 \sin\phi\,\mathrm{Im}(a)$. Thus

$$\begin{aligned} \bar{J}_A &= \lambda^2 f_0 \frac{k_0}{q_0} q_0 \sin\phi\,\mathrm{Im}(a) \\ &= -2\lambda^2 k_0 f_0^2 \frac{\Omega \sin\phi(1-\cos\phi)}{\Omega^2 + 4f_0^2(1-\cos\phi)^2}. \end{aligned} \tag{26}$$

Again, using Eq.(12)

$$\begin{aligned} \bar{J}_B &= -\frac{\lambda^2 f_0}{2N\tau}\left(1 - \frac{2k_0}{q_0}\right) \sum_{s=1}^{N} \int_0^\tau dt\,(u_s - u_{s-1})(\rho_{s-1}^{(1)} + \rho_s^{(1)}) \\ &= -\frac{\lambda^2 f_0}{2N}\left(1 - \frac{2k_0}{q_0}\right) \sum_{s=1}^{N} 2\,\mathrm{Re}\left[(\eta_s^* - \eta_{s-1}^*)(A_{s-1}^{(1)} + A_s^{(1)})\right]. \end{aligned} \tag{27}$$

The relation $(\eta_s^* - \eta_{s-1}^*)(A_{s-1}^{(1)} + A_s^{(1)}) = (q_0/2)\sin\phi\,(ia)$, leading to $2\,\mathrm{Re}\left[(\eta_s^* - \eta_{s-1}^*)(A_{s-1}^{(1)} + A_s^{(1)})\right] = q_0 \sin\phi\,\mathrm{Im}(a)$. Thus one gets

$$\begin{aligned} \bar{J}_B &= -\frac{\lambda^2 f_0}{2}\left(1 - \frac{2k_0}{q_0}\right) q_0 \sin\phi\,\mathrm{Im}(a) \\ &= \lambda^2(q_0 - 2k_0) f_0^2 \frac{\Omega \sin\phi(1-\cos\phi)}{\Omega^2 + 4f_0^2(1-\cos\phi)^2}. \end{aligned} \tag{28}$$

4. Simulation

A detailed numerical simulation for model A was presented earlier in Ref. [14]. Here we perform Monte-Carlo simulations of the models B and C and present density dependence of directed current $\bar{J}$ in Fig. 1. In the stochastic simulation, we randomly choose a lattice site s with uniform probability and perform a trial move with rate $w_{s,s\pm1}$. The trial move is accepted if the new site $s \pm 1$ is empty, else it is rejected. A sweep of n trial moves for a system having n particles is considered as one Monte-Carlo step. We use periodic boundary condition. Note that in simulations we do not use the linearized versions of hopping rates, unlike in theory with small potential strength λ. Instead we use the full non-linear forms. In all our simulations we keep $\lambda = 0.5$, unlike the perturbation theory where we assumed $\lambda \ll 1$. At the time-periodic steady state, the current is measured on each bond and then averaged over all bonds in the system, and several time-periods. We also average over many initial conditions to obtain better statistics. For details of the parameter values used in simulations, see figure caption of Fig. 1. All the simulation data show good agreement with predictions presented in Eq.(14).

For all the three variants of the model current vanishes as packing fraction $\rho \to 0$ and $\rho \to 1$ (close pack) limits. The first vanishing is due to absence of particles to carry current, and the second one is due to complete jamming. If all the lattice sites are occupied, within the discrete lattice random sequential dynamics, particles can not move. However, the detailed density dependence of current $\bar{J}$ shows three very different form for the three variants of the model considered. While for model A, $\bar{J}_A \sim \rho^2(1-\rho)$, model B shows a dramatic effect of current reversal with changing density. For model B, the density dependence is $\bar{J}_B \sim \rho(1-\rho)(1-2\rho)$, with a new zero in current appearing at the half filling $\rho = 1/2$. This particular model has a symmetry under the exchange of particles with holes together with swapping direction from right to left. Thus a phase factor ϕ that leads to free particle motion towards right, which is the dominant mode at low densities, will lead to *free* hole motion to right at high densities. Therefore, particle current changes direction from near $\rho = 0$ to near $\rho = 1$. At $\rho = 1/2$ the particle and hole currents cancel each other leading to $\bar{J}_B = 0$. We performed simulations for model C as well, and present the numerically obtained $\bar{J}_C$ in Fig. 1. Our simulation results for model B and C agree well with theoretical predictions (see Fig. 1). Note that, for model C, theory predicts a density dependence $\bar{J}_C \sim \rho(1-\rho)^2$. In Fig. 1 we have also plotted the theoretical prediction for model A, for comparison.

5. Outlook

We presented a discrete pump model in which an external traveling wave potential leads to average directed motion of particles interacting via exclusion process. We discussed three possible choices of external potential dependent hopping rates, all of which obey the microscopic time-reversal symmetry. We studied how a resultant directed current depends on average density of particles. Using a perturbative expansion for small strength of external potential with respect to thermal noise, we obtained analytic expressions for directed current, and compared our results with direct Monte-Carlo simulations to find good agreement. While the time evolution of first order perturbation in local density and correlation functions, are independent of specific choice of the three variants of the lattice model discussed here, the expressions for directed current depend on the choice of local hopping rates. The dependence of average directed current on frequency and phase is the same across all the three choices of hopping rates [see Eq.(12)]. However, it is important to note that the detailed density dependence is very different in the three choices of models (hopping rates) discussed – while model A predicts $\bar{J}_A \sim \rho^2(1-\rho)$, model B and C predict $\bar{J}_B \sim \rho(1-\rho)(1-2\rho)$ and $\bar{J}_C \sim \rho(1-\rho)^2$ respectively [mean field limits of Eq.(12)]. The hopping rates chosen for model-A fails to generate any directed current in absence of particle exclusion – density correlation turns out to be necessary. On the other hand, models B and C allows for driving of directed current for non-interacting particles, also. In Ref. [16] we argued that the model B is a natural choice, if one starts from the corresponding Langevin equation and discretize its dynamics. This model shows a curious current reversal with increasing density of particles.

However, later studies by us in continuum model showed very different density dependence [18], in particular, absence of current reversal predicted by the discrete exclusion process in model-B. In order to discuss the continuum limit of our calculation, let us now use the lattice parameter b explicitly. In the continuum limit, one has to take the lattice parameter $b/L \to 0$, with packing fraction $\rho b \ll 1$. Since b defines the length scale of inter-particle repulsion as well, in this limit, one obtains results valid for non-interacting continuum dynamics. However, in the real continuum system the space is continuum, but the hard core particles have finite size and repel each other. Thus the continuum limit of the discrete exclusion process used here,

Journal of Physics: Conference Series **638** (2015) 012011 doi:10.1088/1742-6596/638/1/012011

fails to capture the behavior of hard core particles moving in continuum space under stochastic thermal force, and time-oscillatory external potential. A correct discrete model would require the length scale of exclusion process to be defined as a new variable $\sigma = \nu b$, such that in the continuum limit, $b/L \to 0$ with $\nu \to \infty$ keeping σ constant. Experimental realization of the model presented here looks possible, using colloidal particles confined in narrow channels driven by traveling wave potential. This would provide better insight into driven many-body dynamics, and could have potential applications.

Acknowledgments
The author thanks Abhishek Dhar for numerous discussions, and a collaboration on related topics which led to the publication of Refs. [16, 18].

References

[1] Jülicher F, Ajdari A and Prost J 1997 *Rev. Mod. Phys.* **69** 1269
[2] Reimann P 2002 *Phys. Rep.* **361** 57–265
[3] Reimann P and Hänggi P 2002 *Appl. Phys. A* **75** 169–178
[4] Gadsby D, Takeuchi A, Artigas P and Reyes N 2009 *Phil. Trans. R. Soc. B* **364** 229
[5] Astumian R D 2003 *Phys. Rev. Lett.* **91** 118102
[6] Brouwer P 1998 *Phys. Rev. B* **58** R10135
[7] Sela E and Oreg Y 2006 *Phys. Rev. Lett.* **96** 166802
[8] Cavaliere F, Governale M and König J 2009 *Phys. Rev. Lett.* **103** 136801
[9] Strass M, Hänggi P and Kohler S 2005 *Phys. Rev. Lett.* **95** 130601
[10] Derényi I and Vicsek T 1995 *Phys. Rev. Lett.* **75** 374
[11] Derenyi I and Ajdari A 1996 *Phys. Rev. E* **54** R5
[12] Aghababaie Y, Menon G I and Plischke M 1999 *Phys. Rev. E* **59** 2578
[13] Savel'ev S, Marchesoni F and Nori F 2004 *Phys. Rev. E* **70**(6) 061107
[14] Jain K, Marathe R, Chaudhuri A and Dhar A 2007 *Phys. Rev. Lett.* **99**(19) 190601
[15] Marathe R, Jain K and Dhar A 2008 *J. Stat. Mech. Theor. Exp.* **2008** P11014
[16] Chaudhuri D and Dhar A 2011 *Europhys. Lett.* **94** 30006
[17] Schütz G M 2000 *Phase transitions and critical phenomena* vol 19 ed Domb C and Lebowitz J (London: Academic Press) pp 3 – 242
[18] Chaudhuri D, Raju A and Dhar A 2015 *Phys. Rev. E* **91** 050103

Collective effects in force generation by multiple cytoskeletal filaments pushing an obstacle

Aparna J S[1], Dipjyoti Das[2], Ranjith Padinhateeri[3] and Dibyendu Das[2]

[1] Centre for Research in Nanotechnology and Science, Indian Institute of Technology Bombay, Mumbai 400076, India.
[2] Department of Physics, Indian Institute of Technology Bombay, Mumbai 400076, India.
[3] Department of Biosciences and Bioengineering, Indian Institute of Technology Bombay, Mumbai 400076, India.

E-mail: dibyendu@phy.iitb.ac.in

Abstract. We report here recent findings that multiple cytoskeletal filaments (assumed rigid) pushing an obstacle typically generate more force than just the sum of the forces due to individual ones. This interesting phenomenon, due to the hydrolysis process being out of equilibrium, escaped attention in previous experimental and theoretical literature. We first demonstrate this numerically within a constant force ensemble, for a well known model of cytoskeletal filament dynamics with random mechanism of hydrolysis. Two methods of detecting the departure from additivity of the collective stall force, namely from the force-velocity curve in the growing phase, and from the average collapse time versus force curve in the bounded phase, is discussed. Since experiments have already been done for a similar system of multiple microtubules in a harmonic optical trap, we study the problem theoretically under harmonic force. We show that within the varying harmonic force ensemble too, the mean collective stall force of N filaments is greater than N times the mean stall force due to a single filament; the actual extent of departure is a function of the monomer concentration.

1. Introduction

Cytoskeletal filaments are dynamic, long protein-polymers, which form the structural components of cells. Two major types of cytoskeletal filaments are microtubules and actin filaments. Actin filaments are made of G-actin monomers and microtubules are made of tubulin monomers. These monomers are typically bound to ATP/GTP molecules when they polymerize, and they hydrolyze to ADP/GDP molecules with passing time [1]. In addition to maintaining structural integrity, they also aid in several functions, such as cell motility, movement of chromosomes during cell division, and transportation within cells [2, 3]. These functional properties are aided by their ability to generate forces against barriers they encounter, by regulating their structure based on microscopic polymerization, depolymerization and hydrolysis processes. A single filament or a bundle of filaments can grow on an average, against the barrier, if the collective force generated by the system is greater than the force exerted by the barrier. If the compressive force increases, the average growth velocity of the system decreases until it vanishes at the 'stall force' of the system. This is the maximum force that can be generated by the bundle of filaments, at which the bundle, on an average, stalls [4, 5, 6]. The highly

Journal of Physics: Conference Series **638** (2015) 012012 doi:10.1088/1742-6596/638/1/012012

dynamic nature of cytoskeletal filaments are observed at high external forces or low free monomer concentrations, in a process termed as dynamic instability, which is characterised by large length fluctuations of the filaments. During this, the filament can shrink to very short length by undergoing 'catastrophes' and then grow back in events called 'rescues' [7, 8]. Understanding these dynamic features of cytoskeletal filaments and their force generation capabilities are important to understand their contribution in cellular functions.

In the absence of irreversible hydrolysis of ATP/GTP-bound monomer to ADP/GDP-bound monomer in the filament, thermodynamic arguments can deduce the stall force of the system as

$$f_s = \frac{k_B T}{d} log(\frac{u_0}{w_0}) \tag{1}$$

where w_0 is the depolymerization rate and u_0 is the concentration dependent polymerization rate [9, 10, 4, 5]. For a N-filament system, these theories obtain a stall force that is proportional to N, and the constant of proportionality is equal to the stall force of a single filament:

$$f_s^{(N)} = N f_s^{(1)}. \tag{2}$$

Introduction of hydrolysis into the system introduces two different depolymerization rates for the filament, depending on the tip of the filament being in either ATP/GTP-bound or ADP/GDP-bound state. Detailed studies, taking into account the monomer level polymerization/depolymerization and hydrolysis, but on single filaments, have been carried out theoretically in the past [6, 11, 12, 13, 14]. These models incorporate the mechanism of hydrolysis as either sequential or random. In the former, an ATP/GTP subunit can hydrolyse only at its interface with an ADP/GDP subunit, thereby maintaining a 'cap' of ATP/GTP-bound monomers at the end of the filament. In random hydrolysis model, hydrolysis can happen at any random position along the polymer. Recently, we have shown that [15, 8], in the presence of hydrolysis, the collective stall force of a bundle of N filaments, $f_s^{(N)} \neq N f_s^{(1)}$, and for typical experimental parameters, $f_s^{(N)} > N f_s^{(1)}$.

Experiments have studied the dynamics of single filament in presence of a barrier (exerting force) [16], and multiple filaments in harmonic traps [17, 18]. Laan *et al.* [18] in particular, studied multiple microtubules in the presence of a harmonic force (an optical trap), as opposed to a constant applied force, as considered in the papers discussed above. They observed a linear variation of the collective stall force of the filament bundle with the number of filaments.

In this paper, we discuss the cooperativity of a bundle of N filaments growing against a constant compressive force similar to what is discussed in [15]. This result is further corroborated by studying catastrophe timescales of the filaments, under high forces [8]. We then proceed to present some new results for filaments in the presence of a harmonic force. Even in the latter case, we find that, additivity does not hold for average forces generated by the bundle.

2. The model

The model we study consist of N parallel filaments, growing against a barrier, which exerts a compressive force against their growth. First we would consider a constant force as in Figure 1 [15], and then a harmonic force as in Figure 4. A filament here is supposed to represent an entire microtubule with all 13 protofilaments – thus with the addition of each monomer, the system effectively moves a distance equal to 0.6 nm (=8 nm/13), where 8 nm is the actual subunit length of a tubulin monomer [6, 14]. The filaments are independent, except that they interact with each other through their contact with the barrier. Filaments are allowed to grow only from one end, and the other end acts as a reflecting boundary. Filaments, in contact with the barrier, grow with a polymerization rate $u = k_0 C exp(-\frac{fd}{k_B T})$, in accordance with Kramer's theory[15, 19], where k_0 is the intrinsic polymerization rate and C is the free

Table 1: Rates corresponding to microtubules.

Parameter	k_0 ($\mu M^{-1}s^{-1}$)	w_T (s^{-1})	w_D (s^{-1})	r (s^{-1})
Value	3.2	24	290	0.2

GTP-bound monomer concentration in the solution, k_B is the Boltzmann constant and T is the absolute temperature. Filaments away from the barrier polymerizes with a force independent rate $u_0 = k_0C$. The GTP-bound monomer can undergo random hydrolysis, i.e., a GTP-monomer at any random position along the filament can undergo hydrolysis with rate 'r'. This 'chemical switching' of monomers from the GTP-bound to GDP-bound state is irreversible in nature, forcing the dynamics to be out of equilibrium. The tip of the filament can now be in either GTP or GDP-bound state, accordingly there are different depolymerization rates, w_T and w_D, respectively, for GTP-monomer and GDP-monomer. The GDP-monomer is found to have a much higher depolymerization rate than their GTP counterpart [1], .i.e., $w_D \gg w_T$. This suggests that the filament is more prone to depolymerization when the tip is in the GDP-bound state rather than the GTP-bound state. Hence, a filament with a cap structure (successive T-monomers at the tip) is more stable against depolymerization than those without a cap. The depolymerization rates w_T and w_D are considered to be force independent.

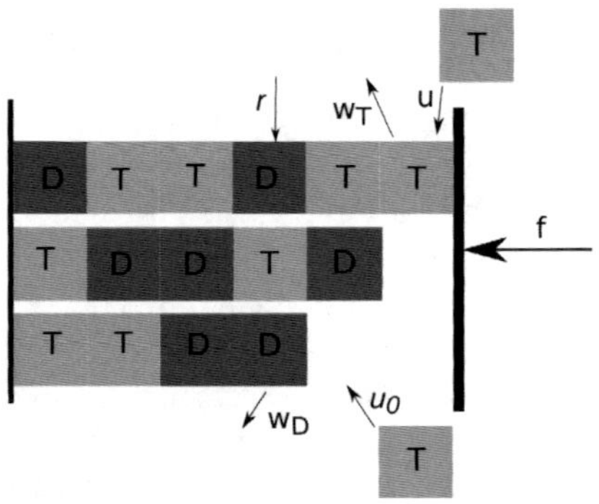

Figure 1: Schematic diagram of a multifilament system of parallel filaments growing against a constant force. Filaments, in contact with the barrier, grow with a polymerization rate $u = k_0Cexp(-\frac{fd}{k_BT})$, where as those away from the barrier polymerize with a rate $u_0 = k_0C$, where k_0 is the intrinsic polymerization rate and C is the free GTP-bound monomer concentration in the solution. Depolymerization rates are w_T and w_D, depending on whether the tip of the filament is in GTP-bound and GDP-bound states, respectively. At any position along the filament, a GTP-bound monomer can hydrolyse into a GDP bound monomer with rate r.

3. Results under constant force

We present here, the results on the collective behaviour of a system of N filaments growing against a constant force. The graphs presented are for microtubules, although similar results can be obtained for a system of actin filaments as well [15]. The parameters used for simulations are listed in Table 1.

3.1. Stall force measurement using growth velocity

Using kinetic Monte-Carlo, we have simulated the growth trajectories of a 2-filament system, in order to study their growth dynamics. The system of filaments grow on an average, as long

Journal of Physics: Conference Series **638** (2015) 012012 doi:10.1088/1742-6596/638/1/012012

as the force exerted is less than the stall force that can be generated by them. As the applied force increases, the average growth velocity of the filament decreases, until it vanishes at the stall force of the system. Figure 2 shows various traces of filament length (equivalently, position of the wall from the reflecting boundary) as a function of time, for a system of two microtubule filaments. The top curves, corresponding to various realisations, represent the situation where we apply a force equal to $2f_s^{(1)} = 14.48pN$. It can be seen that the two filament system does not stall at this force, but grows with a positive average velocity. The force at which the filament growth could be stalled occurs at a slightly higher force (= 15.45 pN), for which the resulting growth trajectories are shown in the bottom curves. Hence, the stall force, when the average filament velocity vanishes, is greater than $2f_s^{(1)}$. Similar results can be obtained for higher number N of filaments (data available in [15]), and it is observed that the excess stall force $\Delta^{(N)} = f_s^{(N)} - Nf_s^{(1)}$, increases almost linearly with the number N.

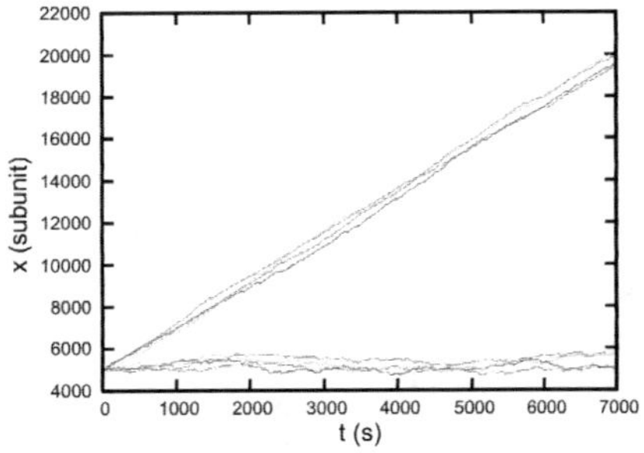

Figure 2: Different traces of the length of a two filament system, growing against a constant compressive force (f), is plotted as a function of time. The top and bottom curves correspond to $f = 2f_s^{(1)}$ and $f = f_s^{(2)}$, respectively. Concentration of GTP-bound monomers in the solution is 25μM.

3.2. Stall force measurement using collapse times of catastrophes

If the system of filaments is subjected to a force larger than its stall force, the steady state system velocity would be negative, but due to the reflecting boundary wall, the system stays bounded in length [6]. In the bounded state, the filaments undergo repeated length fluctuations known as catastrophes and rescues. Stall force of a multi-filament system can be measured by studying these catastrophes and rescues, at high forces [8]. As the applied compressive force decreases, filaments undergo catastrophes with decreasing frequencies. This can be quantified using average collapse time (T_{coll}) as discussed in [8], which is defined as the average time taken for the bundle of filaments to shrink from a maximum length to zero length during a cycle of rescue and catastrophe. As the catastrophes occur less frequently, the collapse time increases with decreasing force. With force approaching the stall force from above, the collapse time diverges. The collective behaviour of multiple filaments can be inferred by studying the collapse time of the system as one approaches the stall force. Figure 3 shows the average collapse time of microtubules, for two, three and four filaments, as a function of the ratio of applied force to the stall force of a single filament. The two curves in each panel correspond to the case with no hydrolysis (red) and with random hydrolysis (blue). Red curves diverge as the force ratio becomes equal to the number of filaments, indicating that the stall force of the N-filament system is equal to $Nf_s^{(1)}$. But, it is evident from the plot that the blue curves show divergence before the applied force is reduced to $Nf_s^{(1)}$. This point towards a cooperative behaviour of the N-filament system, where the stall force of N-filaments $f_s^{(N)}$ is greater than $Nf_s^{(1)}$, corroborating the analysis via force-velocity relation, as discussed in the previous section.

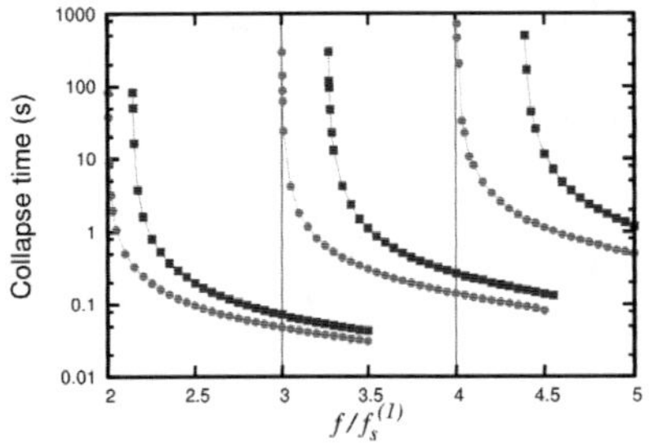

Figure 3: Average collapse time as a function of scaled force $f/f_s^{(1)}$ for a system of two, three and four microtubules, at a free monomer concentration of 25μM. The curves with symbols ● and ■ correspond to the cases without hydrolysis and with hydrolysis, respectively.

4. Results under a harmonic force

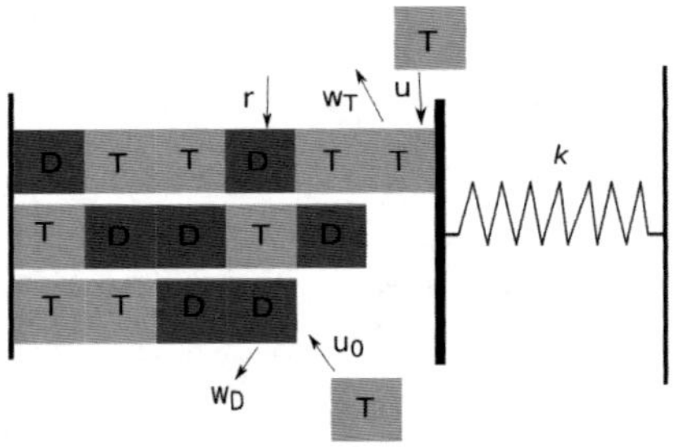

Figure 4: Schematic diagram of a multifilament system of parallel filaments growing against a harmonic force. The kinetic rates here are similar to the ones mentioned in Figure 1.

In vitro experiments [18], conducted on multiple parallel microtubules, growing against a harmonic force, have indicated an additive nature of stall force. These results are contrary to the theme we have built up so far, for constant force. So, this requires careful theoretical study. In the experiment, multiple filaments were allowed to grow from seeds, and one end the filament bundle was attached to a bead in an optical trap. A rigid barrier was held at the growing end of the bundle of filaments. Thus, as the filaments grow against the barrier, the position of the beam in the trap changes, which results in a restoring force, which is a linear function of the filament length. It was observed that the most probable forces produced by the multi-filament system occur at integral multiples of some unit.

Comparison of a theoretical model with the above in vitro study is needed. In this section, we theoretically model the growth of a system of N-filaments against a harmonic force, similar to that in [18]. Here, multiple filaments grow against a barrier which is attached to a spring with a stiffness constant k, as shown in Figure 4. The other end of the filament acts as a reflecting boundary as before. As the filament grows, it pushes against the spring, which in turn applies a restoring force on the filament, which is proportional to the length of the filament in contact with the spring. The important difference with the previous section is that, force now varies, and we would have a distribution of forces.

The forces generated by the system are obtained from filament lengths (x), as $f = -k(x-x_0)$, where x_0 is the reference length. The distribution of forces, related to the distribution of lengths x, is plotted in the insets of Figure 5 (a) and 5 (b), for two different free monomer concentrations C. It is evident from the main figures (see caption of Figure 5 (a) and 5 (b)) that the system

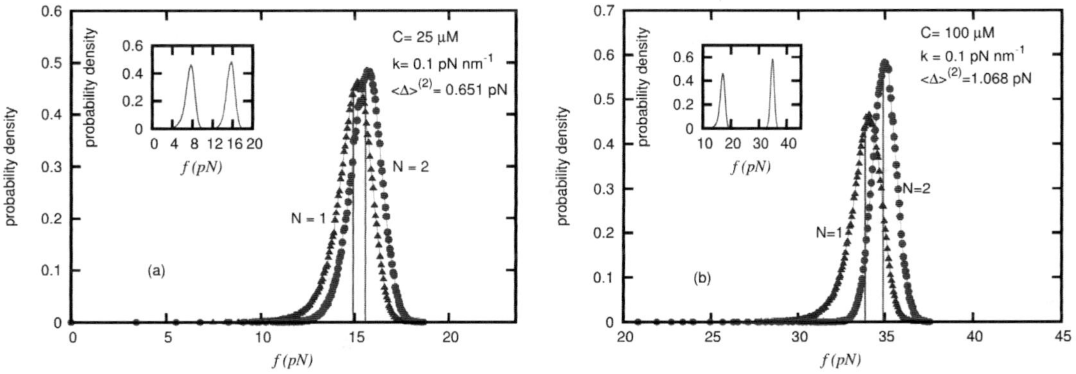

Figure 5: The two insets show the distributions of forces (f) generated by a single microtubule (at lower forces) and two microtubules (at higher forces), at two values of $C = 25\mu$M (a), and 100μM (b). The stiffness constant of the trap is k=0.1 pN nm^{-1}. In the main figures of (a) and (b), the distribution for the single-filament system is plotted against $f + \langle f_s^{(1)} \rangle$ (curves with ▲), for comparison with the distribution of the 2-filament system plotted against f (curves with ●). The point is that they do not coincide, implying that the mean stall forces shown by the vertical lines follow the relation $2\langle f_s^{(1)} \rangle \neq \langle f_s^{(2)} \rangle$.

of two microtubules generate a mean force $\langle f_s^{(2)} \rangle$, greater than twice the individual mean stall force $\langle f_s^{(1)} \rangle$:

$$\langle f_s^{(2)} \rangle > 2\langle f_s^{(1)} \rangle. \quad (3)$$

For $C = 25\mu M$, average excess force, $\langle \Delta \rangle^{(2)} = \langle f_s^{(2)} \rangle - 2\langle f_s^{(1)} \rangle$=0.651 pN, and for $C = 100\mu M$, $\langle \Delta \rangle^{(2)}$=1.068 pN. The deviations are small, but if looked for carefully, forces of ~ 1 pN may be detected in experiments.

5. Conclusion

Cytoskeletal filaments show highly dynamic behaviour, constantly regulating their structures, and in the presence of an obstacle, can generate resistive forces [3, 16]. Force generation capabilities of cytoskeletal filaments have been studied by various theoretical models. But the cooperative behaviour of a bundle of filaments are not well understood, since previous studies were done on either single filament or on multiple filaments in the absence of hydrolysis. The role of irreversible hydrolysis of ATP/GTP-monomer to ADP/GDP monomer has been carefully probed in [15]. It is observed that by introducing hydrolysis, the resulting non-equilibrium system has an enhanced collective stall force, as compared to the sum of individual stall forces, i.e. in a N-filament system, $f_s^{(N)} > N f_s^{(1)}$. The robustness of this result has been checked using a variety of models (with both sequential and random hydrolysis), over broad experimentally relevant parameter regimes, and for an analytically solvable toy model. The relation of this result with the violation of detailed balance is discussed in [15]. This work was further extended in [8], where the fluctuation properties of multiple filaments in the bound phase were studied. The study demonstrates a cooperation among the filaments against catastrophes, quantified by an average collapse time, "T_{coll}"– it is found to increase with the number of filaments, rendering more stability to multi-filament systems as compared to single filaments, against an applied force. Variation of T_{coll} as a function of scaled force $f/f_s^{(1)}$ shows that the forces at which T_{coll} diverges, which are the stall forces of the corresponding system, are not integral multiples of the single filament stall force.

Experimentally, Laan *et al.*[18] inferred that a force dependent collective instability exists in microtubules, by considering the growth of multiple microtubules in a harmonic force. They observe that the most probable forces that the system of microtubules produce are integral multiples of some unit. In order to probe whether the additivity of stall forces shown in [18] is a particularity of harmonic force, we have extended the multifilament model of [15, 8], by replacing the constant force ensemble by a harmonic force ensemble. In this paper, a comparison of force distributions of single filaments with those of multiple filaments within our model show that the mean force exerted by multiple filaments is greater than the sum of mean force of individual filaments. The deviation is small, and likely to be ignored within experimental errors, but strictly speaking, is non-zero. Hence, even in the presence of harmonic force, as opposed to constant force, we see that the cooperativity observed in [15] and [8] still hold. A further detailed study of multiple filaments in the presence of harmonic force, by considering various reaction rates and other parameters is necessary, to understand how the interesting deviation from additivity may be enhanced and brought within detectable limits of experiments.

Acknowledgments
DD would like to thank CSIR(India) (grant No. 03(1326)/14/EMR-II) for financial support.

References

[1] Desai A and Mitchison T J 1997 *Annu. Rev. Cell Dev. Biol.* **13** 83–117
[2] Alberts B, Johnson A, Lewis J, Raff M, Roberts K and Walter P 2002 *Molecular Biology of the Cell* 4th ed (New York: Garland Science)
[3] Howard J 2001 *Mechanics of Motor Proteins and the Cytoskeleton* (Massachusetts: Sinauer Associates, Inc.)
[4] van Doorn G S, Tanase C, Mulder B M and Dogterom M 2000 *Eur. Biophys. J.* **29** 2–6
[5] Tsekouras K, Lacoste D, Mallick K and Joanny J F 2011 *New J. Phys.* **13** 103032
[6] Ranjith P, Lacoste D, Mallick K and Joanny J F 2009 *Biophys. J.* **96** 2146–59
[7] Zelinski B and Kierfeld J 2013 *Phys. Rev. E* **87** 012703
[8] Das D, Das D and Padinhateeri R 2014 *PLOS ONE* **9** e114014
[9] Hill T L 1981 *Proc. Natl. Acad. Sci. U.S.A.* **78** 5613–5617
[10] Krawczyk J and Kierfeld J 2011 *Europhys. Lett.* **93** 28006
[11] Vavylonis D, Yang Q and O'Shaughnessy B 2005 *Proc. Natl. Acad. Sci. U.S.A.* **102** 8543–8548
[12] Stukalin E B and Kolomeisky A B 2005 *Biophys. J.* **90** 2673–2685
[13] Ranjith P, Mallick K, Joanny J F and Lacoste D 2010 *Biophys. J.* **98** 1418–27
[14] Padinhateeri R, Kolomeisky A B and Lacoste D 2012 *Biophys. J.* 1274–83
[15] Das D, Das D and Padinhateeri R 2014 *New J. Phys.* **16** 63032
[16] Dogterom M 1997 *Science* **278** 856–860
[17] Footer M J, Kerssemakers J W J, Theriot J a and Dogterom M 2007 *Proc. Natl. Acad. Sci. U.S.A.* **104** 2181–6
[18] Laan L, Husson J, Munteanu E L, Kerssemakers J W J and Dogterom M 2008 *Proc. Natl. Acad. Sci. U.S.A.* **105** 8920–5
[19] Peskin C S, Odell G M and Oster G F 1993 *Biophys. J.* **65** 316–24

Journal of Physics: Conference Series **638** (2015) 012013 doi:10.1088/1742-6596/638/1/012013

Conformational thermodynamics of biomolecular complexes: The histogram-based method

Amit Das[1#], Samapan Sikdar[1], Mahua Ghosh[*1] and J. Chakrabarti[*1,2]

[1]Department of Chemical, Biological And Macromolecular Sciences, S. N. Bose National Center for Basic Sciences, Sector III, Block JD, Salt lake City Kolkata 700098, INDIA

[2]Also at Unit of Nanoscience and Technology-II and The Thematic Unit of Excellence on Computational Materials Science, S. N. Bose National Centre for Basic Sciences, Block JD, Sector III, Salt Lake, Kolkata-700098, INDIA

[#]Presently at National Centre for Biological Sciences (TIFR), Bellary Road, Bangalore 560 065, INDIA

[*]Email: jaydeb@bose.res.in and mahuaghosh@bose.res.in

Abstract. Conformational changes in biomacromolecules govern majority of biological processes. Complete characterization of conformational contributions to thermodynamics of complexation of biomacromolecules has been challenging. Although, advances in NMR relaxation experiments and several computational studies have revealed important aspects of conformational entropy changes, efficient and large-scale estimations still remain an intriguing facet. Recent histogram-based method (HBM) offers a simple yet rigorous route to estimate both conformational entropy and free energy changes from same set of histograms in an efficient manner. The HBM utilizes the power of histograms which can be generated as accurately as desired from an arbitrarily large sample space from atomistic simulation trajectories. Here we discuss some recent applications of the HBM, using dihedral angles of amino acid residues as conformational variables, which provide good measure of conformational thermodynamics of several protein-peptide complexes, obtained from NMR, metal-ion binding to an important metalloprotein, interfacial changes in protein-protein complex and insight to protein function, coupled with conformational changes. We conclude the paper with a few future directions worth pursuing.

Journal of Physics: Conference Series **638** (2015) 012013 doi:10.1088/1742-6596/638/1/012013

1. Introduction

Bio-macromolecules are large flexible molecules which exist in numerous possible conformations in biological environments. Fluctuations among these conformations are of fundamental importance in their structure and function. One classic example of such conformational fluctuation is the folding of proteins into its native conformation [1]. The biomacromolecules need to adopt suitable conformations in general to participate in molecular recognition, signal transduction, gene expression and so on [2,3]. Furthermore, subtle conformational changes of different biomacromolecules have been utilized in various nano-biotechnological applications [4]. Thus, microscopic understanding of conformational fluctuations in biomacromolecules is important from both biological and technological standpoints.

The most important aspect is the thermodynamic characterization of conformational fluctuations. Experimentally, the isothermal titration calorimetry (ITC) measurements yield the binding enthalpy and the binding free energy for bio-macromolecular complexes [5] but not the thermodynamics of the conformational changes. The NMR relaxation experiments provide measure of the conformational entropy costs in bio-macromolecular complexes from the bond order parameter [2]. Such studies indicate that the conformational entropy changes constitute an important contribution of the thermodynamic changes in the complex. The computer simulations [6,7] often estimate the conformational entropy from the normal modes associated with atomic vibrations. However, the atomic Cartesian coordinates often are not suitable to capture all possible bond rotations, thus providing poor estimates [8] of conformational entropy. The approaches based on purely statistical scoring functions from the crystal structure databases [9] are devoid of microscopic details. The protein dihedral angles, defined by the angle between different atomic planes in the molecule, constitute a convenient set of microscopic conformational variable to define a conformational state of a protein. Multidimensional histograms of the dihedral angle are shown to provide the estimation of the conformational entropy [10] for small molecules. Such studies further show that nearly ~80% of the conformational entropy for different small molecules, like hydrogen peroxide, methanol and some alkanes, is accounted for by neglecting all sorts of correlations among the dihedral angles [11]. In biomacromolecules the dihedral correlations have been generally found [12,13] to be short-ranged. These observations highlight the importance of completely reduced one-dimensional histograms of the independent dihedral angles [14,15]. Accurate and efficient estimation of conformational free energy is challenging as well. The free energy difference between two conformational states of small biomolecules has been measured by the population ratios of the two states via UV resonance Raman measurements [16]. Computational methodologies for conformational free energies are of two types: (1) estimating the absolute free energies of a conformational state [17]; and (2) evaluating the free energy differences between two states [18-20]. However, these approaches are inefficient to study medium to large bio-macromolecular systems.

It is thus important to have efficient and accurate method for evaluating the conformational thermodynamics for large bio-macromolecule, like proteins. If the dihedral correlations can be ignored, the histogram-based method (HBM) [21] utilizing the one-dimensional histograms of the dihedral angles to extract the thermodynamics of conformational changes, including the entropy and free energy, by simply adding all the dihedral angle contributions of a protein. The basic idea can be understood as follows: Suppose that the normalized histograms of a given dihedral angle are computed in two conformational states. One can compare the entropies in the two states using the Gibb's formula and the free energy by comparing the peaks, relating the logarithm of the ratio to the free energy. Thus, unlike the existing methodologies, the HBM can yield both the information simultaneously from a single set of simulations. Moreover, the HBM is capable of yielding the conformational thermodynamics data even for the individual residues.

The histograms of dihedral fluctuations about different conformational states of a protein can be sampled from the equilibrium trajectories from all-atom MD simulations. In this article we focus on

Journal of Physics: Conference Series **638** (2015) 012013 doi:10.1088/1742-6596/638/1/012013

a couple of interesting aspects of the thermodynamics of conformational changes in a protein: (1) The conformational entropy estimated from the one-dimensional histograms can reproduce the conformational entropy obtained for large protein molecules from NMR experiments. This validates the HBM as a viable technique to estimate the conformational thermodynamics from microscopic conformational variables for bio-macromolecules. (2) The changes in the conformational thermodynamics are different for different residues. Even if a given conformational state is stabilized in thermodynamic sense, all the residues are not necessarily stabilized. The destabilized residues participate in binding events which gives clue for functional activities of different residues in different conformational states. The paper is organized as follows: We discuss the theoretical basis of the HBM in Sec. 2. We discuss the extraction of conformational thermodynamics for a system where the experimental data is available in Sec 3. We illustrate the method for a couple of cases, like ion binding to a protein in Sec. 4 and protein-protein interface in Sec. 5. We show the application of HBM in understanding protein function in Sec 6 and conclude with some future directions in Sec. 7.

2. The histogram-based method (HBM)

Let us consider a system with a set of conformational variables $\{\xi_i\}$. The normalized probability distribution at a temperature *T*, is given by:

$$P(\{\xi_i\}) = \frac{1}{Z}\exp[-\mathcal{H}(\{\xi_i\})/k_B T] \quad (1)$$

where k_B is the Boltzmann constant, $\mathcal{H}(\{\xi_i\})$, the Hamiltonian and *Z*, the partition function in the conformational space of the system which normalizes the probability distribution. Let us fix one of the conformational variables $\xi_i = \xi$. By integrating over the other variables in Eq. 1:

$$P(\xi_i = \xi) = (1/Z)\int \exp[-\mathcal{H}(\{\xi_i\})/k_B T]\delta(\xi_i - \xi)\prod_i d\xi_i = (1/Z)\exp[-G(\xi)/k_B T] \quad (2)$$

we get the effective free energy $G(\xi)$ or the potential of mean force associated with a given conformational variable ξ.

Let us specifically consider the case of a protein complexed with a ligand molecule. For dihedral angles θ of protein *p* and dihedral angles τ of ligand, one expresses $\xi \equiv (\theta, \tau)$. Subsequently, the subscripts *p+l*, *p* and *l* indicate quantities associated with the complex, the protein and the ligand respectively, while the superscripts *c* and *f* denote the bound and free state respectively. Then Eq. 2 defines the effective free energies as:

$$P^c_{p+l}(\xi) = \frac{1}{Z}\exp[-G^c_{p+l}(\xi)/k_B T],$$

$$P^f_p(\theta) = \frac{1}{Z}\exp[-G^f_p(\theta)/k_B T]$$

$$\text{and } P^f_l(\tau) = \frac{1}{Z}\exp[-G^f_l(\tau)/k_B T], \quad (3)$$

Therefore, the free energy change for ξ due to complexation is

$$\Delta G_{conf}(\xi) = -k_B T \ln\left[\frac{P^c_{p+l}(\xi)}{P^f_p(\theta)P^f_l(\tau)}\right]. \quad (4)$$

Journal of Physics: Conference Series **638** (2015) 012013 doi:10.1088/1742-6596/638/1/012013

If the conformational variables are independent, and $P^c_{p+l}(\xi) = P^c_p(\theta)\cdot P^c_l(\tau)$. Equation 4 then gives

$$\Delta G_{conf}(\xi) = -k_B T \ln\left[\frac{P^c_p(\theta)}{P^f_p(\theta)}\right] - k_B T \ln\left[\frac{P^c_l(\tau)}{P^f_l(\tau)}\right], \qquad (5)$$

which indicates that the thermodynamics is given separately in terms of the individual dihedrals. Summing over all the dihedral angles of protein and ligand gives the total conformational free energy change from Eq.5:

$$\Delta G^{tot}_{conf} = -k_B T \sum_{\theta} \ln \frac{P^c_p(\theta)}{P^f_p(\theta)} - k_B T \sum_{\tau} \ln \frac{P^c_l(\tau)}{P^f_l(\tau)} = \Delta G^{prot}_{conf} + \Delta G^{lig}_{conf}. \qquad (6)$$

Eqs. (5) and (6) may be simplified if the probability distributions are sharply peaked. Let us denote the normalized histograms $H^c_p(\theta)$ and $H^f_p(\theta)$ for a protein dihedral, and $H^c_l(\tau)$ and $H^f_l(\tau)$ for a ligand dihedral in the bound and the free states, respectively. The peak of the histogram defines the equilibrium value of the relevant dihedrals. Then the equilibrium conformational free energy cost associated with any protein dihedral θ is

$$\Delta G^{eq}_{conf}(\theta) = -k_B T \ln[H^c_{p,\max}(\theta) / H^f_{p,\max}(\theta)] \qquad (7)$$

where the subscript 'max' denotes the maximum of histogram. For multi-modal histograms the free energies can be estimated by taking average, weighted by the maximum values of the peaks. For a particular dihedral ξ with multi-modal histograms in both free and complex states, the free energy cost is $\sum_{i,j} c_{ij}\Delta G_{ij}$ where $\Delta G_{ij} = -k_B T \ln(H^c_{\max,j} / H^f_{\max,i})$, the free energy cost for transition from *i*th peak in free state to *j*th peak in the complex state and $c_{ij} = H^f_{\max,i} H^c_{\max,j} / \left(\sum_i H^f_{\max,i} + \sum_j H^c_{\max,j}\right)$, the respective weights.

The conformational entropy for a particular dihedral ξ can be estimated directly using the Gibbs entropy formula [22]: $S_{conf}(\xi) = -k_B \sum_i H_i(\xi) \ln H_i(\xi)$, where the sum is taken over the histogram bins *i* with a non-zero value of $H_i(\xi)$. Therefore, the conformational entropy change for the dihedral is

$$\Delta S_{conf}(\xi) = -k_B\left[\sum_i H^c_i(\xi) \ln H^c_i(\xi) - \sum_i H^f_i(\xi) \ln H^f_i(\xi)\right]. \qquad (8)$$

The HBM can be extended to estimate the free energy and entropy costs associated with the changes of distribution of any quantity. For instance, the interfacial water molecules undergo changes in the distribution over the interface compared to the free components. Let us denote the distribution of the interfacial water (*iw*) molecules within 5 Å of the interfacial residues in the complex by $g^c(r)$, and those within 5 Å of the binding regions of the free components by $g^f_N(r)$) and $g^f_l(r)$ respectively. Here *r* is the separation between the O-atoms of two water molecules. The free energy cost is given by straightforward extension of Eq. 7:

$$\Delta G_{iw} = -k_B T \ln\left[g^{c}_{\max}(r) / \left(g^{f}_{N,\max}(r) g^{f}_{I,\max}(r)\right)\right] \tag{9}$$

where the peak values in the respective $g(r)$ are used. The corresponding entropy cost ($T\Delta S_{iw}$) is obtained using the Gibbs formula as in Eq. 8:

$$\Delta S_{iw} = -k_B [\sum_i g^c_i \ln g^c_i - \sum_i g^f_{N,i} \ln g^f_{N,i} + \sum_i g^f_{I,i} \ln g^f_{I,i}] \tag{10}$$

where the sum has been taken over non-zero bins marked by the index i.

3. Application to Calmodulin-Peptide complexes

The HBM has been applied to the Calmodulin-peptide complexes. Calmodulin (CaM) is an ubiquitous eukaryotic protein which in calcium (Ca^{2+}) loaded (holo-) state acts as a primary mediator of activities of various cellular proteins responding to changes in intracellular Ca^{2+} levels [27]. CaM belongs to the family of proteins containing EF-hand [28], a helix-loop-helix structural motif that binds to Ca^{2+} ions via the loop. CaM has two globular domains: N-terminal (N-domain) and the C-terminal domain (C-domain), each domain having two EF-hands, I and II in the N-domain and III and IV in the C-domain (Figure 1a). Two domains are connected via a linker region consisting of 29 residues (residues 68-92). Metal-free (apo) CaM, upon Ca^{2+}-saturation, undergoes subtle conformational changes [29,30] in the metal-bound (holo) form (Figure 1a) with the linker becoming a long helix and thus exposing the target-binding hydrophobic faces of CaM for target peptides [5]. Moreover, the linker gets deformed to wrap around the peptides (Figure 1b) which are CaM-binding sequences of large number of proteins including several regulatory enzymes *e.g.* protein kinases, phosphodiesterases, cyclases etc. Five peptides have been considered: CaM-binding sequences of smooth muscle myosin light chain kinase (smMLCK) [31], the neuronal and endothelial nitric oxide synthases (nNOS and eNOS respectively) [32], the calmodulin kinase I (CaMKI)[5] and the calmodulin kinase kinase (CaMKK) [33]. For all these CaM-peptide complexes the ITC data and conformational entropy changes (ΔS_{conf}) measured via NMR relaxation experiments are known [2]. According to observations, the total changes in conformational entropy (ΔS^{tot}_{conf}) is linearly correlated with the total binding entropy (ΔS^{tot}_{bind}) for the complexes. Recent work [2] also shows that ΔS_{conf} for CaM and the target peptides are linearly related to $\langle \Delta S^2_{Me} \rangle$, the average changes in residue-weighted methyl group order-parameters S^2_{Me} which describes the rotational freedom of the methyl group.

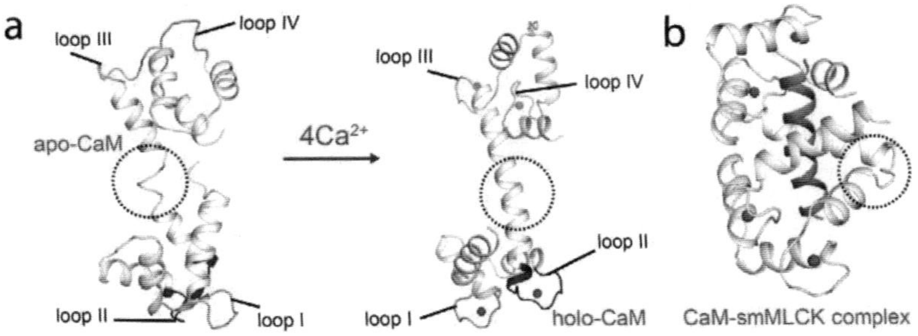

Figure 1: Binding events of CaM. (a) Apo- to holo- transition of CaM using cartoon representations. The EF-hand loops are coloured: I (purple), II (yellow), III (pink) and IV (cyan). The part of linker that undergo loop to helix transformation is marked by circle. (b) CaM bound to a target peptide (blue), smMLCK.

The dihedral angles have been calculated from the atomic coordinates in all atom simulation trajectories. The all-atom MD simulations have been performed for the free protein, metal bound protein, free peptide and the complex in explicitly electroneutral solvent. Simulations are done with the NAMD program [23] at 308K and 1 atm pressure in isothermal-isobaric (*NPT*) ensemble under standard protocols [24], using the CHARMM27 force-field [25], periodic boundary conditions and 1 femtosecond time-step. The number of total particles including water, pressure and temperature are kept fixed for each case to make the simulated ensembles equivalent. The initial configurations are chosen from following protein data bank (PDB) entries: 1CDL (smMLCK), 1NIW (eNOS), 2O60 (nNOS), 2L7L (CaMKI), 1CKK (CaMKK), 1CLL (metal bound CaM) and 1CFD (metal free CaM). The peptide coordinates taken out from the complex structures are the initial configurations of the free peptide simulations. The 50 ns long simulations capture most of the protein motions and peptide motions relevant for binding. The equilibration, which takes about 10 ns, is ensured in any run by monitoring the root mean square deviation. The dihedral angles (backbone dihedrals ϕ, ψ and the side-chain dihedrals χ_1, χ_2, χ_3, χ_4 and χ_5) are calculated from angle between the relevant atomic planes [26] given by the Cartesian coordinates of the associated atoms.

Figures 2a-c show the equilibrium correlations (Eq. 5) among different dihedrals. The correlation data among different dihedrals in CaM (Figure 2a) and peptides (Figure 2b) in various complexes and the cross-correlations between CaM and peptide dihedrals (Figure 2c) indicate only short-ranged correlations as found in earlier studies [13]. In absence of significant correlations, the histograms of the individual dihedrals are taken for the calculation of thermodynamics. Figure 2d shows three set of histograms in free and complex states for the dihedral angle ϕ of CaM-residue Glu45 in smMLCK-complex calculated from 1000 equilibrated configurations from different parts of the trajectory. All the backbone dihedrals (Figure 2e) exhibit sharp single-peaked histograms with the maxima around the equilibrium values in the initial configuration (PDB coordinates). Multi-modal histograms have been mostly observed for the side-chain dihedrals, as illustrated for χ_2 of CaM residue Leu69 (Figure 2f) and for χ_2 of CaM residue Asp118 (Figure 2g), indicating different rotameric states. We use Eqs. 8 and 9 to compute the changes in the conformational thermodynamics of the complex with respect to the free components. The similarities of these histograms in Figure 2d indicate the convergence of thermodynamic quantities calculated based on them, for instance, conformational entropies (Figure 3). Due to equivalence of the samples, the thermodynamic quantities are computed via a flat average over the entire equilibrium trajectory.

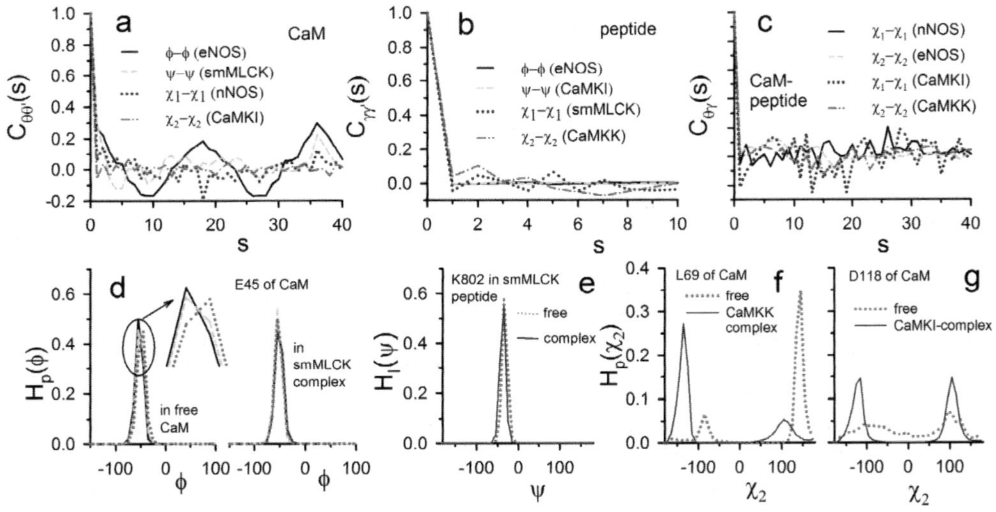

Figure 2: Equilibrium dihedral correlations and histograms. (a-c) ξ - ξ' correlations $C_{\xi\xi'}(s)$ (Eq. 5) where ξ and ξ' are dihedral angles from two residues and s is the difference between their locations. All correlation data have been normalized to $C_{\xi\xi'}(s=0)$ (a) Dihedrals of CaM in different complexes (b) Dihedrals of peptides in different complexes (c) Cross-correlations among side-chain dihedrals ξ from protein and ξ' from peptide. Representative histograms of (d) ϕ of Glu45 of CaM in free and bound form in smMLCK-complex. The inset shows the near-peak region for the free case. In each case three convergent histograms are shown in solid, dashed and dotted lines sampled from different parts of the MD-trajectory; (e) ψ of Lys802 of smMLCK peptide in free (dotted) and CaM-bound form (solid). Multi-modal histograms are shown for (f) χ_2 of CaM residue Leu69 in free and bound form in CaMKK complex and (g) χ_2 of CaM residue Asp118 in free and bound form in CaMKI complex.

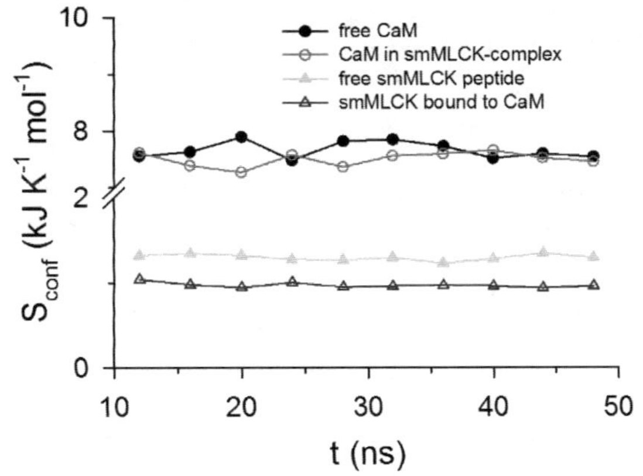

Figure 3: The convergence of simulated conformational entropy is shown for the smMLCK complex chosen as a representative case. The convergence demonstrates the robustness of the HBM. The errors in the data are very small: 0.05 kJ K^{-1} mol^{-1} for free CaM, 0.04 kJ K^{-1} mol^{-1} for CaM in the complex and nearly zero for the peptide.

Journal of Physics: Conference Series **638** (2015) 012013 doi:10.1088/1742-6596/638/1/012013

In Figure 4a, the total conformational entropies ΔS^{tot}_{conf} of the complex obtained from HBM [21] has been compared with the NMR results [2]. Both the theoretical and experimental ΔS^{tot}_{conf} are plotted against the corresponding ΔS^{tot}_{bind} from ITC [2], showing linear correlations between ΔS^{tot}_{conf} and ΔS^{tot}_{bind}. The HBM provide an overall good estimate for all cases, with the correlation line (slope m = 1.3 and linear correlation coefficient R = 0.95) [21] agreeing quite well with the experimental data (m = 0.95 and R = 0.75) [2]. The ΔS_{conf} (in kJ K^{-1} mol^{-1}) for all the side-chain dihedrals in the complexes showing multi-peak histograms contribute more than 60% of the total conformational entropy stabilizations of the complexes. This indicates the importance of the redistributions of populations among various side-chain rotamers in the binding. In Figure 4b, the contributions of the CaM and peptide are reported separately in the complexes. The CaM contributions show good agreement between the HBM and the experimental data. The best fit theoretical line (dashed, m = 1.1, R = 0.88) almost quantitatively matches the experimental correlation line (solid, m = 1.0, R = 0.94). CaM in CaMKI has been observed to be most ordered and least ordered in nNOS which support the same experimental observations [2]. For the peptide, similar ordering has been observed for all the cases which explain the experimental trend [2].

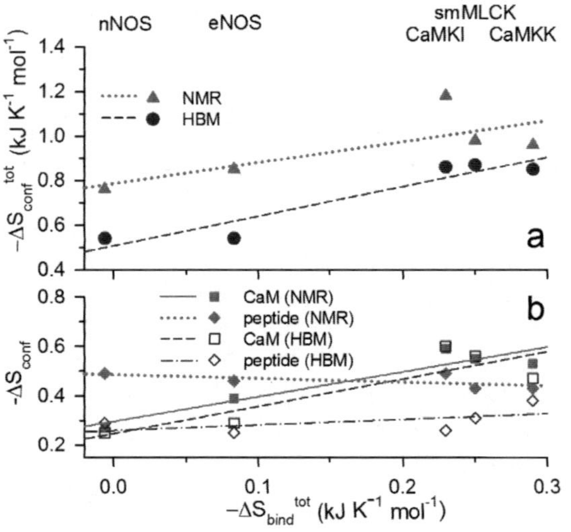

Figure 4: Comparison of theoretical and experimental (NMR [2]) conformational entropy data for CaM-peptide binding plotted against the experimental ΔS^{tot}_{bind} from ITC measurements [2]: (a) Calculated (circles) and experimental (triangles) ΔS^{tot}_{conf}. (b) The conformational entropy contributions of the components obtained from HBM: CaM (open squares) and peptide (open diamonds) contributions. The corresponding experimental data [2] for CaM contributions (filled squares) and peptides (filled diamonds) are also shown. The lines represent the best linear fits to the respective data.

The total conformational free energy changes ΔG_{conf}^{tot} and the contributions of the components are shown in Figure 5a along with the experimental total binding free energy cost [2] ΔG_{bind}^{tot}. The $-\Delta G_{bind}^{tot}$ values fall in a very narrow window (45-52 kJ mol^{-1}). The estimated $-\Delta G_{conf}^{tot}$ lie in a similar range (27-47 kJ mol^{-1}) except CaMKK where the extent of stabilization is nearly double. Although the ΔG_{bind}^{tot}-data seem to indicate that all the peptides exhibit similar affinity to CaM, the conformational contributions of the components bring out a different picture. The protein is conformationally most stabilized in CaMKK-complex, while least stabilized in eNOS-complex. This separates out CaMKK from the others which may be a signature of its opposite binding orientation compared to other four, namely, the N-terminal of four of the peptides interact with the C-terminal of CaM, except CaMKK for where the C-terminal of the peptide interacts with the N-terminal of the protein [33].

A negative change in conformational free energy ΔG_i^{conf} indicates conformational stability in the complex with respect to the free components, whereas that in $T\Delta S_i^{conf}$ represents conformational ordering. The individual residues undergo different degree of changes in the complex with respect to the free states. The thermodynamic changes at each protein residue are shown by surface representations in Figure 5b-c for the most and least stabilized complexes, both free-energetically and entropically. In Figure 5b, CaMKK and eNOS complexes are shown, where the green residues are stabilized, the red residues are destabilized and the white residues undergo a marginal change in conformational free energy. Figure 5c shows the entropy changes for CaMKI and nNOS complexes where the green residues are ordered while the red residues are disordered in the bound protein compared to the free-state.

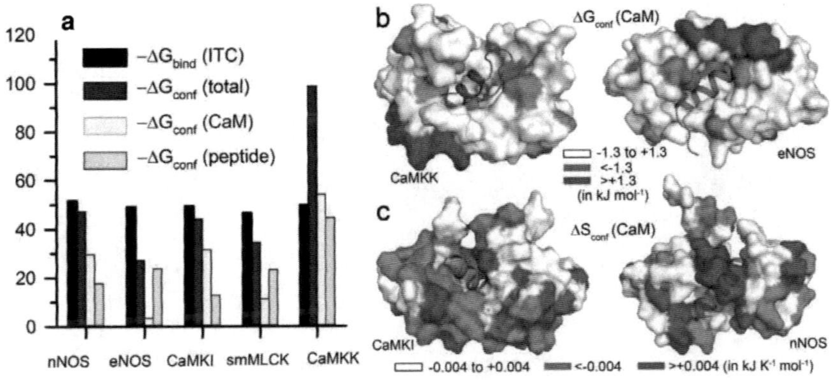

Figure 5: (a) Calculated ΔG_{conf}^{tot} and the individual protein and peptide contributions along with the experimental ΔG_{bind}^{tot}. All data are in kJ mol^{-1}. Surface representations of CaM-peptide complexes showing the residue-wise (b) ΔG_{conf} of CaM for the cases where the protein is most stabilized (CaMKK) and least stabilized (eNOS); (c) ΔS_{conf} of CaM in most ordered (CaMKI) and disordered (nNOS) complexes. The stabilized residues are green and destabilized ones are red. The residues undergoing minor changes are white. Peptide is in violet cartoon representation.

The HBM is applied to make predictions on the conformational thermodynamics of binding of a target peptide from death associated protein kinase (DAPK2) [34] to CaM (PDB ID 1ZUZ) and known ITC data ($\Delta G^{tot}_{bind} = -39.5$ kJ mol^{-1} and $\Delta S^{tot}_{bind} = -0.28$ kJ K^{-1} mol^{-1} at 300K). However, no experimental data is known regarding the conformational entropies. Here $\Delta S^{tot}_{conf} = -1.29$ kJ K^{-1} mol^{-1} using the Gibbs formula where the CaM and peptide contributions are –0.98 and –0.31 kJ K^{-1} mol^{-1} respectively. The ΔS^{tot}_{conf} value follow the same linear scaling with $\langle \Delta S^2_{Me} \rangle$ as the other complexes as shown by the closed circle in Fig.4a. We get $\Delta G^{tot}_{conf} = -38.1$ kJ mol^{-1} with CaM and peptide contributions being –15.4 and –22.7 kJ mol^{-1}, respectively. These free energy values are very similar to the case of smMLCK. Residues 77-83 constitute the maximum destabilized PBR in DAPK2-complex: ΔG_{conf}= +21.2 kJ mol^{-1} and ΔS_{conf}= +0.02 kJ K^{-1} mol^{-1}.

The agreement between the results on ΔS_{conf} from HBM and those of Marlow et al.[2] has got a strong implication. Marlow et al. [2] connects the NMR data on $\langle \Delta S^2_{Me} \rangle$, the average changes of residue weighted S^2_{Me}, to ΔS_{conf} for several CaM-peptide complexes. By definition, $\langle \Delta S^2_{Me} \rangle = n_{CaM} \langle \Delta S^2_{Me} \rangle^{\mathbf{CaM}} + n_{pep} \langle \Delta S^2_{Me} \rangle^{\mathbf{pep}}$ where n_{CaM} and n_{pep} are the numbers of residues in CaM and peptide respectively. $\langle \Delta S^2_{Me} \rangle^{\mathrm{CaM}} = \langle S^2_{Me} \rangle^{\mathrm{CaM}}_c - \langle S^2_{Me} \rangle^{\mathrm{CaM}}_f$ and $\langle \Delta S^2_{Me} \rangle^{\mathbf{pep}} = \langle S^2_{Me} \rangle^{\mathbf{pep}}_c - \langle S^2_{Me} \rangle^{\mathbf{pep}}_f$ where the average is taken over the available methyl groups in the respective system. A microscopic justification for the linear connection described by Marlow et al. [2] between ΔS_{conf} and $\langle \Delta S^2_{Me} \rangle$ is provided by the HBM. Figure 6 shows the linearity between the estimated ΔS^{tot}_{conf} from HBM and $\langle \Delta S^2_{Me} \rangle$ directly obtained from simulations. The slope (m = –0.050) of this linear fit matches quite well with those reported in Marlow et al. (m ~ –0.04) [2]. CaMKK has been an outlier in both the calculated [21] as well as the experimental data [2]. The conformational enthalpy changes $\Delta H^{tot}_{conf} (= \Delta G^{tot}_{conf} + T\Delta S^{tot}_{conf})$ should also have the same linearity with $\langle \Delta S^2_{Me} \rangle$ as ΔS^{tot}_{conf} for thermodynamic consistency. This is shown also in Figure 6 via the plot of $\Delta H^{tot}_{conf} / T$ values of the complexes, excluding the outlier CaMKK.

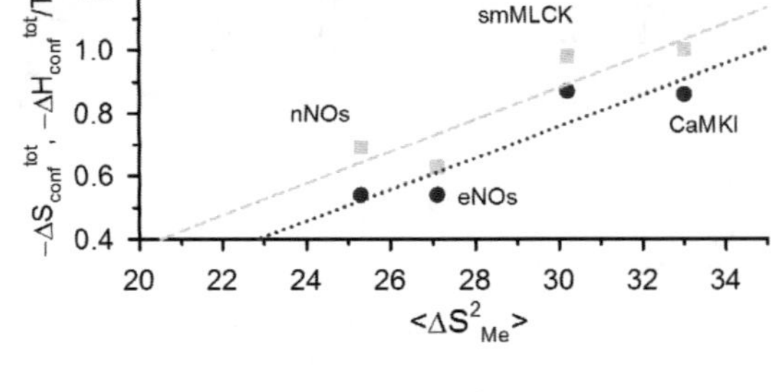

Figure 6: Plot of $-\Delta S^{tot}_{conf}$ (circles) and $-\Delta H^{tot}_{conf}/T$ (squares) vs. $\langle \Delta S^2_{Me} \rangle$ from HBM excluding CaMKK, along with the best fit lines (dotted for $-\Delta S^{tot}_{conf}$ and dashed for $-\Delta H^{tot}_{conf}/T$).

4. Conformational thermodynamics for Ca^{2+}-ion binding to CaM

The HBM has been applied to study the Ca^{2+}-ion binding to apo-CaM [35]. Although some experimental [36-38] and theoretical studies [39,40] exist, the conformational thermodynamics of CaM upon Ca^{2+}-ion binding has remained largely unexplored. The conformational thermodynamic changes have been calculated for the holo-state with respect to the apo-state. Here, $\Delta G^{tot}_{conf} = -54.9$ kJ mol^{-1} for binding of four Ca^{2+}, nearly 30% of the experimentally measured binding free energy [36]. The corresponding $T\Delta S^{tot}_{conf} = -128.7$ kJ mol^{-1} indicating substantial drop in conformational flexibility due to Ca^{2+} binding. In absence of any experimental data, an alternative verification of the estimated $T\Delta S^{tot}_{conf}$ can be obtained from $\langle \Delta S^2_{Me} \rangle$ calculated from the simulations. Using the theoretical value of the slope (−0.05) of the linear dependence between $T\Delta S^{tot}_{conf}$ and $\langle \Delta S^2_{Me} \rangle$ [21], $T\Delta S^{tot}_{conf} = -169$ kJ mol^{-1} from estimated $\langle \Delta S^2_{Me} \rangle$ (=10.95) which is close to the estimated $T\Delta S^{tot}_{conf}$ from dihedral histograms.

Figure 7 shows the contributions of different type of residues in these loops towards the respective $T\Delta S^{L}_{conf}$ (Figure 7a) and ΔG^{L}_{conf} (Figure 7b). The trends are very similar in both the panels for all the loops. In all cases except loop III, the main change comes from the acidic residues, nearly all coordinating to Ca^{2+}. However, the ordering and stabilization of the acidic residues in loop III is nearly half of those observed in loops I and IV. The basic residues in loops I are ordered and stabilized, while that in loop III remains almost unchanged. The residues with hydrophobic side-chains and those with polar side-chains order and stabilize amply in loops I, II and IV.

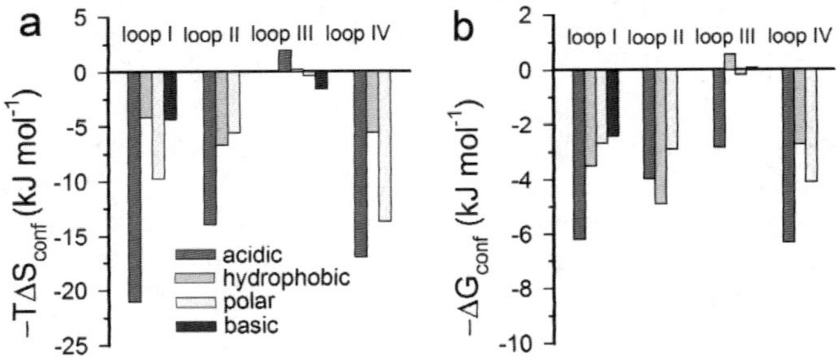

Figure 7: Data for the four EF-hand loops. (a) $T\Delta S_{conf}$ and (b) ΔG_{conf} contributions of different types of residues in the loops. The loop descriptions are: loop I - residues 20-31 (amino acid sequence DKDGDGTITTKE), loop II - residues 56-67 (sequence DADGNGTIDFPE), loop III - residues 93-104 (sequence DKDGNGYISAAE) and loop IV - residues 129-140 (sequence DIDGDGQVNYEE).

The most striking conformational change upon Ca^{2+} binding to apo-CaM occurs in the linker (residues 64-92), connecting the two domains, where a loop (the encircled region in Figure 1a) becomes helical in holo-state helping CaM to expose its hydrophobic surface for target-binding. The metal-induced changes of the linker helix in CaM are quite substantial: $T\Delta S_{conf} = -44.2$ and $\Delta G_{conf} = -13.3$ kJ mol^{-1} for all 29 residues in the linker. These values account for ~34% of $T\Delta S^{tot}_{conf}$ and ~25% of ΔG^{tot}_{conf} for the whole protein. The majority of the changes in linker helix come from the loop region that becomes helical upon Ca^{2+}-binding. $T\Delta S_{conf} = -20.5$ and $\Delta G_{conf} = -10.2$ kJ mol^{-1} for these six residues 76-81 account for ~46% of $T\Delta S_{conf}$ and ~77% of ΔG_{conf} estimated for linker helix.

5. Thermodynamics of conformational changes at protein-protein interface

Recent experiments with bio-macromolecular complexes suggest that structural modifications at the interfaces are vital for stability of the complexes. Although several qualitative aspects about such interfaces are known from structural data, quantification of the interfacial changes is lacking. The thermodynamic changes at the interface in the complex have been estimated using the HBM [41]. A structurally well-characterized protein-protein complex [42], comprising of an enzyme Nuclease A, NucA [43] and its specific inhibitor NuiA [44], a protein has been considered. NucA, a sugar non-specific nuclease from *Anabaena sp.*, is one of the most active nucleases known. These nucleases, having the capacity to hydrolyze nucleic acids without any base-preference, are important for their roles in host cellular mechanisms [45-54]. These enzymes require divalent metal-ion cofactor (like Mn^{2+}, Mg^{2+}) at the active-site to function. NuiA, an intrinsic inhibitor, deactivates NucA specifically via formation of a 1:1 complex with a high stability constant ($> 10^9$ M^{-1}) [47, 48]. The crystal structure of NucA-NuiA complex is particularly important, since isolation of any substrate-bound NucA has been unsuccessful so far due to its high activity. This complex throws light on the plausible nuclease activity of NucA. The crystal structure reveals the conformational changes of NucA and NuiA compared to their free forms and indicate that the electrostatic interactions between basic and the acidic residues dominate the interfacial changes [42, 53]. The crystal structure also indicates the presence of water molecule mediated interactions at the interface. A quantitative analysis of the contributions of water molecules to the interfacial thermodynamic changes is important.

The HBM has been applied on the dihedral angles calculated from the simulated conformations of the entire complex and the free components. Figs. 7(a) and (b) show the conformational entropy and free energy changes of the different types of residues in the interface of the complex with respect to those in the free components. The basic residues of NucA and acidic residues of NuiA undergo maximum stabilization and ordering out of all interfacial residues. This quantitatively confirms the suggestion of the crystal structure data the enzyme-inhibitor interaction is predominantly electrostatic. The stabilizations and ordering of the polar side-chain containing residues at the interface are also significant.

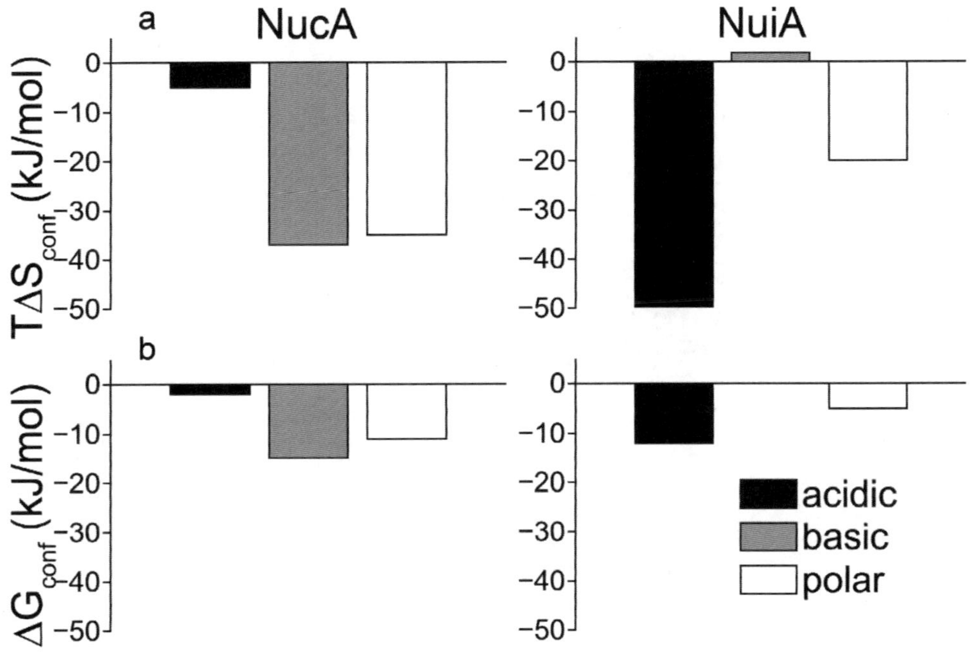

Figure 7: Thermodynamic changes at interface. (a) The conformational entropy and (b) conformational free energy changes of different types of the interfacial residues of NucA and NuiA are shown using a bar diagram.

The HBM has also been extended to the distributions of water molecules present in the NucA-NuiA interface as indicated by Eqs. 9 and 10, using the pair distribution function $g(r)$ [55] for the water molecules which is the probability of finding the centers of mass of a pair of molecules at given separation r. The $g(r)$ for the interfacial water molecules is shown in Figure 8 along with those for the waters present around the binding surfaces of free NucA and free NuiA. The gross behavior of these water distributions is the same in all three cases, except the differences in the peak value. The maximum peak value is observed in the complex, indicating high localization of the water molecules at the interface due to tight inhibitor-binding. Such localization leads to a substantial drop of entropy ($T\Delta S_{iw} \sim -10.2$ kJ mol^{-1} for about 300 interfacial waters) with respect to the free components. However, the free energy cost turns out to be negligible ($\Delta G_{iw} \sim -0.6$ kJ mol^{-1}), indicating entropy dominated contribution of water to the interfacial thermodynamics.

Journal of Physics: Conference Series **638** (2015) 012013
doi:10.1088/1742-6596/638/1/012013

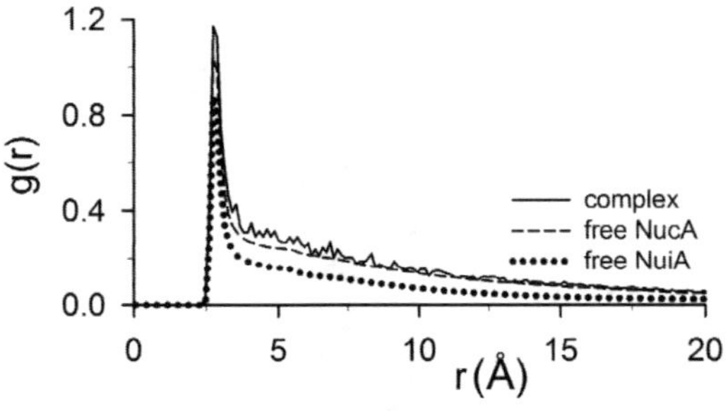

Figure 8: The oxygen-oxygen $g(r)$ of the interfacial water molecules in the complex and the waters around the binding regions of NucA and NuiA.

6. Prediction of protein function

The connection between conformational states and the functionalities of metalloproteins has been illustrated through microscopic conformational thermodynamics [56]. A specific instance is considered where metal ions induce conformational changes in α-Lactalbumin (aLA), a Ca^{2+} ion binding metalloprotein, expressed in the mammary glands during lactation [57]. It participates in a variety of cellular functions, like lactose synthesis [58] and fatty acid binding [59].

As per the crystal structure data [60], the bovine aLA comprises of an α-helical domain and a small β-sheet domain separated by a cleft [61]. The metal ion free native-apo structure of aLA is stable under neutral pH and physiological salt concentration. Calorimetric measurements [62, 63] suggest that apo-aLA binds to physiologically abundant Ca^{2+} at the metal binding loop connecting the two domains, resulting in metal ion bound holo structure. Fluorescence and Circular Dichroism studies show that the metal ions provide stability to native-apo aLA against heat or other denaturing agents [64-67] along with change in conformational state by narrowing of the interfacial cleft. The Ca^{2+}-aLA forms complex with β-1,4-galactosyltransferase (β4GalT), to promote lactose synthesis [58]. The crystal structure of aLA complexed with β4GalT has been reported [58, 68]. Another interesting property of aLA is its ability to form molten globule (MG) conformation [61] in acidic pH upon removal of Ca^{2+}. The MG state has fluctuating tertiary structure and binds to various fatty acids, like oleic acid. It has been identified that the oleic acid is the cytotoxic factor of XAMLET, whereas the MG-aLA acts as its carrier [69, 70]. The binding sites for fatty acid are largely unknown, since the MG state cannot be crystallized. Although the native-aLA cannot bind to oleic acid, it is still inconclusive whether divalent metal ion bound native Ca^{2+}-aLA can bind to oleic acid [70-74]. Since at physiological condition, the pH is neutral and there is abundance of Ca^{2+}, it is of paramount interest to understand the role of these ions in fatty acid binding to aLA, particularly in the context of cyto-toxic activities.

The histograms of dihedral angles of the protein, generated from all-atom molecular dynamics simulations, are used to compare the conformational thermodynamics between the holo Ca^{2+}-aLA and native apo-aLA. We relate the thermodynamics to the functionality of the protein: The residues having instability in conformational free energy ($\Delta G_i^{conf} > 0$) in the holo form compared to the native-apo state of the protein, are likely to participate in further binding event to attain thermodynamic stability by reducing the free energy. The reduction in free energy is further aided by disordering ($T\Delta S_i^{conf} >$ 0). We test this approach for the case of β4GalT binding to Ca^{2+}-aLA where the binding residues are known from crystal structure data. Among the binding residues, the C-terminal residues like aspartate (D) 116, glutamine (Q) 117, tryptophan (W) 118 and leucine (L) 119 are destabilized and disordered and can dock β4GalT onto Ca^{2+}-aLA.

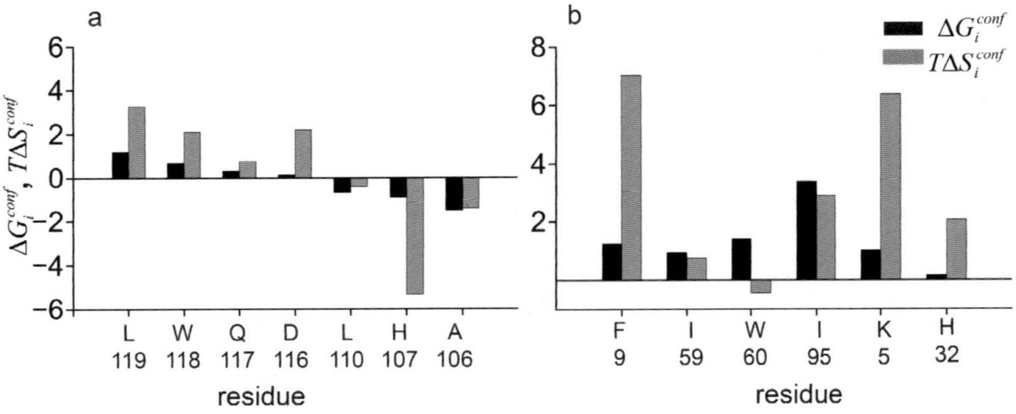

Figure 9: The conformational thermodynamic changes (a) at the β4GalT binding residues and (b) the basic and the hydrophobic residues which can aid in oleic acid binding to Ca^{2+}-aLA compared to the native state.

Fatty acids being amphipathic in nature with negatively charged carboxylate (COO^-) head-groups and long hydrophobic tail, the complexation of fatty acids with aLA is driven by both electrostatic and hydrophobic interactions [69, 75]. We apply similar analysis to oleic acid binding and predict that the Ca^{2+}-aLA complex can bind to oleic acid through the basic histidine (H) 32 of A2 helix and the hydrophobic residues, namely, isoleucine (I) 59, W60 and I95 of the interfacial cleft. We find by docking analysis that the negatively charged head group of oleic acid is stabilized through electrostatic interactions with the side chain imidazole ring of the basic H32 (separation ~ 5.0 Å). The long hydrophobic tail of oleic acid extends towards the interfacial cleft. This long tail is further stabilized through hydrophobic interactions with W104, I95, I59 and W60. The functional sites are often conserved through evolution in a protein family. Multiple sequence alignment of thirty-six different species of aLA shows that the basic H32 along with the hydrophobic residues like W60, I95 and W104 are highly conserved across the aLA sequences.

7. Conclusion

To summarize, the thermodynamic changes in biomacromolecular conformations can be extracted from the distributions of the dihedral angles using the HBM from a single set of simulations in the relevant conformational states. This method is sufficiently fast and accurate to be applicable large macromolecular complex like protein-protein, protein-DNA or protein-ligand complexes. The detailed thermodynamic analysis of the binding regions would enable us to identify the prime spots of binding. The functional activities coupled to the protein conformations can be extracted from the destabilized residues due to the changes in conformation. The HBM may be useful to: (1) compare the relative stability and ordering between different states and thus selecting thermodynamically plausible state to supplement homology modelling in structurally unknown protein[76]; (2) provide functional insight following changes in conformational states and hence, aiding in rational drug designing [77] and a variety of nano-biotechnological applications [78]. The dynamical information extracted from the time series of the dihedral fluctuations may provide useful insight to dynamic phenomena associated with bio-macromolecular interactions, including signal transduction [79].

Acknowledgements
AD thanks Council of Scientific and Industrial Research, India and SS thanks the University Grants Commission, India for research fellowships. MG and JC thanks the Department of Science and Technology, India for funding.

References

[1] Thirumalai D, Liu Z, O'Brien EP, Reddy G, *Curr Opin Struct Biol*, 2013, **23**, 22–29.
[2] Marlow MS, Dogan J, Frederick KK, Valentine KG, Wand AJ, *Nature Chem Biol*, 2010, **6**, 352-358.
[3] Soukup JK, Soukup GA, *Curr Opin Struct Biol*, 2004, **14**, 344–349.
[4] Hall WP, Modica J, Anker J, Lin Y, Mrksich M, et al., *Nano Lett*, 2011, **11**, 1098–1105.
[5] Brokx RD, Lopez MM, Vogel HJ, Makhatadze GI, *J Biol Chem*, 2001, **276**, 14083–14091.
[6] Schafer H, Mark AE, Gunsteren WFv, *J Chem Phys*, 2000, **113**, 7809-7817.
[7] Baron R, Hu¨nenberger PH, McCammon JA, *J Chem Theory Comput*, 2009, **5**, 3150–3160.
[8] Chang C-E, Chen W, Gilson MK, *J Chem Theory Comput*, 2005, **1**, 1017-1028.
[9] Gromiha MM, Fukui K, J, *Chem Inf Model*, 2011, **51**, 721–729.
[10] Trbovic N, Cho J-H, Abel R, Friesner RA, Rance M, et al., *J Am Chem Soc*, 2009, **131**, 615–622.
[11] Killian BJ, Kravitz JY, Gilson MK , *J Chem Phys*, 2007, **127**, 024107
[12] Mendez R, Bastolla U, *Phys Rev Lett*, 2010, **104**, 228103
[13] Li D-W, Meng D, Bru¨schweiler R, *J Am Chem Soc*, 2009, **131**, 14610–14611.
[14] Li D-W, Bru¨schweiler R, *Phys Rev Lett*, 2009, **102**, 118108
[15] Li D-W, Showalter SA, Bru¨schweiler R, *J Phys Chem B*, 2010, **114**, 16036–16044.
[16] Ma L, Ahmed Z, Mikhonin AV, Asher SA, *J Phys Chem B*, 2007, **111,** 7675-7680.
[17] Ytreberg FM, Zuckerman DM, *J Chem Phys*, 2006, **124,** 104105.
[18] Cecchini M, Krivov SV, Spichty M, Karplus M, *J Phys Chem B*, 2009, **113**, 9728–9740.
[19] Park S, Lau AY, Roux B, *J Chem Phys*, 2008, **129**, 134102.
[20] Spichty M, Cecchini) M, Karplus M, *J Phys Chem Lett*, 2010, **1**, 1922–1926.
[21] Das A, Chakrabarti J, Ghosh M, *Biophys J*, 2013, **104**, 1274-1284.
[22] Reif F, *Fundamentals of statistical and thermal physic,* 1965 (Auckland: McGraw-Hill).
[23] Phillips JC, Braun R, Wang W, Gumbart J, Tajkhorshid E, et al., *J Comp Chem*, 2005, **26**, 1781-1802.
[24] Humphrey W, Dalke A, Schulten K, *J Molec Graphics*, 1996, **14**, 33-38.
[25] Brooks BR, III CLB, Mackerell AD, Nilsson L, Petrella RJ, et al., *J Comp Chem*, 2009, **30**, 1545-1615.
[26] Branden C, Tooze J, *Introduction to Protein Structure,* 1999 (New York: Garland Publishing, Inc.).
[27] Babu YS, Sack JS, Greenhough TJ, Bugg CE, Means AR, et al., *Nature*, 1985, **315**, 37-40.
[28] Nelson MR, Thulin E, Fagan PA, Forsén S, Chazin WJ, Protein Sci, 2002, **11**, 198–205.
[29] Kuboniwa H, Tjandra N, Grzesiek S, Ren H, Klee C, et al., *Nat Struct Biol*, 1995, **2**, 768-776.
[30] Zhang M, Tanaka T, Ikura M, *Nature Struct Biol*, 1995, **2**, 758 - 767.
[31] Wintrode PL, Privalov PL, *J Mol Biol*, 1997, **266**, 1050-1062.
[32] Zhang M, Vogel HJ, *J Biol Chem*, 1994, **269**, 981-985.
[33] Osawa M, Tokumitsu H, Swindells MB, Kurihara H, Orita M, et al., *Nature struct Mol Biol*, 1999, **6**, 819 - 824.
[34] Kuczera K, Kursula P , *J Biomol Struct Dyn*, 2012, **30**, 45-61.
[35] Das A, Chakrabarti J, Ghosh M, *Chem Phys Lett*, 2013, **581**, 91-95.
[36] Linse S, Helmersson A, Forsen S, *J Biol Chem*, 1991, **266**, 8050-8054.

[37] Wu G, Gao Z, Dong A, Yu S, *Int J Biol Macromol*, 2012, **50**, 1011– 1017.
[38] Ye Y, Lee H-W, Yang W, Shealy S, Yang JJ, *J Am Chem Soc*, 2005, **127**, 3743-3750.
[39] Kobayashi C, Takada S., *Biophys J*, 2006, **90**, 3043–3051.
[40] Lepsik M, Field MJ, *J Phys Chem B*, 2007, **111**, 10012-10022.
[41] Das, A., Chakrabarti, J. and Ghosh M, *Mol Biosyst*, 2014, **10**, 437-445.
[42] Ghosh, M.; Meiss, G.; Pingoud, A.; London, R.E and Pedersen, L.C, *J. Biol. Chem.*, 2007, **282**, 5682–5690.
[43] Muro-Pastor, A. M., Flores, E., Herrero, A. & Wolk, C. P., *Mol. Microbiol.*, 1992, **6**, 3021-3030.
[44] Muro-Pastor, A. M., Herrero, A., Flores, E., *J. Mol. Biol.*, 1997, **268**, 589-598.
[45] Meiss, G., Franke, I., Gimadutdinow, O., Urbanke, C., and Pingoud, A., *Eur. J. Biochem.* , 1998, **251**, 924–934.
[46] Rangarajan, E.S. and Shankar, V., *FEMS Microbiol. Rev.*, 2001, **25**, 583-613.
[47] Hsia, K.C., Li, C.L. and Yuan, H.S., *Curr. Opin. Struct. Biol.* , 2055, **15**, 126-134.
[48] Fox K.R. and Waring, M.J., *Biochim. Biophys. Acta.*, 1987, **909**, 145-155.
[49] Krupp G. and Gross, H.J., *Nucleic Acids Res.*, 1979, **6**, 3481-3490.
[50] Varecha, M., Amrichova, J., Zimmermann, M., Ulman, V., Lukasova, E. and Kozubek, M., *Apoptosis* 2007, **12**, 1155–1171.
[51] Korn, C.; Meiss, G.; Gast, F.U.; Gimadutdinow, O.; Urbanke, C. and Pingoud, A., *Gene*, 2000, **253**, 221–229.
[52] Meiss, G., Gimadutdinow, O., Haberland, B., Pingoud, A., *J. Mol. Biol.*, 2000, **297**, 521-534.
[53] Ghosh, M.; Meiss, G.; Pingoud, A.; London, R.E and Pedersen, L.C, *J. Biol. Chem.*, 2005, **280**, 27990–27997.
[54] Kirby, T.W.; Mueller, G.A.; DeRose, E.F.; Lebetkin, M.S.; Meiss, G.; Pingoud, A. and London, R.E, *J. Mol. Biol.*, 2002, **320**, 771–782.
[55] Hansen, J. P.; and McDonald, I. R. *Theory of Simple Liquids*, 2006 (Academic: San Diego).
[56] Sikdar, S.; Chakrabarti, J.; Ghosh, M. *Mol Biosyst,* 2014, **10**, 3280-3289.
[57] Anderson, P. J.; Brooks, C. L.; Berliner, L. J. *Biochemistry,* 1997, **36**, 11648-11654.
[58] Ramakrishnan, B.; Shah, P. S.; Qasba, P. K. *J Biol Chem,* 2001, **276**, 37665-37671.
[59] Cawthern, K. M.; Narayan, M.; Chaudhuri, D.; Permyakov, E. A.; Berliner, L. J. *J Biol Chem,* 1997, **272**, 30812-30816.
[60] Chrysina, E. D.; Brew, K.; Acharya, K. R. *J Biol Chem,* 2000, **275**, 37021-37029.
[61] Permyakov, E. A.; Berliner, L. J. *FEBS Lett,* 2000, **473**, 269-274.
[62] Griko, Y. V.; Remeta, D. P. *Protein Sci,* 1999, **8**, 554-561.
[63] Permyakov, S. E.; Khokhlova, T. I.; Uversky, V. N.; Permyakov, E. A. *Proteins,* 2010, **78**, 2609-2624.
[64] Ostrovsky, A. V.; Kalinichenko, L. P.; Emelyanenko, V. I.; Klimanov, A. V.; Permyakov, E. A. *Biophys Chem,* 1988, **30**, 105-112.
[65] Permyakov, E. A.; Kalinichenko, L. P.; Morozova, L. A.; Yarmolenko, V. V.; Burstein, E. A. *Biochem Biophys Res Commun,* 1981, **102**, 1-7.
[66] Hiraoka, Y.; Sugai, S. *Int J Pept Protein Res,* 1984, **23**, 535-542.
[67] Hiraoka, Y.; Sugai, S. *Int J Pept Protein Res,* 1985, **26**, 252-261.
[68] Ramakrishnan, B.; Qasba, P. K. *J Mol Biol,* 2001, **310**, 205-218.
[69] Spolaore, B.; Pinato, O.; Canton, M.; Zambonin, M.; Polverino de Laureto, P.; Fontana, A. *Biochemistry,* 2010, **49**, 8658-8667.
[70] Svensson, M.; Fast, J.; Mossberg, A. K.; Duringer, C.; Gustafsson, L.; Hallgren, O.; Brooks, C. L.; Berliner, L.; Linse, S.; Svanborg, C. *Protein Sci,* 2003, **12**, 2794-2804.
[71] Stanciuc, N.; Aprodu, I.; Rapeanu, G.; Bahrim, G. *Eur Food Res Technol,* 2013, **236**, 257-266.
[72] Barbana, C.; Perez, M. D.; Sanchez, L.; Dalgalarrondo, M.; Chobert, J. M.; Haertle, T. *Int Dairy J,* 2006, **16**, 18-25.

[73] Knyazeva, E. L.; Grishchenko, V. M.; Fadeev, R. S.; Akatov, V. S.; Permyakov, S. E.; Permyakov, E. A. *Biochemistry,* 2008, **47**, 13127-13137.

[74] Svanborg, C.; Agerstam, H.; Aronson, A.; Bjerkvig, R.; Duringer, C.; Fischer, W.; Gustafsson, L.; Hallgren, O.; Leijonhuvud, I.; Linse, S.; Mossberg, A. K.; Nilsson, H.; Pettersson, J.; Svensson, M. *Adv Cancer Res,* 2003, **88**, 1-29.

[75] Tolin, S.; De Franceschi, G.; Spolaore, B.; Frare, E.; Canton, M.; Polverino de Laureto, P.; Fontana, A. *FEBS J,* 2010, **277**, 163-173.

[76] Punta, M.; Coggill, P. C.; Eberhardt, R. Y.; Mistry, J.; Tate, J.; Boursnell, C.; Pang, N.; Forslund, K.; Ceric, G.; Clements, J.; Heger, A.; Holm, L.; Sonnhammer, E. L.; Eddy, S. R.; Bateman, A.; Finn, R. D. *Nucleic Acids Res,* 2012, **40**, D290.

[77] Mavromoustakos, T.; Durdagi, S.; Koukoulitsa, C.; Simcic, M.; Papadopoulos, M. G.; Hodoscek, M.; Grdadolnik, S. G. *Curr Med Chem,* 2011, **18**, 2517-2530.

[78] Salata, O. *J Nanobiotechnology,* 2004, **2**, 3.

[79] Berg J.M., Tymoczko J.L., Stryer L., Signal-Transduction Pathways: An Introduction to Information Metabolism, *Biochemistry* 2002 (New York: W H Freeman) Chapter 15

Journal of Physics: Conference Series **638** (2015) 012014 doi:10.1088/1742-6596/638/1/012014

Social inequality: from data to statistical physics modeling

Arnab Chatterjee[1], Asim Ghosh[1,2], Jun-ichi Inoue[3]*, Bikas K. Chakrabarti[1,4]

[1]Condensed Matter Physics Division, Saha Institute of Nuclear Physics, 1/AF Bidhannagar, Kolkata 700064, India.

[2]Complex Systems Unit, Aalto University School of Science, P.O. Box 12200, FI-00076, Finland.

[3]Graduate School of Information Science and Technology, Hokkaido University, N14-W9, Kita-ku, Sapporo 060-0814, Japan.

[4]Economic Research Unit, Indian Statistical Institute, 203 B. T. Road, Kolkata 700108, India.

E-mail: arnabchat@gmail.com

Abstract. Social inequality is a topic of interest since ages, and has attracted researchers across disciplines to ponder over it origin, manifestation, characteristics, consequences, and finally, the question of how to cope with it. It is manifested across different strata of human existence, and is quantified in several ways. In this review we discuss the origins of social inequality, the historical and commonly used non-entropic measures such as Lorenz curve, Gini index and the recently introduced k index. We also discuss some analytical tools that aid in understanding and characterizing them. Finally, we argue how statistical physics modeling helps in reproducing the results and interpreting them.

1. Introduction

Repeated social interactions produce spontaneous variations manifested as inequalities at various levels. With the availability of huge amount of empirical data for a plethora of measures of human social interactions makes it possible to uncover the patterns and look for the reasons behind socio-economic inequalities. Using tools of statistical physics, researchers are bringing in knowledge and techniques from various other disciplines [1], e.g., statistics, applied mathematics, information theory and computer science to have a better understanding of the precise nature (both spatial and temporal) and the origin of socio-economic inequalities that is prevalent in our society.

Socio-economic inequality [2, 3, 4, 5, 6] is concerned with the existence of unequal opportunities and rewards for various social positions or statuses in a society. Structured and recurrent patterns of unequal distributions of goods, wealth, opportunities, and even rewards and punishments are the key features, and measured as *inequality of conditions*, and *inequality of opportunities*. The first one refers to the unequal distribution of income, wealth and material goods, while the latter refers to the unequal distribution of 'life chances' across individuals, as

* Dedicated to the memory of Prof. J.-I. Inoue

is reflected in education, health status, treatment by the criminal justice system, etc. Socio-economic inequality is held responsible for conflict, war, crisis, oppression, criminal activity, political unrest and instability, and affects economic growth indirectly [7]. Usually, economic inequalities have been studied in the context of income and wealth [8, 9, 10]. However, it is also measured for a plethora of quantities, including energy consumption [11]. The inequality in society [12, 13, 14, 15] is an issue of current focus and immediate global interest, bringing together researchers across several disciplines – economics and finance, sociology, demography, statistics along with theoretical physics (See e.g., Ref. [9, 16, 17]).

Socio-economic inequalities are quantified in numerous ways. The most popular measures are absolute, as quited with a single number, in terms of indices, e.g., Gini [18], Theil [19], Pietra [20] indices. The alternative measure approach is relative, using probability distributions of various quantities, but the most of the previously mentioned indices can be computed once one has the knowledge of the distributions. What is usually observed is that most quantities display broad distributions, usually lognormals, power-laws or their combinations. For example, the distribution of income is usually an exponential followed by a power law [9, 21].

In one of the popular methods of measuring inequality, one has to consider the Lorenz curve [22], which is a function that represents the cumulative proportion X of ordered (from lowest to highest) individuals in terms of the cumulative proportion of their sizes Y. X can represent income or wealth of individuals. But it can as well be citation, votes, city population etc. of articles, candidates, cities etc. respectively. The Gini index (g) is defined as the ratio between the area enclosed between the Lorenz curve and the equality line, to that below the equality line. If the area between (i) the Lorenz curve and the equality line is represented as A, and (ii) that below the Lorenz curve as B (See Fig. 1), the Gini index is $g = A/(A+B)$. It is an useful measure to quantify socio-economic inequality. Ghosh et al. [23] recently introduced the 'k index' (where 'k' symbolizes for the extreme nature of social inequalities in Kolkata), which is defined as the fraction k such that $(1-k)$ fraction of people or papers possess k fraction of income or citations respectively [24].

When the probability distribution is described using an appropriate parametric function, one can derive these inequality measures as a function of those parameters analytically. In fact, several empirical evidence have been reported to show that the distributions can be put into a finite number of types. Most of them turn out to be a of mixture of two distinct parametric distributions with a single crossover point [24].

This review is organized as follows: Sec. 2 discusses the evolutionary view of socio-economic inequalities and Sec. 3 discusses the basic and most popular quantities which are used for measuring inequalities. Next, in Sec. 4 we discuss the most realistic scenario in which probability distributions have to fitted to more than a single function, and how to measure inequalities from them. Further, we discuss if inequalities are natural, in context of the recent works of Piketty in Sec. 5, followed by a section on statistical physics modeling in Sec. 6. Finally, we summarize in Sec. 7.

2. Evolutionary view of socio-economic inequality

The human species is known to have lived their life as hunter gathers for more than 90% of its existence. It is widely believed that those early societies were egalitarian, still seen in the lifestyle of various tribes like the !Kung people of the Kalahari [25]. Early hunter-gatherer societies are even believed to have championed sex equality [26]. Traditionally having a few possessions, they were semi nomadic in the sense they were moving periodically. Having hardly mastered farming and lived as small groups, the mere survival instinct was driving them to overlook individual interests. They were sharing what they had, so that all of their group members were healthy and strong, be it food, weapons, property, or territory [27]. With the advent of agricultural societies, elaborate hierarchies were created, with less stable leadership in course of time. These

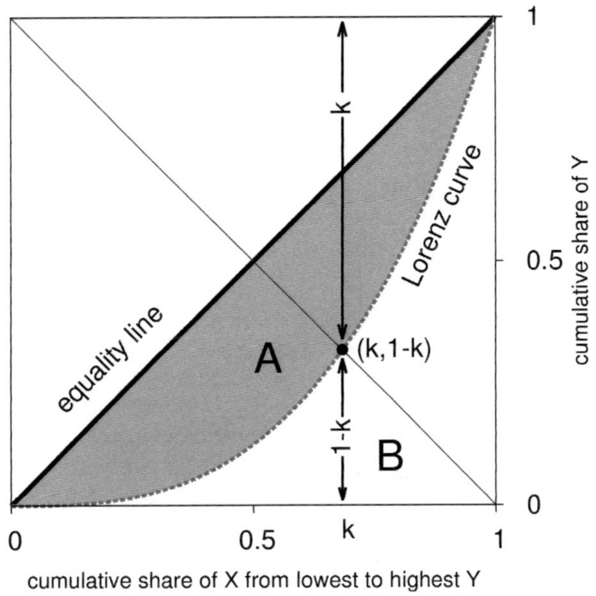

Figure 1. Schematic representations for Lorenz curve, Gini index g and k index. The dashed line depicts the Lorenz curve and the solid line represents perfect equality. The area enclosed between the equality line and the Lorenz curve is A and that below the Lorenz curve is B. The Gini index is given by $g = A/(A+B)$. The k index is given by the abscissa of the intersection point of the Lorenz curve and $Y = 1 - X$ (adapted from Ref. [24]).

evolved into clans or groups led mostly by family lines, which eventually developed as kingdoms. In these complex scenarios, new strategies for hoarding surplus produce of agriculture or goods were adapted by the chiefs or kings, predominantly for survival in times of need, and this lead to the concentration of wealth and power (see Ref. [28] for models with savings). Along with the advancement of technologies, intermediate mechanisms helped in wealth multiplication. This completed the transition from egalitarianism to societies having competition and the inequality paved way for the growth of chiefdoms, states and industrial empires [29, 6].

3. Basics and generic properties of inequality measures

In this section, we formally introduce the measures to quantify the degree of social inequality, namely, Lorenz curve, Gini index and k index.

The Lorenz curve shows the relationship between the cumulative distribution and the cumulative first moment of $P(m)$:

$$X(r) = \int_{m_0}^{r} P(m)dm, \quad Y(r) = \frac{\int_{m_0}^{r} mP(m)dm}{\int_{m_0}^{\infty} mP(m)dm}. \tag{1}$$

The set $(X(r), Y(r))$ defines the Lorenz curve, assuming $P(m)$ to be defined in $[m_0, \infty)$. Fig. 1 shows the typical behavior of Lorenz curve. The Lorenz curve gives the cumulative proportion X of ordered individuals (from lowest to highest) holding the cumulative proportion Y of wealth. Lorenz curve, Gini index etc. were historically introduced in the context of income/wealth. So, let us call X as individuals and Y as 'wealth', but in principle the attributes X and Y can be any of the combinations like article/citations, candidate/vote, city/population, student/marks, company/ employee etc. Hence, when all individuals take the same amount of wealth, say m', we have $P(m) = \delta(m - m')$, with $m_0 < m' < \infty$, and one obtains

$$X(r) = \int_{m_0}^{r} \delta(m - m')dm = \Theta(r - m'), \;\; Y(r) = \frac{\int_{m_0}^{r} m\delta(m - m')dm}{\int_{m_0}^{\infty} m\delta(r - m')dm} = \frac{m'\Theta(r - m')}{m'} = X(r). \tag{2}$$

where $\Theta(x)$ is a step function defined by

$$\Theta(x) = \begin{cases} 1, & x \geq 1, \\ 0, & x < 1. \end{cases} \tag{3}$$

Thus $Y = X$ is the 'perfect equality line' (see Fig. 1), where X fraction of people takes X fraction of wealth in the society. On the other extreme, if the total wealth in the society of N persons is concentrated to a few persons, $P(m) = (1 - \varepsilon)\delta_{m,0} + \varepsilon\delta_{m,1}$, with $\varepsilon \sim \mathcal{O}(1/N)$ and with the total amount of wealth is normalized to 1, we get $X(r) = 1 - \varepsilon + \varepsilon\delta_{r,1}$ and $Y(r) = \delta_{r,1}$. Hence, $Y = 1$ iff $X = r = 1$ and $Y = 0$ otherwise, and the Lorenz curve is given as 'perfect inequality line' $Y = \delta_{X,1}$ where $\delta_{x,y}$ is a Kronecker's delta (see Fig. 1).

For a given Lorenz curve, the Gini index is defined by twice of area between the curve $(X(r), Y(r))$ and perfect equality line $Y = X$. (shaded part 'A' in Fig. 1). It reads

$$g = 2\int_0^1 (X - Y)dX = 2\int_{r_0}^{\infty} (X(r) - Y(r))\frac{dX}{dr}dr, \tag{4}$$

where $X^{-1}(0) = r_0, X^{-1}(1) = \infty$ should hold. Graphically (see Fig. 1), the Gini index is the ratio of the two areas ('A' and 'B'), $g = \mathrm{A}/(\mathrm{A}+\mathrm{B})$. Thus, the Gini index g is zero for perfect equality and unity for perfect inequality. The Gini index may be evaluated analytically when the distribution of population is obtained in a parametric way.

The recently introduced k index is the value of X-axis for the intersection between the Lorenz curve and a straight line $Y = 1 - X$. For the solution of equation $X(r) + Y(r) = 1$, say $r_* = Z^{-1}(1), Z(r) \equiv X(r) + Y(r)$, the k index is given by $k = X(r_*)$. Thus, the value of k index indicates that k fraction of people shares totally $(1 - k)$ fraction of the wealth. Hence, the k index equals 1/2 for perfectly equal society, and 1 for perfectly unequal society. This is obviously easier to estimate by eyes in comparison with the Gini index (shaded area A in Fig. 1).

Apart from these indices, Pietra's p index [20] and m or median index [30] has been used as inequality measures, and can be derived from the Lorenz curve. The p index is defined by the maximal vertical distance between the Lorenz curve and the line of perfect equality $Y = X$, while the m index is given by $2m - 1$ for the solution of $Y(m) = 1/2$.

4. Results for mixture of distributions

It is rather easy to perform analytic calculations for the g and k indices when the distribution of population are described by parametric distributions such as a uniform, power law and lognormal distributions. It is quite common to find that the probability distributions of quantities like wealth, income, votes, citations etc. fit to more than one theoretical function depending on the range. Formally, $P(m) = F_1(m)\theta(m, m_\times) + F_2(m)\Theta(m - m_\times)$, with $\theta(m, m_\times) \equiv \Theta(m) - \Theta(m - m_\times)$, where $m_\times$ is the crossover point. The functions $F_1(m)$ and $F_2(m)$ are suitably normalized and computed for their continuity at $m_\times$. In such cases, one can also develop a framework [24], using which it is reasonably straightforward to calculate Lorenz curve, Gini and k-index.

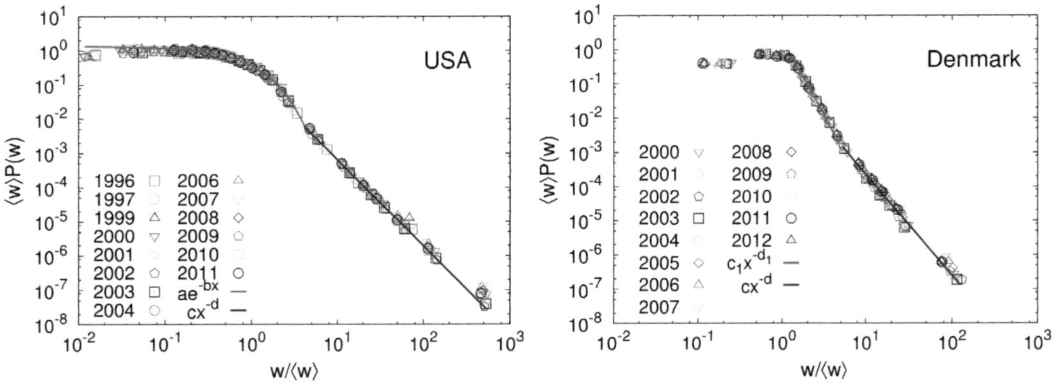

Figure 2. Left: Income distribution $p(w)$ for USA [31] for a few years, the data is rescaled by the average income $\langle w \rangle$. The lower income range fits to an exponential form $a\exp(-bw)$ with $b = 1.3$ while the high income range fits to a power law decay cw^{-d} with exponent $d = 2.52$. Right: Same for Denmark [32]. The middle and upper ranges fit to different power laws cw^{-d} with $d = 2.96$ and $c_1 w^{-d_1}$ with $d_1 = 4.41$. Taken from Ref. [24].

4.1. Empirical data

The most well studied data in the context of socio-economic inequality is that of income. Incomes are re-calculated from the income tax data reported in the Internal Revenue Service (IRS) [31] of USA for 1996-2011. This data is used to compute the probability distribution $P(w)$ of income w for each year. Similar data from Denmark were used from the years 2000-2012 [32]. The g index is found to be around $0.54 - 0.60$ for USA and $0.34 - 0.38$ for Denmark, while the k-index is around $0.69 - 0.71$ for USA and $0.65 - 0.69$ for Denmark [24]. Fig. 2 shows both these data sets.

The data related to several socio-economic measures were analyzed. The (i) voting data of proportional elections from a few countries [33])., (ii) citations of different science journals and institutions, collected from ISI Web of Science [34], and (iii) population data for city sizes for Brasil [35], municipalities of Spain [36] and Japan [37] were analyzed. The data showed broad distributions of the above quantities, which well-fitted to (see Ref. [24] for details) (a) a single lognormal, (b) a single lognormal with a power law tail, (c) uniform with a power law tail, (d) uniform with a lognormal tail, (e) a mixture of power laws, (f) a single power law with a lognormal tail.

The inequality indices computed using a combination of fitting functions were compared with that computed directly from the empirical data.

5. Is wealth and income inequality natural?

One is often left wondering whether inequalities in wealth and income are natural. It has been shown using models [38] and their dynamics that certain minimal dynamics over a completely random exchange process and subsequent entropy maximization produces broad distributions. Piketty [39] argued recently that inequality in wealth distribution is quite natural. He pointed out that before the great wars of the early 20th century, the strong skewness of wealth distribution was prevailing as a result of a certain 'natural' mechanisms. The Great Depression that followed after the two great World Wars have helped in dispersion of wealth. This in turn brought the prevailing extreme inequality under check and subsequently gave rise to a sizable middle class. After analyzing very accurate data, he concluded that the world is currently

Journal of Physics: Conference Series **638** (2015) 012014 doi:10.1088/1742-6596/638/1/012014

'recovering' back to this 'natural state', which is happening due to capital ownership driven growth of finance [12], which has been dominant over a labor economy, and this is simply the result of the type of institution and policies that are adopted by the society. His work raises quite fundamental issues concerning bot only economic theory but also the future of capitalism. It points out the large increases in the wealth/output ratio. According to standard economic theory, such increases are attributed to the decrease in the return to capital and an increase in the wages. However, the return to capital has not diminished, while the wages have. He has also prescribed the following: higher capital-gains and inheritance taxes, higher spending towards access to education, tight enforcement of anti-trust laws, corporate-governance reforms that restrict pay for the executives, and finally, the financial regulations which have been an instrument for banks to exploit the society. It is anticipated that all of these might be able to help reduce inequality and increase equality of opportunity. There is further speculation that this might be able to restore the shared and quick economic growth that characterized the middle-class societies in the mid-twentieth century.

6. Statistical physics modeling

One of the most efficient ways to model evolution of systems to broad distributions showing strong inequality, is by using the toolbox of statistical physics. Microscopic and macroscopic modeling helps in imitating real socio-economic systems.

There is a whole body of empirical evidence supporting the fact that a number of social phenomena are characterized by emergent behavior out of the interactions of many individual social components. Recently, the growing community of researchers have analyzed large-scale social dynamics to uncover certain 'universal patterns'. There has also been an attempt to propose simple microscopic models to describe them, in the same spirit as the minimalistic models commonly used in statistical physics [16, 17].

6.1. Income & wealth distribution

In case of wealth distributions, the popular models are chemical kinetics motivated Lotka-Volterra models [40, 41, 42], polymer physics inspired models [43] and most importantly, models inspired by kinetic theory of gases [44, 45, 46, 28] (see Ref. [9] for details). The two-class structure [8] of the income distribution (exponential dominated low income and power law tail in the high incomes) is well understood to be a result of very different dynamics of the two classes. The bulk is described by a process which is more of a random kinetic exchange [45, 44], producing a distribution dominated by an exponential functional form. The dynamics is very simple, as described in the kinetic theory of gases [47]. The minimal modifications that one can introduce are additive or multiplicative terms.

Processes creating inequality involving uniform retention rates [48] or equivalently, savings [45] produce Gamma-like distributions. These models are defined as a microcanonical ensemble, with fixed number of agents and wealth. Here, the wealth exchanging *agent* retains a certain fraction (termed as 'saving propensity') of what they had before each trading process and randomly exchanges the rest of the wealth. When agents are assigned with the same value of the 'saving propensity' (as in Ref. [45]), it could not produce a broad distribution of wealth. What is important to note here is that the richest follow a different dynamic from the poor and thus heterogeneity in the saving behavior plays a crucial role. So, to obtain the power law distribution of wealth for the richest, one needs to simply consider that each agent is different in terms of how much fraction of wealth they will save in each trading [46], which is a very natural ingredient to assume, because it is quite likely that agents in a trading market think very differently from one another. In fact, with this small modification, one can explain the entire range of the wealth distribution [28]. These models, moreover, can show interesting characteristics if the exchange processes and flows are made asymmetric, e.g., put on directed networks [49]. A plethora of variants of these models, results and analyses find possible applications in a variety of trading processes [9].

6.2. Cities & firms

City [50] and firm sizes [51] consistently exhibit broad distributions with power law tails for the largest sizes, commonly known to be Zipf's law.

Gabaix [52] showed that if cities grow randomly at the same expected growth rate and the same variance (Gibrat's law [53]), the limiting distribution converges to the Zipf's law. He proposed that growth 'shocks' are random and they impact utilities in both positive and negative way. A similar approach resulted in diffusion and multiplicative processes [54]. Shocks were also used to immitate sudden migration [55]. Simple economics arguments demonstrated that expected urban growth rates were identical irrespective of city sizes and variations were random normal deviates, resulting Zipf law with exponent unity.

6.3. Consensus

Consensus in social systems is an interesting topic, due to its dynamics. The dynamics of agreement and disagreement in a 'society' is complex, and statistical physicists working on opinion dynamics have been brave enough to model opinion states in a population and their dynamics that determine the transitions within such states. A huge body of old and recent literature [17, 16] discusses models that explain various social phenomena and the observed inequality in such instances of consensus formation.

6.4. Bibliometrics

The increasing amount of data produced from bibliometric tools have led to a better understanding of how researchers and their publications 'interact' with one another in a 'social system' consisting of articles and researchers. The patterns of citation distribution and growth are now well studied, and some of the most successful models have used statistical physics [56].

Statistical physics tools have aided in formulating these microscopic models, which are simple enough yet rich in terms of socio-economic ingredients. Toy models help in understanding the basic mechanism at play, and demonstrate the crucial elements that are responsible for the emergent distributions of income and wealth. A variety of models, ranging from zero-intelligence variants to the more complex agent based models (including those incorporating game theory) have been proposed over years and are found to be successful in interpreting the empirical results [9]. Simple modeling is also effective in understanding how entropy maximization produce distributions which are dominated by exponentials, and also explaining the reasons for aggregation at the high range of wealth, including the power law Pareto tail [8, 9].

7. Summary and discussions

Social inequalities are manifested in several forms, and are recorded well in history, being the reason of unrest, crisis, wars and revolutions. Traditionally a subject of study of social sciences, though scholars from different fields have been investigating the causes and effects from a sociological perspective, and trying to understand its consequence on the prevailing economic system. Reality is not as simple and pointing out the causes and the effects are much more complex.

Imagine a the world which is very equal, where it would have been difficult to compare the extremes, differentiate the good from the bad, hardly any leadership people will look up to, will lack stable ruling governments if there were almost equal number of political competitors, etc.

The recent concern about the increase of inequality in income and wealth, as pointed out from different measurements [39] has renewed the interest on this topic among the leading social scientists across the globe. Society always had classes, and climbing up and down the social ladder [57, 58] is quite difficult to track, until recent surveys which provide some insight into the dynamics. Several deeper and important issues of our society [4] still need attention in terms of inequality research, and this can only be achieved by uncovering hidden patterns on further analysis of the available data.

Measurement of inequalities in society can be as simple as measuring measuring zeroth order quantities as Gini index, to finding exact probability distributions. The complexity of the underlying problems have inspired researchers to propose multi-dimensional inequality indices [59], which serve well in explaining a lot of factors in a compact form.

As physicists, our interests are mostly concentrated on subjects which are amenable to modeling using macroscopic or microscopic frameworks. Tools of statistical physics can very well explain the emergence of broad distributions which are signatures of inequalities. The literature already developed, contains serious attempts to understand socio-economic phenomena, under *Econophysics* and *Sociophysics* [60]. The physics perspective brings alternative ideas and a fresh outlook compared to the traditional approach taken by social scientists, and is reflected in the increasing collaborations between researchers across disciplines [1].

Acknowledgments
A.C. thanks V.M. Yakovenko for discussions, in the context of a more extensive material which is being written up. A.C. and B.K.C. acknowledges support from B.K.C.'s J. C. Bose Fellowship and Research Grant. J.I. was financially supported by Grant-in-Aid for Scientific Research (C) of Japan Society for the Promotion of Science (JSPS) No. 2533027803 and Grant-in-Aid for Scientific Research (B) of 26282089, Grant-in-Aid for Scientific Research on Innovative Area No. 2512001313. He also thanks Saha Institute of Nuclear Physics for their hospitality during his stay in Kolkata.

References

[1] Lazer D, Pentland A, Adamic L, Aral S, Barabási A L, Brewer D, Christakis N, Contractor N, Fowler J, Gutmann M, Jebara T, King G, Macy M, Roy D and Van Alstyne M 2009 *Science* **323** 721–723
[2] Arrow K J, Bowles S and Durlauf S N 2000 *Meritocracy and economic inequality* (Princeton: Princeton Univ. Press)
[3] Stiglitz J E 2012 *The price of inequality: How today's divided society endangers our future* (New York: WW Norton & Company)
[4] Neckerman K 2004 *Social Inequality* (New York: Russell Sage Foundation)
[5] Goldthorpe J H 2010 *Eur. Sociological Rev.* **26** 731–744
[6] Chatterjee A 2015 Socio-economic inequalities: a statistical physics perspective *Econophysics and Data Driven Modelling of Market Dynamics, Eds. F Abergel, H. Aoyama, B.K. Chakrabarti, A. Chakraborti, A. Ghosh,* (Milan: Springer) pp 287–324
[7] Hurst C E 1995 *Social Inequality: Forms, Causes, and Consequences* (Boston: Allyn and Bacon)
[8] Yakovenko V M and Barkley Rosser Jr J 2009 *Rev. Mod. Phys.* **81** 1703
[9] Chakrabarti B K, Chakraborti A, Chakravarty S R and Chatterjee A 2013 *Econophysics of income and wealth distributions* (Cambridge: Cambridge Univ. Press)
[10] Aoyama H, Fujiwara Y and Ikeda Y 2010 *Econophysics and companies: statistical life and death in complex business networks* (Cambridge: Cambridge Univ. Press)
[11] Lawrence S, Liu Q and Yakovenko V M 2013 *Entropy* **15** 5565–5579
[12] Piketty T and Saez E 2014 *Science* **344** 838–843
[13] Cho A 2014 *Science* **344** 828
[14] Chin G and Culotta E 2014 *Science* **344** 818–821
[15] Xie Y 2014 *Science* **344** 809–810
[16] Castellano C, Fortunato S and Loreto V 2009 *Rev. Mod. Phys.* **81** 591–646
[17] Sen P and Chakrabarti B K 2013 *Sociophysics: An Introduction* (Oxford: Oxford University Press)
[18] Gini C 1921 *Econ. J.* **31** 124–126
[19] Theil H 1967 *Economics and information theory* vol 7 (Amsterdam: North-Holland)
[20] Eliazar I I and Sokolov I M 2010 *Physica A* **389** 117–125

Journal of Physics: Conference Series **638** (2015) 012014 doi:10.1088/1742-6596/638/1/012014

[21] Drăgulescu A A and Yakovenko V M 2001 *Physica A* **299** 213–221
[22] Lorenz M O 1905 *Am. Stat. Assoc.* **9** 209–219
[23] Ghosh A, Chattopadhyay N and Chakrabarti B K 2014 *Physica A* **410** 30–34
[24] Inoue J I, Ghosh A, Chatterjee A and Chakrabarti B K 2015 *Physica A* **429** 184–204
[25] Pennisi E 2014 *Science* **344** 824–825
[26] Dyble M, Salali G D, Chaudhary N, Page A, Smith D, Thompson J, Vinicius L, Mace R and Migliano A B 2015 *Science* **348** 796–798
[27] Buchanan M 2014 *Nature Phys.* **10** 471–471
[28] Chatterjee A and Chakrabarti B K 2007 *Eur. Phys. J. B* **60** 135–149
[29] Pringle H 2014 *Science* **344** 822–825
[30] Eliazar I I and Cohen M H 2014 *Physica A* **401** 148 – 158
[31] Internal Revenue Service (IRS) – Individual Tax Tables, retreived June, 2014 http://www.irs.gov/uac/SOI-Tax-Stats—Individual-Statistical-Tables-by-Size-of-Adjusted-Gross-Income
[32] Statistics Denmark, retreived June, 2014 http://www.dst.dk
[33] Chatterjee A, Mitrović M and Fortunato S 2013 *Sci. Reports* **3** 1049
[34] ISI Web of Science, retreived April, 2014 http://portal.isiknowledge.com
[35] DATASUS, retreived August, 2013 http://www2.datasus.gov.br/DATASUS/
[36] Población de España - datos y mapas, retreived August, 2013 http://alarcos.inf-cr.uclm.es/per/fruiz/pobesp/
[37] e-Stat: Official Statistics of Japan, retreived March, 2014 https://www.e-stat.go.jp/SG1/chiiki/Welcome.do?lang=02
[38] Chatterjee A, Sinha S and Chakrabarti B K 2007 *Current Science* **92** 1383–1389
[39] Piketty T 2014 *Capital in the Twenty-first Century* (Harvard: Harvard Univ. Press)
[40] Levy M and Solomon S 1997 *Physica A* **242** 90–94
[41] Solomon S and Richmond P 2002 *Eur. Phys. J. B* **27** 257–261
[42] Richmond P and Solomon S 2001 *Int. J. Mod. Phys. C* **12** 333–343
[43] Bouchaud J P and Mézard M 2000 *Physica A* **282** 536–545
[44] Drăgulescu A and Yakovenko V 2000 *Eur. Phys. J. B* **17** 723–729
[45] Chakraborti A and Chakrabarti B K 2000 *Eur. Phys. J. B* **17** 167–170
[46] Chatterjee A, Chakrabarti B K and Manna S S 2004 *Physica A* **335** 155–163
[47] Saha M N and Srivastava B N 1931 *A treatise on heat* (Allahabad: Indian Press)
[48] Angle J 2006 *Physica A* **367** 388–414
[49] Chatterjee A 2009 *Eur. Phys. J. B* **67** 593–598
[50] Zipf G K 1949 *Human Behaviour and the Principle of Least-Effort* (Cambridge: Addison-Wesley)
[51] Axtell R L 2001 *Science* **293** 1818–1820
[52] Gabaix X 1999 *Am. Econ. Rev.* **89** 129–132
[53] Gibrat R 1931 *Les inégalités économiques* (Paris: Sirey)
[54] Zanette D H and Manrubia S C 1997 *Phys. Rev. Lett.* **79** 523
[55] Marsili M and Zhang Y C 1998 *Phys. Rev. Lett.* **80** 2741
[56] Newman M E J, Barabási A L and Watts D J 2006 *The structure and dynamics of networks* (Princeton: Princeton Univ. Press)
[57] Mervis J 2014 *Science* **344** 836–837
[58] Bardoscia M, De Luca G, Livan G, Marsili M and Tessone C J 2013 *J. Stat. Phys.* **151** 440–457
[59] Sen A 1992 *Inequality reexamined* (Oxford: Oxford University Press)
[60] Chakrabarti B K, Chakraborti A and Chatterjee A (eds) 2007 *Econophysics and Sociophysics: Trends and Perspectives* (Weinheim: Wiley-VCH)

Journal of Physics: Conference Series **638** (2015) 012015 doi:10.1088/1742-6596/638/1/012015

To join or not to join: collective foraging strategies

Kunal Bhattacharya[1] and Tamás Vicsek[2]

[1]Aalto University School of Science, P.O. Box 12200, FI-00076, Finland

[2]Department of Biological Physics, Eötvös University, Pázmány Péter setány 1A, 1117 Budapest, Hungary

E-mail: kunal.bhattacharya@aalto.fi

Abstract. We present a minimal model to address the issue of how the interaction between foragers influence their encounter rates with resources. We consider a two-dimensional lattice model to study the optimal strategy for a group of foragers (walkers) searching for targets distributed heterogeneously. A forager who has not detected any target has the choice of, either joining other foragers in the neighbourhood, who have detected targets, or continue searching independently. The optimal strategy appears to be a mixture of these choices. Remarkably, we observe that scale-free walks appear for strategies far from optimality. In general, we investigate the behaviour of the model with a parameter characterizing the propensity of the foragers to aggregate and the radius of interaction characterizing the neighbourhood of a forager.

1. Introduction

The phenomena of foraging or searching is ubiquitous in nature. Animals are found to forage for food or shelter. The efficiency with which an individual animal can forage becomes a key to its survival. Conventionally, the subject of behavioural ecology has tried to understand, how the foraging behaviour determines the foraging efficiency in different species [1, 2]. However, the problem has been found to be relevant in several other contexts ranging from military search procedures [3], to proteins searching for target sites on the DNA [4]. Interestingly, theoretical investigations based on ideas in statistical physics have provided a paradigm for the analysis of the complexity of the phenomena [5, 6]. The frameworks of random walks and diffusion equations have been extensively used for modelling [7, 8]. The possibility of recording trajectories of animals searching for food have further facilitated such studies [9, 10].

To model the scale-free walk-length distributions of animals that forage alone, Lévy walks have been proposed [11, 12, 13]. Studies have shown that under certain conditions Lévy walks are optimal, but deviations greatly reduce the advantage of Lévy walks in favour of ballistic or Brownian motion [12, 14, 15, 16]. For a single animal looking for resources, the foraging efficiency is dependent on its searching strategy and the distribution of the targets [8, 11]. However, when in a group, the interactions between the foragers become important in determining the success rate. An animal which is unsuccessful in locating food by searching on its own, is usually found to locate other distant members who are successful and join them [1, 17, 18]. When food is abundant there is hardly any benefit for individuals to join others to increase the chance of encountering food. However, when the distribution of food in space is heterogeneous, or a temporal variation is present, interactions between the foragers may become important.

Several field studies and theoretical models have investigated how collective effects influence the search behaviour [19, 20]. For example, a stochastic environment was shown to initiate cooperation between foragers [21]. Social behaviour in primates has been shown to give rise to scale-free move lengths [22]. Very recently, the influence of communication on the foraging pattern of gazelles was investigated [23]. The study showed that communication over intermediate length scales result in faster searches. Remarkably, the movement of gazelles was found to be Brownian.

In this paper we study the problem of collective foraging. In a recent investigation [24], we developed a minimal model to investigate the effect of interactions on the search efficiency, and the spatio-temporal scalings that emerged in the movement of foragers. In [24] we assumed a forager to be informed about the state of all other foragers at any instant of time. In the present investigation we assume a finite interaction radius for the foragers. In our model, the foragers are walkers on a lattice and a forager executes a random walk when it is independently searching for targets. We characterize the propensity of a forager to approach the nearest site where targets have been discovered, through a parameter. Unlike the foragers, the targets are immobile but regenerative. We study the resulting forager-target interactions by varying this parameter and the radius.

2. The model

Below, we provide a brief description of the model. We consider a two dimensional square lattice of size L with the periodic boundary condition (PBC). The PBC is applied with regard to the movement of the foragers as well as the regeneration of the targets. There are N_F foragers which are initially distributed randomly on this lattice. The targets are regenerative such that at every time step the total number is N_T. We define the neighbourhood of a forager as a circular region of radius R centred on it. We assume that at any instant of time, a forager is able to determine its neighbours that have been successful in identifying and consuming food. However, a forager is only able to detect targets at the site where it has arrived. A forager which is searching for targets moves towards the nearest site in the neighbourhood, where, targets have been discovered, with probability, $p = \exp(-\alpha d)$, where d is the distance to the site ($d < R$), or, continues executing a random walk (RW), with probability $1 - p$. We term the movement in the former case as a targeted walk (TW). A forager which is already executing a TW continues with it, unless the targeted site becomes empty of targets. In the latter case the forager reverts back to executing a RW. In the limit $\alpha \to \infty$, the foragers are not following each other and the trajectories of are essentially RWs. The tendency of foragers to travel towards distant sites, increases with decrease in α. In general, the trajectories are a mixture of RWs and TWs. The detailed rules pertaining to the behaviour of foragers can be found in [24].

The rule pertaining to the regeneration of targets at every instant is as follows. Let $\Delta N(t)$ be the number of targets that have been consumed at any instant of time t. We randomly select one of the remaining $N_T - \Delta N(t)$ targets. At a distance d ($1 \leq d \leq L/\sqrt{2}$) from the location of this target a new target is placed, in a random direction. The distance d is chosen from a distribution $P(d) \sim d^{-\gamma}$, where γ is a parameter. This process is repeated another $\Delta N(t) - 1$ times. Thus, the total number of targets at the beginning of the time step $t + 1$, again becomes equal to N_T. The distribution $P(d)$ generates targets distributed in well separated clusters with the parameter γ effectively characterizing the degree of heterogeneity of the distribution. The method effectively creates a foraging environment with a high degree of spatio-temporal heterogeneity [21, 25, 26], and allows to preserve the overall statistical nature of the environment during the time evolution of the model. We also take, $N_F = N_T = N$, which would correspond to the case of targets being rather scarce, and we expect that collective effects are relevant when there is a scarcity of targets.

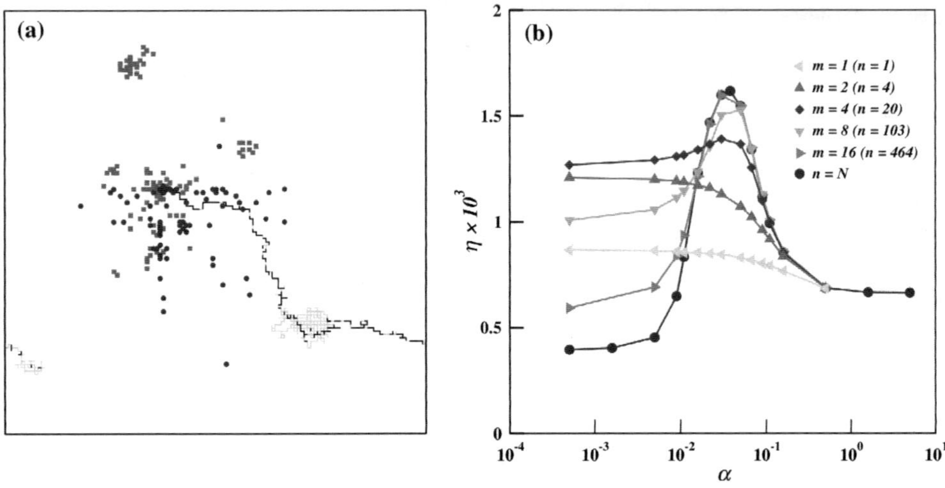

Figure 1. (a) A typical snapshot during the time evolution of the model for $N = 128$, $L = 128$, $\gamma = 2.5$ and $\alpha = 0$. The foragers are shown with circles (blue) and targets are shown with squares (green). The path of a randomly chosen forager is drawn. The grey steps belong to random walks and the black steps belong to targeted walks. (b) Variation of the efficiency, η, with α for $L = 512$, $N = 512$ and $\gamma = 2.5$. The different symbols pertain to different values of m (and hence, R) and are shown in the legend.

3. Results

We associate a radius (r_0) with the area per particle, so that $r_0 = L/\sqrt{N\pi}$. We choose to measure R in units of r_0 such that $R = m.r_0$, where, m is a positive integer. Having $m < (L/r_0)$ implies that at any instant, on the average, a forager remains informed about the state of n other foragers with $n < N$. In [24] we found that, $n \sim m^{\nu}$, where, $\nu = 2.23$. The Fig. 1(a) shows a typical snapshot during the time evolution of the model for $N = 128$, $L = 128$, $\gamma = 2.5$, $\alpha = 0.0$ and $n = N$ (m is rather large). The position of the foragers and the targets are shown with different symbols. The figure reveals the spatial clustering of the targets. The snapshot corresponds to a case when the foragers are increasingly aggregating in a region where the density of the targets is high. This event also becomes apparent by looking at the trajectory of a typical forager. The trajectory is a mixture of RWs (marked in grey) and TWs (marked in black). However, the dynamics depicted in the figure is actually deterministic owing to the fact that α is zero. A forager, searching for targets, is always drawn towards the nearest site where targets have been discovered, independent of the distance of separation.

For the rest of the study we take, $N = 512$, $L = 512$ and $\gamma = 2.5$. We define the efficiency of foraging, η as the ratio of the total number of targets consumed by all the N foragers, to the total distance travelled by all the foragers, in a given amount of time. We first allow the system to reach a stationary state and we collect the statistics over a period of 10^6 time steps starting from 200 arbitrary initial conditions. In Fig. 1(b) we plot η as function of α for different values of m. We note that the distance dependence of the decision to approach other foragers actually results from the span of the neighbourhood as well as the value of α. When $\alpha \to \infty$ the foragers search independently through RWs, the collective effects are absent and the efficiency is the same for all values of m. The efficiency is found to increase when α is lowered. When targets are clustered in space, we expect that a forager travelling to a region where targets have already been discovered increases its own chance of encountering a target. For very small values of m, the initial rise in efficiency is rapid. The parameter α being the inverse of a length scale, for a given R, we expect the rapid rise in η to occur when $\alpha \sim 1/R$. For lower values of α, the

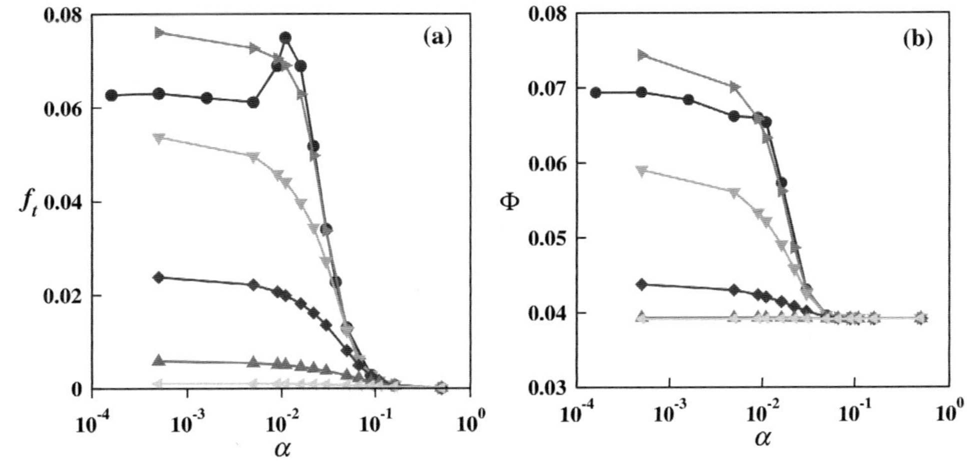

Figure 2. Plots of the fraction of walkers executing targeted walks at any instant, f_t (a) and the flocking order parameter, Φ (b) against α for $L = 512$, $N = 512$ and $\gamma = 2.5$. The different symbols correspond to different values of m and are the same as that in the Fig. 1(b).

increase is slow. This is because a forager does not detect other foragers at distances larger than R. Therefore, when the neighbourhood is small, following other successful foragers is always beneficial. For larger values of m, a maximum is seen to appear in η. This corresponds to the most efficient strategy in terms of α. This strategy pertains to joining others when clusters are discovered nearby and opting to search independently, otherwise. Lesser values of α makes the foragers follow other foragers indiscriminately. When a cluster is discovered, a large fraction of the foragers are drawn towards it through TWs. This delays the discovery of other clusters in space. As a result the efficiency drops. We call the inverse of the optimal α as d_{opt}. In [24] we showed that, $d_{\text{opt}} \sim 1/N^{\zeta}$, for a given L, with $\zeta = 0.65$. Additionally, as m increases further, the efficiency in $\alpha \to 0$ limit is found to become lesser than the $\alpha \to \infty$ limit. This crossover is seen to occur at around $m = 4 = m_c$, which suggests $m_c r_0 \sim d_{\text{opt}}$.

The TWs are signature of the collective effects and we measured the fraction of walkers (f_t) that execute TWs at any instant (Fig. 2)(a). In general, for a given value of m, when α is decreased, we find f_t to increase rapidly at the beginning and then slowly, later. The reason is very similar to what was discussed in the previous paragraph. Lowering of α causes foragers to follow neighbours through TWs but because of the finite size of the neighbourhood the increase in f_t is slowed down. However, for a given α, f_t is found to increase with m. This is because the larger the neighbourhood, the greater is the chance that a forager detects another successful forager. However, in the extreme case when R encompasses the whole of the lattice ($n = N$), we find there is a maximum in f_t. This is related to the drop in efficiency, η. In this particular case, when $\alpha \to 0$, all the foragers get involved in TWs on the discovery of a cluster, the simultaneous discovery of a new cluster is prevented, and as such less number of TWs take place on the average at any instant.

We found that on the simultaneous discovery of clusters foragers execute TWs from one cluster to another. To characterize this collective movement we used the flocking order parameter [27], modified for the lattice, $\Phi = \left\langle \sqrt{(f_{x+} - f_{x-})^2 + (f_{y+} - f_{y-})^2} \right\rangle$, where f_{x+} denotes the fraction of foragers moving towards the positive x-direction at any instant, and likewise. A large value value of Φ would imply that, on the average, the foragers are moving in the same direction, at any instant of time. In Fig. 2(b) the plot Φ versus α is shown. The dependence of Φ on α is

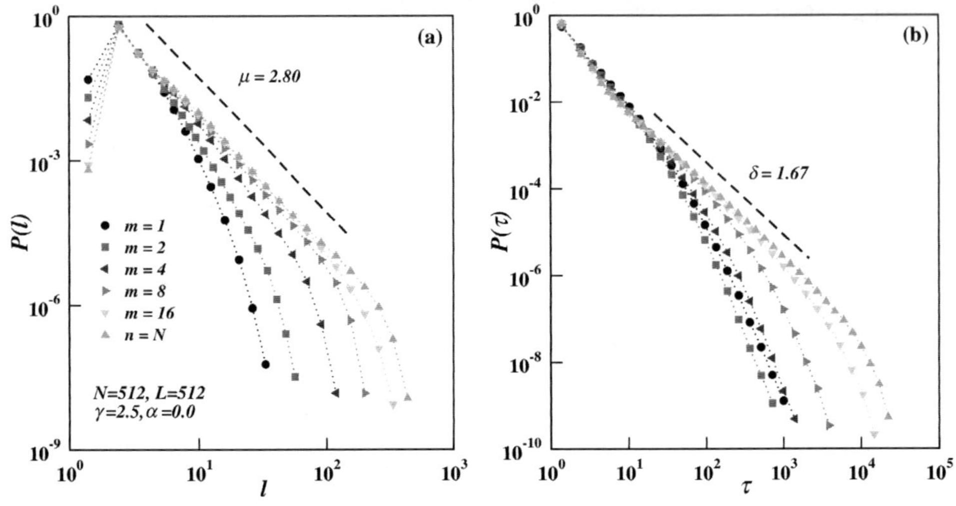

Figure 3. (a) The plots of probability density function, $P(l)$, of the length of targeted walks, for $L = 512$, $N = 512$, $\gamma = 2.5$ and $\alpha = 0.0$. The different curves correspond to different values of m (or n) as indicated in the legend. The dashed line is a guide to the eye and indicates the power-law nature in the region for the curve with $n = N$. Regression fit with $P(l) \sim l^{-\mu}$ gives the value of $\mu = 2.80$. (b) The plots of probability density function, $P(\tau)$, of waiting times between forager-target encounters, for $L = 512$, $N = 512$, $\gamma = 2.5$ and $\alpha = 0.0$. The different curves correspond to the same legend a in (a). The dashed line is a guide to the eye and indicates the power-law nature in the region for the curve with $n = N$. Regression fits with $P(\tau) \sim \tau^{-\delta}$ gives the value of $\delta = 1.67$.

very similar to that of η and f_t. With the decrease in α there is a transition from a regime where the value of Φ is dominated by fluctuations, due to random movement of the foragers, to higher values of Φ. The actual values of Φ are low because of the movements being restricted to the lattice. When $n = N$, the simultaneous discoveries are less and after the consumption of one cluster the foragers mostly execute RWs until a new cluster is detected. This causes the decrease in the value Φ.

We analyse the collective effects on the individual foragers from the distribution of the length of the TWs, $P(l)$. Long TWs become more probable as α is reduced. We plot $P(l)$ against l, for different values of m in Fig. 3(a), in the limit, $\alpha \to 0$. The plot reveals that for $m \geq m_c$, the walks become scale-free. For $n = N$, we find $P(l) \sim l^{-\mu}$ with $\mu = 2.80$ (using regression analysis). This also shows that for the emergence of scale-free walks, it is not necessary for foragers to have information about all the other foragers in the lattice. A TW initiated towards a targeted site, may get diverted towards a new site, when the targets at the original site are consumed, or when targets are discovered in nearer locations. A sequence of such diversions gives rise to very long walks.

We characterize the temporal heterogeneity in the behaviour of the foragers through the inter-event time distributions, $P(\tau)$, where, τ is the time between successive encounters with targets by the group, as a whole. As with $P(l)$, we evaluate $P(\tau)$ in the limit, $\alpha \to 0$. We observe from Fig. 3(b) that the states of the foragers become increasingly correlated as the size of the neighbourhood increases. The distribution $P(\tau)$ becomes fat tailed with increase in m. For $n = N$, we have, $P(\tau) \sim \tau^{-\delta}$, with $\delta = 1.67$.

4. Discussion

Our model shows that when the target distribution is heterogeneous, collective effects among foragers can help enhance the efficiency of searching. The maximum efficiency is found to be in the case when each forager is equipped with the information about all the other foragers, but adopts a probabilistic strategy to follow only nearby foragers. However, in practice the radius of interaction would be finite and, at any instant of time, a forager would have information about only a limited number of other foragers. Our present investigation enumerates the efficient strategies when the neighbourhood is restricted to a finite radius. If the radius is small, then following other successful agents is always beneficial for a searcher. However, we find an optimal bound for this radius, dependent on the density of agents. Exchange of information above this length scale causes unnecessary aggregation at clusters of targets; and reduction in efficiency. In our model scale-free walks emerge as a result of the collective behaviour of the foragers. The presence of correlations in the behaviour of the foragers is also reflected in the scale-free inter-event time distribution. Overall, our model demonstrates the complexity that emerges when collective effects are incorporated into the behaviour of a set of agents which are involved in searching for targets. We believe that our model is representative of the foraging behaviour seen is animal groups and human populations [28, 29, 30, 31]. The model can also be of help from the viewpoint of design of artificial swarms [32, 33].

Acknowledgments

KB acknowledges helpful discussions with Vishwesha Guttal (IISc, Bangalore) and Anjan Nandi (IISER, Kolkata).

References

[1] Giraldeau L A and Beauchamp, G 1999 Food exploitation: searching for the optimal joining policy *Trends Ecol. Evol.* **14** 102
[2] Vickery W L, Giraldeau L A, Templeton J J, Kramer D L and Chapman C A 1991 Producers, scroungers, and group foraging *Am. Nat.* **137** 847
[3] Champagne L, Carl RG and Hill R 2003 Search theory, agent-based simulation, and u-boats in the Bay of Biscay *Proceedings of The 2003 Winter Simulation Conference* **1** 991
[4] Coppey M, Bénichou O, Voituriez R and Moreau M 2004 Kinetics of target site localization of a protein on DNA: a stochastic approach *Biophysical Journal* **87** 1640
[5] Bénichou O, Loverdo C, Moreau M and Voituriez R 2011 Intermittent search strategies *Rev. Mod. Phys.* **83** 81
[6] Viswanathan G M, da Luz, M G E, Raposo E P and Stanley H E 2011 *The physics of foraging* (Cambridge, UK: Cambridge University Press)
[7] Turchin P 1998 *Quantitative analysis of movement: measuring and modelling population redistribution in animals and plants* (Sunderland, USA: Sinauer Associates)
[8] Bartumeus F and Catalan J 2009 Optimal search behaviour and classic foraging theory *J. Phys. A: Math. Theor.* **42** 434002
[9] Viswanathan G M, Afanasyev V, Buldyrev S V, Murphy E J, Prince P A and Stanley H E 1996 Lévy flight search patterns of wandering albatrosses *Nature* **381** 413
[10] Sims D W *et al.* 2008 Scaling laws of marine predator search behaviour *Nature* **451** 1098
[11] Viswanathan G M, Buldyrev S V, Havlin S, da Luz M G E, Raposo E P and Stanley H E 1999 Optimizing the success of random searches *Nature* **401** 911
[12] Bartumeus F, Catalan J, Fulco U L, Lyra M L and Viswanathan G M 2002 Optimizing the encounter rate in biological interactions: Lévy versus Brownian strategies *Phys. Rev. Lett.* **88** 097901
[13] Bartumeus F, da Luz M G E, Viswanathan G M and Catalan J 2005 Animal search strategies: a quantitative random-walk analysis *Ecology* **86** 3078
[14] James A, Plank M J and Edwards A M 2011 Assessing Lévy walks as models of animal foraging *J. R. Soc. Interface* **8** 1233 and references therein.
[15] Raposo E P, Buldyrev S V, Da Luz M G E, Santos M C, Stanley H E and Viswanathan G M 2003 Dynamical robustness of Lévy search strategies *Phys. Rev. Lett.* **91** 240601
[16] James A, Plank M J and Brown R 2008 Optimizing the encounter rate in biological interactions: ballistic versus Lévy versus Brownian strategies *Phys. Rev. E* **78** 051128

[17] Pitcher T J, Magurran A E and Winfield I J 1982 Fish in larger shoals find food faster *Behav. Ecol. Sociobiol.* **10** 149
[18] Haney J C, Fristrup K M and Lee D S 1992 Geometry of visual recruitment by seabirds to ephemeral foraging flocks *Ornis Scandinavica* **23** 49
[19] Beauchamp G 2008 A spatial model of producing and scrounging *Anim. Behav.* **76** 1935
[20] Tania N, Vanderlei B, Heath J P and Edelstein-Keshet L 2012 Role of social interactions in dynamic patterns of resource patches and forager aggregation *Proc. Natl. Acad. Sci.* **109** 11228
[21] Torney C J, Berdahl A and Couzin I D 2011 Signalling and the evolution of cooperative foraging in dynamic environments *PLoS Comput Biol* **7** e1002194
[22] Boyer D, Miramontes O, Ramos-Fernández G, Mateos J L and Cocho G 2004 Modeling the searching behavior of social monkeys *Physica A* **342** 329
[23] Martínez-García R, Calabrese J M, Mueller T, Olson K A and López C 2013 Optimizing the search for resources by sharing information: Mongolian gazelles as a case study Phys. Rev. Lett. **110** 248106
[24] Bhattacharya K and Vicsek T 2014 Collective foraging in heterogeneous landscapes *J. R. Soc. Interface* **11** 20140674
[25] Boyer D, Ramos-Fernández G, Miramontes O, Mateos J L, Cocho G, Larralde H, Ramos H and Rojas F 2006 Scale-free foraging by primates emerges from their interaction with a complex environment *Proc. R. Soc. Lond. B* **273** 1743
[26] Fauchald P 1999 Foraging in a hierarchical patch system *Am. Nat.* **153** 603
[27] Vicsek T, Czirók A, Ben-Jacob E, Cohen I and Shochet O 1995 Novel type of phase transition in a system of self-driven particles *Phys. Rev. Lett.* **75** 1226
[28] Rand A L 1954 Social Feeding Behavior of Birds *Fieldiana Zool.* **36** 5
[29] Clay Z and Zuberbühler K 2009 Food-associated calling sequences in bonobos *Anim. Behav.* **77** 1387
[30] Galef Jr B G and Giraldeau L A 2001 Social influences on foraging in vertebrates: causal mechanisms and adaptive functions *Anim. Behav.* **61** 3
[31] Mangel M and Clark C W 1983 Uncertainty, search, and information in fisheries *J. Cons. int Explor. Mer* **41** 93
[32] Dai H 2010 *Design of adaptive collective foraging in swarm robotic systems* (Kalamazoo, MI, USA: Western Michigan University)
[33] Winfield A F T 2009 Foraging robots *Encyclopaedia of Complexity and System Science* (New York: Springer) p 3682

Journal of Physics: Conference Series **638** (2015) 012016 doi:10.1088/1742-6596/638/1/012016

Relation of classical non-equilibrium dynamics and quantum annealing

Hidetoshi Nishimori

Department of Physics, Tokyo Institute of Technology, Oh-okayama, Meguro-ku, Tokyo 152-8551, Japan

Abstract. Non-equilibrium dynamics of the Ising model is a classical stochastic process whereas quantum mechanics has no stochastic elements in the classical sense. Nevertheless, it has been known that there exists a close formal relationship between these two processes. We reformulate this relationship and use it to compare the efficiency of simulated annealing that uses classical stochastic processes and quantum annealing to solve combinatorial optimization problems. It is shown that classical dynamics can be efficiently simulated by quantum-mechanical processes whereas the converse is not necessarily true. This may imply that quantum annealing may be regarded as a more powerful tool than simulated annealing for optimization problems.

1. Introduction

Quantum annealing (QA) is a metaheuristic for combinatorial optimization problems. In the words of physics, the task is to find the ground state of the Ising model with generally complicated interactions [1, 2, 3, 4]. QA uses quantum fluctuations realized by transverse fields to search for the solution in the phase space. A restricted version of QA is adiabatic quantum computation, in which the system is supposed to follow the instantaneous ground state of the time-dependent Schrödinger equation [5]. QA may be regarded as the quantum-mechanical alternative to simulated annealing (SA) [6, 7], in which classical thermal fluctuations play the role of quantum fluctuations in QA. It is an interesting problem how QA can be better (or worse) than SA.

Recent years have observed high activities in this area, most of which indicate some kind of supremacy of QA over SA [1, 2, 3, 8]. Experimental approaches have also become available using the D-Wave machine [9]-[26]. Experimental data suggest that quantum effects indeed play roles at least to some extent.

Pioneering examples of numerical data, obtained from direct numerical integration of the Schrödinger equation and quantum Monte Carlo simulations, are found in the thesis of Kadowaki [8]. It has been found that the probability of the system being in the true ground state is larger in QA than in SA for spin glass cases and travelling salesman problems. It has also been shown that the achieved values of the final energy are lower in QA than in SA.

From the analytical side, theorems have been proven that give a sufficient condition for the time dependence of the strength of quantum fluctuations, *i.e.* the annealing schedule, such that the system reaches the ground state in the infinite-time limit [27, 28, 29, 30]. The resulting condition turns out to give a faster reduction of the control parameter in QA than in SA. It should nevertheless be remembered that this result does not mean that a given difficult problem can be solved efficiently by the quantum method. Theoretical supremacy of QA is usually quantitative, not qualitative, as far as the sufficient condition for the worst case is concerned.

One of the interesting questions is whether or not QA can be efficiently simulated on a classical computer. The above-mentioned sufficient condition indicates that the amplitude of the transverse field should be reduced as a function of time in the same manner in the natural Schrödinger dynamics and in the master equation dynamics corresponding to quantum Monte Carlo simulations [27, 28, 29]. This is a surprising result because the Schrödinger dynamics is completely different from the classical stochastic dynamics of Monte Carlo processes. The result may be understood to suggest that QA can be efficiently simulated on a classical computer. We should, however, be very careful because, first, quantum Monte Carlo operates at low but finite temperatures whereas the Schrödinger dynamics is a zero-temperature process. Second, quantum Monte Carlo is a stochastic sampling of the classical master equation. Directly solving the master equation consumes an exponential amount of time and memory on a classical computer, which is the same situation as in solving the Schrödinger equation directly on a classical computer.

In the present paper, we try to shed some light on this problem. Following [31], we show that an ingenious mapping between classical stochastic dynamics and a quantum system allows us to derive a conclusion that simulated annealing may be efficiently simulated by quantum annealing but the converse is not necessarily true.

2. Classical to quantum mapping

Let us first consider the classical dynamics of the Ising model described by the Hamiltonian $H_0(\sigma)$, where σ stands for the set of N Ising spins, $\sigma = \{\sigma_1, \sigma_2, \cdots, \sigma_N\}$. We have added a subscript 0 to the Hamiltonian to distinguish this classical Ising Hamiltonian from the quantum Hamiltonian defined later. The master equation for the probability $P_\sigma(t)$ that the system is in the state σ

$$\frac{dP_\sigma(t)}{dt} = \sum_{\sigma} W_{\sigma\sigma'} P_{\sigma'}(t), \tag{1}$$

where $W_{\sigma\sigma'}$ denotes an element of the transition matrix, defines the classical stochastic dynamics. The transition probability is a function of the temperature, which is implicit in the above notation. The temperature is fixed for the moment. The transition matrix has non-positive right eigenvalues

$$W\psi^{(R,n)} = -\lambda_n \psi^{(R,n)} \tag{2}$$

with $\lambda_0 = 0$ and $0 > -\lambda_1 > -\lambda_2 > \cdots$. The leading eigenvector $\psi^{(R,0)}$ corresponds to the equilibrium state, and other eigenvectors represent relaxing modes. The solution to the master equation (1) has a general expression as

$$P_\sigma(t) = P_\sigma^{\text{eq}} + a_\sigma^{(1)} e^{-\lambda_1 t} + a_\sigma^{(2)} e^{-\lambda_2 t} + \cdots, \tag{3}$$

where P_σ^{eq} is the equilibrium Gibbs-Boltzmann distribution. This expression makes it clear that the relaxation time toward equilibrium is the inverse of the leading non-vanishing eigenvalue,

$$\tau_{\text{relax}} = \frac{1}{\lambda_1}. \tag{4}$$

The relaxation time diverges at a transition point $T = T_c$. If the transition is of second order, the relaxation time at $T = T_c$ diverges polynomially as the system size N increases,

$$\tau_{\text{relax}}\left(= \frac{1}{\lambda_1}\right) \propto N^a \ (a > 0), \tag{5}$$

whereas the divergence is usually exponential at a first-order transition,

$$\tau_{\rm relax}\left(=\frac{1}{\lambda_1}\right)\propto e^{bN}\ (b>0). \tag{6}$$

This means that the energy gap between the leading eigenvalue $\lambda_0 = 0$ and the next eigenvalue λ_1 closes polynomially at a second-order transition point and exponentially at a first-order transition.

If we define a $2^N \times 2^N$ matrix H with element [31, 32, 33]

$$H_{\sigma\sigma'} = -e^{\beta H_0(\sigma)/2}W_{\sigma\sigma'}e^{-\beta H_0(\sigma')/2}, \tag{7}$$

where β is the inverse temperature, then the detailed balance condition $W_{\sigma\sigma'}P^{\rm eq}_{\sigma'} = W_{\sigma'\sigma}P^{\rm eq}_{\sigma}$ imposed on the transition matrix guarantees that H is a symmetric matrix,

$$H_{\sigma\sigma'} = H_{\sigma'\sigma}. \tag{8}$$

Since all matrix elements of H are real, this is a Hermitian matrix and thus can be regarded as the Hamiltonian of a quantum system.

The Hamiltonian H shares the eigenstates and eigenvalues with W, up to trivial factors,

$$H\phi^{(n)} = \lambda_n\phi^{(n)},\ \phi^{(n)} = e^{\beta H_0/2}\psi^{(R,n)} \tag{9}$$

as can be verified from the definition (7). Since the energy gap between the ground and first excited states $\Delta = \lambda_1$ is common to both systems (the classical dynamical system and the quantum system), these two systems share the existence and properties of a phase transition. In particular, the energy gap of the quantum system closes polynomially as a function of the system size at a second-order quantum transition and exponentially at a first-order quantum transition if the same is true for the classical case.

It is to be noticed that this classical-to-quantum mapping preserves the spatial dimension of the system. It is also important to remember that the definition (7) leads to short-range interactions in H if the same is true for H_0 as can be verified from the definition (7). A simple example is the one-dimensional ferromagnetic Ising model

$$H_0(\sigma) = -\sum \sigma_j\sigma_{j+1}, \tag{10}$$

from which the following quantum Hamiltonian is derived under the heat-bath dynamics [31],

$$H = -\frac{1}{2}\sum_j \sigma_j^z\sigma_{j+1}^z - \frac{1}{2\cosh 2\beta}\sum_j(\cosh^2\beta - \sinh^2\beta\ \sigma_{j-1}^z\sigma_{j+1}^z)\sigma_j^x. \tag{11}$$

3. Quantum to classical mapping

The converse mapping from a quantum Hamiltonian to classical dynamics can be formulated as follows [31, 32, 33]. Consider a quantum spin system that can be expressed in the basis of eigenstates of $\{\sigma_j^z\}_j$ as $H_{\sigma\sigma'}$. It is necessary to impose the conditions of negative semi-definiteness of off-diagonal elements

$$H_{\sigma\sigma'} \le 0\ (\sigma \ne \sigma') \tag{12}$$

and positive semi-definiteness of the eigenvalues,

$$H\phi^{(0)} = 0, \tag{13}$$

where $\phi^{(0)}$ is the ground-state eigenvector. The latter condition can always be satisfied by a shift of the energy standard. Under these conditions, the matrix W defined by

$$W_{\sigma\sigma'} = -e^{-H_0(\sigma)/2} H_{\sigma\sigma'} e^{H_0(\sigma')/2} \tag{14}$$

can be regarded as the transition matrix of a classical stochastic process of the Ising model $H_0(\sigma)$ whose elements are defined by

$$H_0(\sigma) = -2 \log \phi_\sigma^{(0)}. \tag{15}$$

This last definition is valid because the ground-state eigenvector $\phi^{(0)}$ has all its elements positive

$$\phi_\sigma^{(0)} > 0 \tag{16}$$

due to the Perron-Frobenius theorem that applies to matrices with the properties described in equation (12). Notice that the prescription (15) has been inspired by the relation

$$\phi_\sigma^{(0)} = e^{-\beta H_0(\sigma)/2} \tag{17}$$

in the classical-to-quantum mapping. It is also interesting that the present quantum-to-classical mapping does not change the spatial dimension in contrast to the Suzuki-Trotter decomposition or the path-integral formulation of quantum mechanics, by which the effective spatial dimension increases by one.

The classical Ising Hamiltonian $H_0(\sigma)$ defined by equation (15) has in general many-body long-range interactions with 2^N coefficients

$$H_0(\sigma) = \text{const} - \sum h_j \sigma_j - \sum J_{ij} \sigma_i \sigma_j - \cdots - J_N \sigma_1 \sigma_2 \cdots \sigma_N. \tag{18}$$

The reason is that equation (15) must be satisfied by all possible values of $\sigma = \{\sigma_1, \sigma_2, \cdots, \sigma_N\}$, and there are 2^N of them. Hence $H_0(\sigma)$ should have 2^N coefficients in front of various products of spin variables as they appear in equation (18). We set up 2^N linear equations for these coefficients, the solution of which gives, in general, non-vanishing values for all coefficients. This is in sharp contrast to the classical-to-quantum mapping, in which short-range interactions are mapped to short-range interactions. Since the many-body long-range interactions in equation (18) are hard to simulate, we may conclude that classical simulation of the quantum system is not efficient.

4. Conclusion

We have reformulated the mapping between classical stochastic dynamics of the Ising model and a quantum system in stationary state. It has been shown that classical stochastic dynamics with fixed temperature can be efficiently simulated by a quantum system whereas quantum systems cannot necessarily be efficiently simulated by classical dynamics as long as the present quantum-to-classical mapping is concerned. The present results indicate that simulated annealing, if the temperature is changed slowly, may be efficiently simulated by quantum annealing if a control parameter in the quantum system, typically the transverse field, is changed slowly or adiabatically. The converse process to simulate quantum annealing with a slow change of a control parameter by simulated annealing with a slow decrease of temperature is not necessarily efficient. It may be premature to conclude only from this result that quantum annealing is superior to simulated annealing because other formulations of classical-quantum correspondence may be possible in which the present conclusion does not hold. It is nevertheless important to have established that the present method yields asymmetric correspondence between classical stochastic dynamics and quantum mechanics.

Acknowledgments
This was done in collaboration with Junichi Tsuda and Sergey Knysh.

References
[1] Kadowaki T and Nishimori H 1998 *Phys. Rev.* E **58** 5355
[2] Santoro G E and Tosatti E 2006 *J. Phys. A: Math. Gen.* **39** R393
[3] Das A and Chakrabarti B K 2008 *Rev. Mod. Phys.* **80** 1061
[4] Suzuki S, Inoue J -i and Chakrabarti B K 2013 *Quantum Ising Phases and Transitions in the Transverse Ising Models* (Heidelberg: Springer), Chapter 8
[5] Farhi E, Goldstone J, Gutmann S, Lapan J, Lundgren A, and Preda D 2001 *Science* **292** 472 (2001)
[6] Kirkpatrick S, Gelatt Jr C D, and Vecchi M P 1983 *Science* **220** 671
[7] Aarts E and Korst J 1989 *Simulated Annealing and Boltzmann Machines* (Hoboken, New Jersey: Wiley)
[8] Kadowaki T 1998 *Thesis*, Tokyo Institute of Technology; arXiv:quant-ph/0205020.
[9] Johnson J W *et al.* 2011 *Nature* **473** 194
[10] Perdomo-Ortiz A, Dickson N, Drew-Brook M, Rose G, and Aspuru-Guzik A 2012 *Scientific Reports* **2** 571
[11] Boixo S, Albash T, Spedalieri F M, N Chancellor and Lidar D 2013 *Nature Commun* **4** 2067
[12] Dickson N G *et al.* 2013 *Nature Commun* **4** 1903
[13] Bian A, Chudak F, Macready W G, Clark L and Gaitan F 2013 *Phys. Rev. Lett.* **111** 130505
[14] Boixo S *et al.* 2014 *Nature Phys.* **10** 218
[15] Pudenz K L, Albash T and Lidar D L 2014 *Nature Comm.* **5** 3243
[16] Ronnow T F *et al.* 2014 *Science* **345** 420
[17] Lanting T *et al.* 2014 *Phys. Rev.* X **4** 021041
[18] Vinci W, Albash T, Mishra A, Warburton P A and Lidar D A 2014 *Preprint* arXiv:1403.4228
[19] Katzgraber H G, Hamze F and Andrist R S 2014 *Phys. Rev.* X **4** 021008
[20] Venturelli D, Mandra S and Knysh S 2014 *Preprint* arXiv:1406.7553
[21] Perdomo-Ortiz A *et al.* 2014 *Preprint* arXiv:1406.7601
[22] Boixo S *et al.* 2014 *Preprint* arXiv:1411.4036
[23] Pudenz K L, Albash T and Lidar D A 2014 *Preprint* arXiv:1408.4382
[24] Hen I *et al.* 2014 *Preprint* arXiv:1502.01663
[25] Martin-Mayor V and Hen I 2015 *Preprint* arXiv:1502.02494
[26] Boixo S *et al.* 2015 *Preprint* arXiv:1502.05754
[27] Morita S and Nishimori H 2006 *J. Phys. A: Math. Gen.* **39** 13903
[28] Morita S and Nishimori H 2007 *J. Phys. Soc. Jpn.* **76** 064002
[29] Morita S and Nishimori H 2008 *J. Math. Phys.* **49** 125210
[30] Morita S 2008 *Thesis*, Tokyo Institute of Technology
[31] Nishimori H, Tsuda J and Knysh S 2015 *Phys. Rev.* E **91** 012104
[32] Henley C 2004 *J. Phys.: Cond. Matt* **16** S891
[33] Castelnovo C, Chamon C, Mudry C and Pujol P 2005 *Ann. Phys.* **318** 316

Journal of Physics: Conference Series **638** (2015) 012017 doi:10.1088/1742-6596/638/1/012017

Some results for k-SAT on trees

Sumedha
National Institute of Science Education and Research, Institute of Physics Campus, Bhubaneswar, Odisha- 751 005

Supriya Krishnamurthy
Department of Physics, Stockholm University, SE- 106 91, Stockholm, Sweden

Abstract. Phase transitions in random k-SAT problems are connected to their computational complexity. While polynomical time algorithms are known to solve the problem for $k = 2$, for $k \geq 3$ the problem is known to be NP-complete. Recently we have studied random k-SAT and many of its variants on regular infinite trees. We find that the solvability threshold for $k = 2$ matches the exact value of the threshold on regular random graphs. For higher k, the values are very close to those predicted using other techniques like cavity method.

1. Introduction

K-satisfiability problems are an important example of constraint satisfaction problems, which have been fundamental in understanding computational complexity. The aim is to find a satisfying assignment for a randomly generated logical expression of M clauses(logical constraints). Each clause is an OR of k Boolean variables, which are chosen randomly from a set of N Boolean variables. For $k \geq 3$ variables, this problem is NP -complete [1], *i.e.*, potential solutions can be verified easily for correctness, but finding a solution can take exponential time in the worst case. In addition, being NP-complete, should a polynomial-time algorithm be found for solving SAT, it is also possible to adapt it to solve any problem in NP in polynomial-time.

As the constraint density ($\alpha = M/N$) increases, the number of satisfying assignments decreases. In the limit of $M \to \infty$ and $N \to \infty$, the system is known to have a sharp threshold in constraint density α_c below which the probability of finding satisfiable assignments approaches 1 and above which it vanishes [2, 3].

The problem is originally defined on a random graph, but because of the presence of loops, this is hard to solve exactly for arbitrary k. Hence the location of a sharp threshold α_c has been known rigorously only for $k = 2$ [4, 5]. For higher k only upper and lower bounds on this threshold are proven[6]. However using non-rigorous but powerful methods from statistical physics, namely the replica and cavity methods, estimates for the threshold are obtained which seem to be very close to the values obtained numerically [7, 8, 9].

The replica and cavity methods also predict that the solvability threshold is only one of many thresholds that exist in the problem, as the number of constraints is increased. Before the solvability transition occurs, it is conjectured that the set of solutions first breaks up into a large number of well separated clusters at the clustering transition α_d [7, 10]. As the number of constraints further increases, it is argued that there is first a condensation transition [11] in which the number of clusters changes from being exponentially numerous to sub-exponential

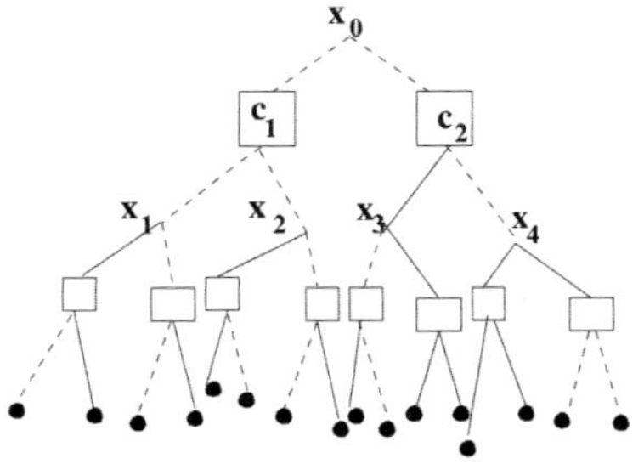

Figure 1. 3-SAT on a rooted tree of depth 2 and $d = 2$. Only the clauses neighboring the root are labelled. Surface variables (or leaves) are depicted by a •. Variable x_0 is at depth 2, variables $x_1 - x_4$ are at depth 1 and the leaves are at depth 0.Dashed/full lines between a variable and a clause indicate that it is negated/non-negated.

and a freezing transition beyond which some variables take the same value in all the solutions of a given cluster [12, 13]. Recently, both the existence of clusters [14] as well as the 1-RSB prediction for the satisfiability threshold has been established for random k-SAT [15] for large k. All the approaches mentioned above for $k \geq 3$ look at the properties of solution clusters.

Recently, we have studied the random k-SAT problem on regular d-ary rooted trees[16, 17, 18]. We study the probability of having a satisfiable assignment as a function of tree depth. In this paper we will review some of these results. We find that the probability of having a satisfiable assignment for an infinite tree gives a value for the solvability threshold that matches the value of the clustering (or dynamic) transition calculated using cavity method for random graphs. On a Bethe lattice we obtain a solvability threshold that matches the solvability threshold on the random graphs. We also present some preliminary results for a $(2+p)$-SAT problem. The plan of the paper is as follows: In Sec. 2, we define the model on a d-ary rooted tree. In Sec. 3 we present the solution of the random k-satisfiability problem and in Sec. 4, we compare our method with the tree reconstruction problem. In Sec. 5 we briefly discuss the $(2+p)$-SAT before concluding in Section 6.

2. The model

We define the k-SAT problem on a rooted tree as follows: Consider a regular d-ary tree T in which every vertex has exactly d descendents. The root of the tree x_0 has degree d and its d edges are connected to function nodes $\{c_1, c_2..., c_d\}$. Each function node has degree k, and each of its $k-1$ descendents $\{x_i = x_1, x_2, ...x_{k-1}\}$ is the root of an independent tree (see Fig. 1). Hence the root has a degree d while all the other vertices on the tree (except the leaves which have a degree $= 1$) have a degree $d+1$. Each vertex can take only two values: -1 or 1. Each function node is associated independently with a clause $\phi(x_0, x_1, ..., x_{k-1}) = \ell_0 \bigvee \ell_1 \bigvee ....\ell_{k-1}$. Here ℓ_i is one of the two literals x_i or $\overline{x_i}$, determined by whether x_i is joined to the function node by a dashed or a solid line (see Fig. 1). An assignment σ of all the variables on the tree is a solution iff $\phi = 1$ for all the clauses on the tree. One configuration of dashed and solid lines on the tree defines a realization R.

We study the probability that a realization has no solution on this tree for a fixed boundary. This can happen if there is even a single variable on the graph, which, whether it takes the value -1 or 1, causes atleast one clause to be unsatisfied. Such a variable then is a variable that can take 0 values by our definition, and a realization that is not solvable has at least one variable of this type.

On the tree graph, we can define the probabilities of a variable taking 0, 1 or 2 values on the corresponding subtree. We define $P_i(0)$ as the conditional probability for a variable x_i to cause a contradiction, in the subtree of which it is the root, given that all the other variables in the subtree can take at least 1 value. Note that because of the tree structure and because of the definition of the specific quantity we are looking at, all variables x_i at depth n will have the same probability $P_n(0)$. The conditional probability that a variable at depth n, can take only one of the two values -1 or 1 is defined to be $P_n(1)$ (the boundary nodes have $P_0(1) = 1$, for example). Similarly the conditional probability that a variable at depth n can take both values is $P_n(2) = 1 - P_n(0) - P_n(1)$. The probability of a realization *having* a solution (or the fraction of realizations that have solutions) is then exactly equal to the product $\Pi_i(1 - P_i(0))$, where the product is over all the variables in the graph. The tree structure also gives us a way to calculate the P_n's via recursions. For the problems we look at, we are interested in the recursions for these quantities deep within the tree, so that we can get rid of boundary effects.

3. Random k-SAT on a regular tree

Let us first calculate $P_{n+1}(0)$ for variable x_0 (assuming it is at depth $n+1$), given these quantities for its descendents. Assume variable x_0 has a degree d and assume it is not negated on d_1 of these clauses. Variable x_0 will not be able to take the value -1 in the case when *at least* one of the d_1 clauses is *not* satisfied by the $k-1$ variables at the other end. In this case there will be at least one unsatisfied clause if x_0 takes the value -1. Similarly, if at least one of the $d - d_1$ clauses which are satisfied by x_0, are also not satisfied by the $k-1$ variables at the other end, then x_0 cannot take the value 1 either.

It is easy to see that averaging over all realizations at depth $n+1$ implies averaging over all values of d_1, as well as averaging over all realizations at depth n. It is important to note however that the realizations at depth $n+1$ are only built up from those realizations at depth n that do have solutions. We define Q_n as the conditional probability that a depth n variable does not satisfy the clause above (to depth $n+1$), given that it has to be able to take at atleast one value (which satisfies the sub tree of which it is the root). In terms of $P_n(0)$ and $P_n(1)$, $Q_n = \frac{P_n(1)}{2(1-P_n(0))}$. The recursion for $P_{n+1}(0)$ is then:

$$\begin{aligned} P_{n+1}(0) &= \frac{1}{2^d}\sum_{d_1=1}^{d_1=d-1}\binom{d}{d_1}[1-(1-Q_n^{k-1})^{d_1})(1-(1-Q_n^{k-1})^{d-d_1}] \\ &= 1+(1-Q_n^{k-1})^d - 2(1-0.5Q_n^{k-1})^d \end{aligned} \quad (1)$$

Similarly one can work out the recursion for $P_{n+1}(1)$. This is:

$$P_{n+1}(1) = 2(1-0.5Q_n^{k-1})^d - 2(1-Q_n^{k-1})^d \quad (2)$$

Eqs. 1 and 2 result in the following relation between Q_{n+1} and Q_n:

$$Q_{n+1} = \frac{[1-0.5Q_n^{k-1}]^d - [1-Q_n^{k-1}]^d}{2[1-0.5Q_n^{k-1}]^d - [1-Q_n^{k-1}]^d} \quad (3)$$

From this equation the threshold at which the fraction of realizations goes to zero exponentially with the depth of the tree may be extracted. This is the solvability threshold for these models on an infinite tree. A fixed point analysis of Eq. 3 predicts a continuous transition for $k = 2$ and a first order transition for $k > 2$(see Fig. 2 and 3). The value of d at which the system undergoes a continuous transition for $k = 2$ can be extracted by expanding to order Q^2 in Eq. 3 at the fixed point. This gives, for $k = 2$:

$$Q_c = \frac{8(d/2-1)}{3(d-1)d} \quad (4)$$

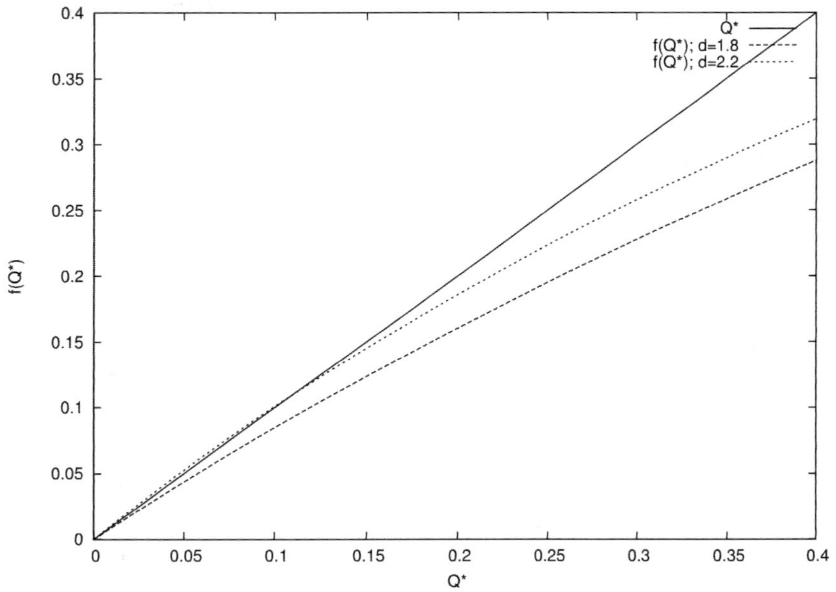

Figure 2. Fixed points for $d < 2$ and $d > 2$ for 2-SAT

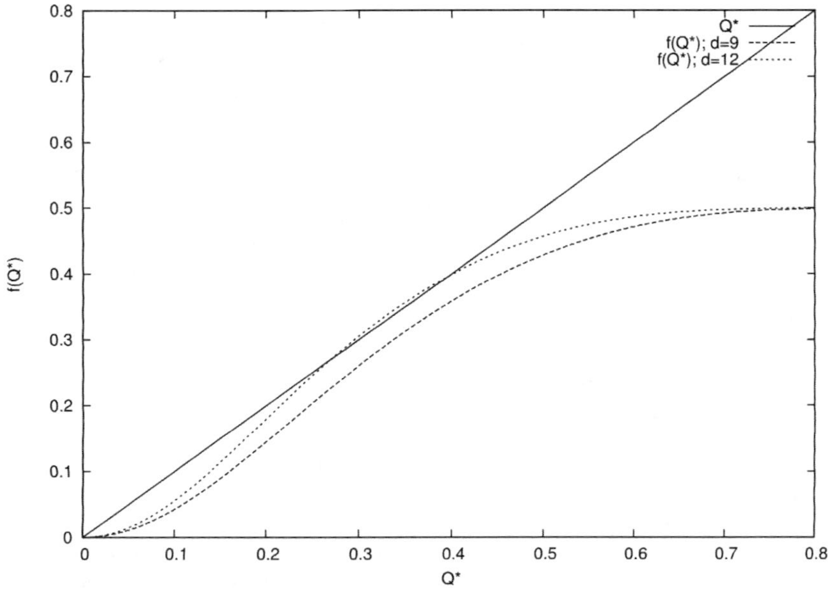

Figure 3. Fixed points for $d < 11.5$ and $d > 11.5$ for 3SAT

Journal of Physics: Conference Series **638** (2015) 012017 doi:10.1088/1742-6596/638/1/012017

Table 1. We compare the values of d_c and d_s obtained from our tree calculations with the clustering (α_d) and SAT-UNSAT(α_s) threshold values known from the cavity calculations [8]. $\alpha = (d+1)/k$ for the tree. The difference in numbers for smaller values of k is due to the difference in the degree distributions. We study the problem on regular graphs, while [8] looks at the problem on poisson graphs.

k	$(d_c+1)/k$	$(d_s+1)/k$	α_d (from [8])	$\alpha_s(from[8])$
2	1.5	1.5	1	1
3	4.166	4.55	3.927	4.267
4	8.4	10.15	8.297	9.93
5	16.2	21.26	16.12	21.12
6	30.5	43.41	30.5	43.37
7	57.28	87.84	57.22	87.79
8	107.13	176.57	107.24	not available

which implies $d_c = 2$. The value of α corresponding to a give value of d should be $(d+1)/k$ [19, 20]. Hence α_c corresponding to d_c is $3/2$, which matches the exact threshold for random 2-SAT on regular random graphs (these are random graphs in which each variable is connected with the same number of clauses)[17, 5].

For $k > 3$ no exact results for random k-SAT on random graphs is known, but a lot of progress has been made using replica and cavity methods. Cavity calculations conjecture the breaking of solution space into many disconnected clusters before the SAT-UNSAT transition. This is the point where belief propagation(BP) iterations lead to contradiction. From the replica point of view this is also the point where the replica symmetry gets broken, and for systems with discontinuous transitions one has to look at 1-RSB solutions. For $k = 3$, on regular random graphs, it has been found that BP iterations stop converging at $d = 12$, which gives the value of clustering transition on a regular graph to be $4 \leq \alpha_d < 4.33$ [21]. On the tree we find for $k = 3$, Eq. 3 starts having non zero values of Q at around $d_c = 11.5$, which gives $\alpha_c = 4.166$ on an infinite tree. Given the fact that BP iterations were done for integer values of the degree on a regular random graph, the values match well within error bars.

In Table 1 we compare the solvability threshold on an infnite tree with the value of α_d obtained via the cavity method. The later have been obtained for random graphs with poisson degree distribution. This also accounts for the difference in the value. As expected the effect of degree distribution goes down with increasing k. We find for $k \geq 6$, that the difference is less than 1% in the two values.

Unlike $k = 2$, for $k \geq 3$, the transition to the UNSAT phase does not happen at the point of first appearence of a new fixed point of Eq. 3. We find that if we study the problem on a graph which is only locally tree-like (Bethe lattice), with all nodes having the same degree $d+1$, the SAT-UNSAT threshold is different for the infinite tree and for the Bethe lattice, for $k \geq 3$. We calculate the fraction of satisfied realizations per node on such a lattice [18, 22] and find that the calculation yields a value of threshold which is larger than α_d for $k > 2$ and is very close to the known 1-RSB static transition threshold (see Table 1).

4. Connection with tree reconstruction problem

The reconstruction problem, as originally defined, is a broadcast model on a tree, such that information is sent from the root to the leaves, across edges which act as noisy channels. The problem then is to understand whether we can recover information about the root from a knowledge of the configuration of the leaves. It has been shown that the recursions developed in the reconstruction context are the same as obtained by other means (such as the replica or

Journal of Physics: Conference Series **638** (2015) 012017 doi:10.1088/1742-6596/638/1/012017

cavity methods) for the dynamical glass transition on a random graph [23] or the clustering transition for k SAT.

In terms of reconstruction, these fixed point recursions are developed for the unconditional probability distribution at the root of the tree to have a certain 'bias' ; namely, the fraction of boundary conditions (out of all boundary conditions that have a non-zero solution set for a fixed instance), weighted by the total number of solutions each of these boundary conditions possesses, that leads to the root taking the value -1 a certain number of times and the value 1 a certain number of times.

Our approach apriori looks very different from the reconstruction problem mainly because the quantities $P_n(1)$ and $P_n(2)$ are the fraction of realizations that have a non zero solution set (for any fixed boundary condition) and not the fraction of boundaries that have a non-zero solution set for any fixed instance. However, we show in [17] that if we define the probability space over boundary conditions instead of realizations, then we can derive the recursion for fraction of boundary conditions that fix the value unambiguously at the root at level n, given this quantity at level $n-1$. These equations turn out to be the same as Eqs. 1 and 2 in Sec. 3. The fact that for k-SAT, cavity recursions for the clustering transition matches the tree reconstruction recursions [23], gives a connection between the solvability transition on infinite trees as calculated in Sec. 3 and clustering transition on random graphs.

5. Variants of Random k-SAT

The method outlined in Sec. 3 can straightforwardly be used to study many variants of random k-SAT. We have used it to successfully predict transitions for biased k-SAT, balanced k-SAT [17], k-NAE SAT and $(2+p)$-SAT [22]. As a simple illustration of the wide applicability of the method we briefly mention the results for $(2+p)$-SAT here.

For $k=2$ the random k-SAT is known to have a unique continuous transition, while for $k=3$ it exhibits multiple transitions. $(2+p)$-SAT interpolates between these two systems as every clause has a probability p of being a 3-clause and $(1-p)$ of being a 2-clause. The problem is NP-complete for any $p>0$ as it contains a subformula of 3-clauses. However using the replica method and numerical simulations [24, 25], it was found that random $(2+p)$-SAT with poisson degree distribution continues to have a continuous transition for $p<0.41$. Also the computational cost of proving a formula SAT or UNSAT stays linear in this regime for all values of α.

We studied the problem on a rooted tree. As before we define Q_n as the conditional probability that a depth n variable does not satisfy the clause above (to depth $n+1$). The recursions for $P_{n+1}(0)$ and $P_{n+1}(1)$ come out to be

$$\begin{aligned} P_{n+1}(0) &= \frac{1}{2^d}\sum_{d_1=1}^{d-1}[1-\{p(1-Q_n^2)+(1-p)(1-Q_n)\}^{d_1}][1-\{p(1-Q_n^2)+(1-p)(1-Q_n)\}^{d-d_1}] \\ &= 1+[p(1-Q_n^2)+(1-p)(1-Q_n)]^d-2\left[\frac{1+p(1-Q_n^2)+(1-p)(1-Q_n)}{2}\right]^d \end{aligned} \quad (5)$$

Similarly for

$$P_{n+1}(1)=2\left[\frac{1+p(1-Q_n^2)+(1-p)(1-Q_n)}{2}\right]^d-2\left[p(1-Q_n^2)+(1-p)(1-Q_n]\right)^d \quad (6)$$

$$Q_{n+1}=\frac{[1+p(1-Q_n^2)+(1-p)(1-Q_n)]^d-[2p(1-Q_n^2)+2(1-p)(1-Q_n)]^d}{2[1+p(1-Q_n^2)+(1-p)(1-Q_n)]^d-[2p(1-Q_n^2)+2(1-p)(1-Q_n)]^d} \quad (7)$$

6. Conclusions

In conclusion, we have looked at the k-satisfiability class of constraint satisfaction problems on a regular tree as well as a Bethe lattice. Our approach directly looks at all realizations giving them weights 0 and 1 if they are unsuccessful or successful respectively. We do not weight them by the number of solutions. Interestingly, even though we do not look at the solution space, the solvability threshold on the infnite tree seems similar to the clustering (or dynamic) transition in solution space obtained via BP iterations.

For $k > 2$ almost for all variants of the satisfiability problem, the clustering transition is known to be well below the solvability transition of random graphs. Solvability transition for random graphs via the cavity approach is estimated by looking at a reduced entropy called complexity, which is defined as the log of the number of clusters in the solution space for a typical realization. Complexity is conjectured to be equivalent to counting different backbones of frozen variables, evaluated using populations of typical realizations [26]. We find [18, 22] that the value of the solvability transition obtained via complexity matches very well with the value of solvability threshold on Bethe lattices. Our method does not make any assumption about the structure of solution space and assigns the same weight to all realizations with solutions.

Inspite of these differences in the two approaches, the expression and threshold obtained by us matches the corresponding expressions and thresholds on regular random graphs, obtained using the cavity method. This intriguing connection need to be explored further to get a better understanding of the problem. Our method can easily be extended to k-SAT problems with non-uniform degree distributions, making it a useful tool even to study the real-world SAT applications.

References

[1] Cook S, 1971 *Proc. 3rd Annual ACM Symp. on Theory of Computing (Shaker Heights, Oh)* (New York: ACM press) p151
[2] Mitchell D, Selman B and Levesque H 1992, *Proc. 10th Nat. Conf. Artif. Intel.*, 459
[3] Kirkpatrick S and Selman B 1994 *Science* **264** 1297
[4] Chvatal V and Reed B, 1992 *33rd FOCS*, 620; Goerdt A, 1996 *J. Comput. System. Sci.* **53** 469
[5] Cooper C, Frieze A and Sorkin G B, 2002 *Proceedings of the* 13th *annual ACM-SIAM symposium on Discrete Algorithms* 316-320; Fernandez de la vega W, 2001 *Theoret. Comput. Sci.* **265** 131
[6] Achlioptas D 2001 *Theoret. Compt. Sci.* **265** 159
[7] Mèzard M, Parisi G and Zecchina R 2002 *Science* **297** 812
[8] Mertens S, Mèzard M and Zecchina R 2006 *Random Structures and Algorithms* **28** 340
[9] Monasson R and Zecchina R 1996 *Phys. Rev. Lett.* **76** 3881
[10] Biroli G, Monasson R and Weigt M 2000 *Eur. Phys. J. B* **14** 551
[11] Krzakala F, Montanari A, Ricci-Tersenghi F, Semerjian G and Zdeborová L 2007 *Proc. Natl. Acad. Sci. U.S.A.* **104** 10318
[12] Semerjian G 2008 *J. Stat. Phys* **130** 251
[13] Zdeborová L and Krzakala F 2007 *Phys. Rev. E* **76** 031131
[14] Achlioptas D and Coja-Oghlan A 2008 *Proc. 49th FOCS* 793
[15] Ding J, Sly A and Sun N 2014 *arXiv:1411.0650*
[16] Krishnamurthy S and Sumedha 2012 *J. Stat. Mech* P05009.
[17] Sumedha, Krishnamurthy S and Sahoo S 2013 *Phys. Rev. E* **87** 042130.
[18] Krishnamurthy S and Sumedha 2014 *arXiv:1412.2460*
[19] Baxter R J 1982 *Exactly solvable models in statistical mechanics* (London: Academic Press)
[20] Gujrati P D 1995 *Phys. Rev. Lett.* **74** 809
[21] Castellana M, Zdeborová L 2011 *J. Stat. Mech.* P03023
[22] Krishnamurthy S and Sumedha, in preparation
[23] Mèzard M and Montanari A 2006 *J. Stat. Phys.* **124** 1317
[24] Monasson R, Zecchina R, Kirkpatrick S, Selman B and Troyansky L 1999 *Random Structure and Algorithms* **15** 414
[25] Monasson R and Zecchina R 1988 *J. Phys. A* **31** 9209
[26] Parisi G 2004 *Lecture Notes in Computer Science* **2919** Springer 203

Journal of Physics: Conference Series **638** (2015) 012018 doi:10.1088/1742-6596/638/1/012018

Learning probability distributions from smooth observables and the maximum entropy principle: some remarks

Tomoyuki Obuchi[1,†], **Rémi Monasson**[2]

[1]Department of Computational Intelligence and Systems Science, Tokyo Institute of Technology, Yokohama, Japan

[2]Laboratoire de Physique Théorique de l'Ecole Normale Supérieure, associé au CNRS et l'Université Pierre et Marie Curie, 75005 Paris, France

E-mail: [†]`obuchi@sp.dis.titech.ac.jp`

Abstract. The maximum entropy principle (MEP) is a very useful working hypothesis in a wide variety of inference problems, ranging from biological to engineering tasks. To better understand the reasons of the success of MEP, we propose a statistical-mechanical formulation to treat the space of probability distributions constrained by the measures of (experimental) observables. In this paper we first review the results of a detailed analysis of the simplest case of randomly chosen observables. In addition, we investigate by numerical and analytical means the case of smooth observables, which is of practical relevance. Our preliminary results are presented and discussed with respect to the efficiency of the MEP.

1. Introduction

The maximum entropy (ME) principle, which was proposed by Jaynes in 1957 to rederive statistical mechanics from an information-theoretic viewpoint [1], recently stands out as a useful working hypothesis in a wide variety of inference problems [2, 3, 4, 5, 6, 7, 8, 9, 10]. However, a better comprehension of why and when ME applies is still needed. In this respect, we have recently introduced a statistical-mechanical formulation, in which we treated the space of all probability distributions satisfying a set of moment-matching conditions [11]. Moment-matching conditions were defined through the measured observables, and provided information about the target distribution of interest. It was assumed in that work that the observables were identically and independently distributed random variables, which enabled us to perform analytical investigations and to get a precise quantitative characterization of the volume of the space of probability distributions and of the distance between certain probability distributions and the target one as the number of moment-matching conditions varies. An interesting, but disappointed outcome of the analysis was that the ME distribution was not 'closer' to the target distribution than most of the other distributions satisfying the conditions. It was however clear that the failure of the ME principle was due to the irrealistic nature of the observable statistics. In practical applications of the ME, indeed, the observables are not chosen randomly, but are designed to represent some underlying features of the target distribution.

In the present paper, we try to relax the above randomization hypothesis. This strategy leads us to introduce a bias in choosing observables. In particular, we consider the case of 'smooth' observables, that is, taking values that cannot differ much between two nearby microscopic configurations of the system. Though we have not been able to extend the detailed analysis of [11] to this (much) more involved case, we report here some analytical and numerical results we have recently obtained.

The organization of this paper is as follows. In the Sec. 2, we present our statistical-physics framework, and discuss different distributions of observables (random and smooth); we also give an overview of our previous results on the random case. In Sec. 3, we report numerical calculations based on a Monte Carlo method in several different situations. In Sec. 4 we report some attempts at solving the problem analytically. Last of all, we discuss our findings are their possible implications for the ME. in the last Section.

2. Formalism and overview of results in the random observable case

2.1. Basic formulation

Let us consider a system described by configurations of N binary spins, $\boldsymbol{s} = \{s_i = \pm 1\}_{i=1}^N$, which we call target system. The probability distribution of the target system, the target distribution, is denoted by $\hat{p}_{\boldsymbol{s}}$. Trial distributions, which we will fit based on the measurements of observables (moments). will be generally written as $p_{\boldsymbol{s}}$.

We assume that a measurement gives a value of a linear combination of some moments of the target distribution. Thus we can write a measurement of an observable $\{v_{\boldsymbol{s}}\}_{\boldsymbol{s}}$ as

$$\sum_{\boldsymbol{s}} v_{\boldsymbol{s}} \hat{p}_{\boldsymbol{s}} = \boldsymbol{v} \cdot \hat{\boldsymbol{p}}. \tag{1}$$

where the summation runs over all the spin configuration and we introduce a vector notation of the 2^N dimension corresponding to all the spin configurations. The moment-matching (MM) conditions for M observables $\{\boldsymbol{v}^\mu\}_{\mu=1}^M$ are then written as

$$\boldsymbol{v}^\mu \cdot \boldsymbol{p} = \boldsymbol{v}^\mu \cdot \hat{\boldsymbol{p}}, \ \forall \mu. \tag{2}$$

Those equations impose that the trial distribution $\boldsymbol{p}$ reproduces the measurements of the average values observable $\boldsymbol{v}^\mu$ over the target distribution.

We are interested in the phase space of trial distributions constrained by the MM conditions (2). To investigate this, we introduce the following probability distribution on trial distributions

$$\rho\left(\boldsymbol{p}|\Gamma, E, \{\boldsymbol{v}^\mu\}_{\mu=1}^M, \hat{\boldsymbol{p}}\right) = \frac{1}{V}\prod_{\boldsymbol{s}} \theta(p_{\boldsymbol{s}})\delta\left(\sum_{\boldsymbol{s}} p_{\boldsymbol{s}} - 1\right) \exp\left\{-\frac{1}{2E}\sum_{\mu=1}^M (\boldsymbol{v}^\mu \cdot (\boldsymbol{p} - \hat{\boldsymbol{p}}))^2 + \Gamma S(\boldsymbol{p})\right\}, \tag{3}$$

where $S(\boldsymbol{p})$ is the Shannon entropy whose mathematical expression is

$$S(\boldsymbol{p}) = -\sum_{\boldsymbol{s}} p_{\boldsymbol{s}} \log p_{\boldsymbol{s}}. \tag{4}$$

The denominator V is the normalization factor, or the partition function,

$$V\left(\Gamma, E, \{\boldsymbol{v}^\mu\}_{\mu=1}^M, \hat{\boldsymbol{p}}\right) = \int_0^\infty \prod_{\boldsymbol{s}} dp_{\boldsymbol{s}}\delta\left(\sum_{\boldsymbol{s}} p_{\boldsymbol{s}} - 1\right) \exp\left\{-\frac{1}{2E}\sum_{\mu=1}^M (\boldsymbol{v}^\mu \cdot (\boldsymbol{p} - \hat{\boldsymbol{p}}))^2 + \Gamma S(\boldsymbol{p})\right\}. \tag{5}$$

The factor $\exp\left(-\frac{1}{2E}\sum_{\mu=1}^M (\boldsymbol{v}^\mu \cdot (\boldsymbol{p} - \hat{\boldsymbol{p}}))^2\right)$ corresponds to the MM conditions and E is the tolerance against the error on the MM conditions. If $\boldsymbol{p}$ satisfies the MM conditions, the corresponding probability value $\rho(\boldsymbol{p})$ is large, and small otherwise. Especially in the $E \to 0$ limit, trial distributions exactly satisfying all the MM conditions are the only ones to 'survive' and to contribute to V. We call those distributions 'good' distributions.

All good distributions have an equal weight value of ρ at $\Gamma = 0$, but for large $\Gamma \gg 0$ good distributions with larger entropies are assigned larger weights. Hence Γ is a bias towards distributions with large entropies. Studying the two extreme cases $\Gamma = 0$ and $\Gamma \to \infty$, for which only the ME distribution survives, we can compare typical good distributions and the ME distribution. Actually, we observe some drastic changes for specific intermediate values of Γ, which correspond to phase transitions; hence, the whole range of $\Gamma > 0$ is of interest.

2.2. Distribution of observables

We now need to specify how the observables $\{\boldsymbol{v}^{\mu}\}_{\mu}$ are chosen. In [11], we assumed the observables were identically independently distributed (i.i.d.) Gaussian random variable

$$P_{\text{rand}}(\boldsymbol{v}) = \prod_{\boldsymbol{s}} \frac{1}{\sqrt{2\pi}} e^{-\frac{1}{2}v_{\boldsymbol{s}}^2}. \tag{6}$$

The analysis of this random case, which is briefly reported in Section sec. 2.3, yields some important notions and order parameters. However, it is unrealistic as the components of observables are uncorrelated with each other. In particular, two configurations $\boldsymbol{s}$ and $\boldsymbol{s}'$ differing by a single spin, say, s_i, have totally uncorrelated observable values, $v_{\boldsymbol{s}}$ and $v_{\boldsymbol{s}'}$. This is clearly not true for most systems: changing one among N spins generically does not affect dramatically the system properties.

An inspiring idea comes from the Fourier transform in the spin-configuration space. Let us consider an arbitrary function of spin configurations $f_{\boldsymbol{s}}$ and the Fourier transform characterized by a wavenumber vector of binary components $\boldsymbol{k} = \{k_i = 0, 1\}_{i=1}^{N}$. The explicit expression of the Fourier transform is

$$f_{\boldsymbol{k}} = \sum_{\boldsymbol{s}} \left(\frac{1}{\sqrt{2^N}} \prod_{i=1}^{N} s_i^{k_i} \right) f_{\boldsymbol{s}} = \boldsymbol{w}^{\boldsymbol{k}} \cdot \boldsymbol{f}, \tag{7}$$

where we defined the $\boldsymbol{s}$-component of $\boldsymbol{w}^{\boldsymbol{k}}$ to be equal to $(\prod_{i=1}^{N} s_i^{k_i})/\sqrt{2^N}$. The modes $\{\boldsymbol{w}^{\boldsymbol{k}}\}_{\boldsymbol{k}}$ constitute a complete orthogonal set, and thus any vector in the phase space can be expanded on this set. Each mode has a physical meaning: a mode with a wavenumber $k = |\boldsymbol{k}| = \sum_i k_i$ corresponds to a k-spin multiplet. In particular, if $f_{\boldsymbol{s}}$ is a probability distribution, $f_{\boldsymbol{k}}$ is the average value of the k-spin multiplet (divided by $1/\sqrt{2^N}$).

If a random vector drawn from eq. (6) is expanded in the Fourier space, the corresponding weights are statistically uniform on all the modes, which implies that the corresponding MM condition includes correlations of very high-order multiplets of spins with relevant weights. This is not natural because low-order multiplets are often considered to be more important than high-order ones in a general situation. Actually in most of recent applications ranging from biological to engineering problems [2, 3, 4, 5, 6, 7, 8, 9, 10], the ME models are constructed based on observations of singlets and doublets, $|\boldsymbol{k}| = 1$ and 2, of spins. Hence, a naive extension from the random case to a more realistic situation is to bias the weights on some (low-$|\boldsymbol{k}|$'s) Fourier modes in generating observables.

A simple choice along the above idea is using the following distribution

$$P(\boldsymbol{v}) = C \exp\left\{ -\frac{1}{2} \sum_{\boldsymbol{s},\boldsymbol{t}} M_{\boldsymbol{s}\boldsymbol{t}} v_{\boldsymbol{s}} v_{\boldsymbol{t}} \right\}, \tag{8}$$

instead of eq. (6). In the distribution above, the matrix M_{st} is chosen such to make sure that components of the observable **v** corresponding to nearby configurations have similar values. In other words, the presence of M should make the observable *smooth* over the configuration space. Alternatively, and perhaps more simply, we may require that the Fourier expression of eq. (8) becomes diagonal and the variance only depends on the absolute value of $k = |\boldsymbol{k}|$

$$P(\boldsymbol{v}) = C \exp\left\{ -\frac{1}{2} \sum_{\boldsymbol{k}} \frac{v_{\boldsymbol{k}}^2}{V_k} \right\}. \tag{9}$$

Expression eq. (9) implies that the matrix entry M_{st} depends on $\boldsymbol{s}$ and $\boldsymbol{t}$ through the overlap $\sum_i s_i t_i$ only. If we choose V_k to be a rapidly-decreasing function of the modulus k, modes with large k will be suppressed. There are many different ways to introduce a k-dependence in V_k. We try several different expressions for V_k in the numerical studies reported in Section sec. 3.

2.3. *Overview of results for random observables*

We now briefly recall the results of [11] corresponding to the random case defined in eq. (6). The average value of logarithm of V in the large-system-size limit can be computed with the replica method (within the replica symmetric (RS) ansatz),

$$F = [\log V]_{\text{rand}} = \lim_{n\to 0} \frac{1}{n} \log [V^n]_{\text{rand}}, \tag{10}$$

where the square brackets $[\cdots]_{\text{rand}}$ denote the average over the observables (6). Our analysis showed that F depends on the target distribution only through the entropy curve

$$\sigma(\omega) = \frac{1}{N} \sum_{\boldsymbol{s}} \delta(\omega - \omega_{\boldsymbol{s}}), \tag{11}$$

where we assume the target distribution obeys an exponential scale for $N \gg 1$, *i.e.* that

$$\omega_{\boldsymbol{s}} \equiv -\frac{1}{N} \log \hat{p}_{\boldsymbol{s}}, \tag{12}$$

has a well-defined limit for $N \to \infty$. This assumption is true for common physical systems. We give a typical shape of $\sigma(\omega)$ as the left panel of Fig. 1, which corresponds to the independent spin model (ISM) with $H = 0.5$ as the target distribution

$$\hat{p}_{\boldsymbol{s}}^{\text{ISM}} = \frac{e^{H \sum_i s_i}}{(2\cosh H)^N}. \tag{13}$$

The characteristic values of ω, ω_k with $k = 0, 1, 2$, are defined by

$$\left.\frac{d\sigma(\omega)}{d\omega}\right|_{\omega=\omega_k} = k, \tag{14}$$

and they are connected to certain phase transitions (see [11] for details of the transitions). These characteristic values of ω separate the entropy curve in three regions (the rightmost region is not relevant), each of which is called I, II, and III as shown in Fig. 1.

The analysis in the random case revealed that the following three order parameters are useful

$$Q(\{\boldsymbol{v}^\mu\}_\mu, \hat{\boldsymbol{p}}) \equiv \sum_{\boldsymbol{s}} \left\langle (p_{\boldsymbol{s}} - \hat{p}_{\boldsymbol{s}})^2 \right\rangle, \tag{15}$$

$$R(\{\boldsymbol{v}^\mu\}_\mu, \hat{\boldsymbol{p}}) \equiv \sum_{\boldsymbol{s}} \left(\langle p_{\boldsymbol{s}} \rangle - \hat{p}_{\boldsymbol{s}} \right)^2, \tag{16}$$

$$D(\{\boldsymbol{v}^\mu\}_\mu, \hat{\boldsymbol{p}}) \equiv Q - R = \sum_{\boldsymbol{s}} \left(\langle p_{\boldsymbol{s}}^2 \rangle - \langle p_{\boldsymbol{s}} \rangle^2 \right), \tag{17}$$

IOP Publishing
doi:10.1088/1742-6596/638/1/012018

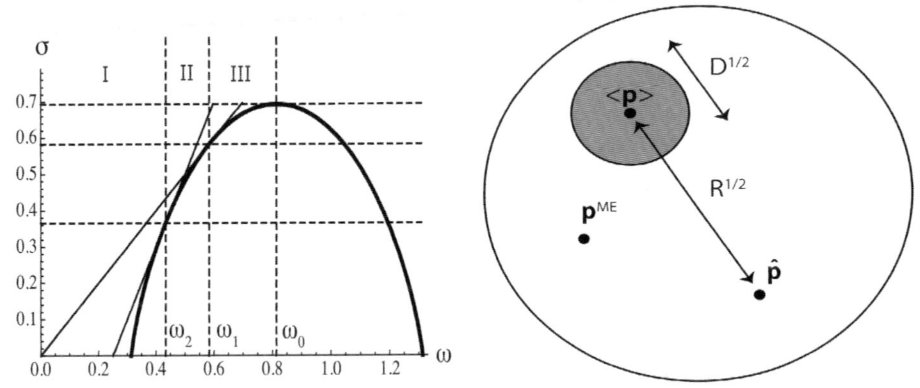

Figure 1. (Left) The entropy curve of the ISM with $H = 0.5$. There are three characteristic values of ω_k with $k = 0, 1, 2$, each of which is defined as a tangential point of the slope k (tangential lines are denoted by solid straight lines). (Right) A schematic diagram of the phase space. The large circle represents the set of all good probability vectors. The shaded area represents the typical fluctuating region of $\boldsymbol{p}$. The target distribution exists apart from $\langle \boldsymbol{p} \rangle$ by $\sqrt{R}$. The ME distribution $\boldsymbol{p}^{\mathrm{ME}}$ also exists somewhere inside the large circle.

where the angular brackets $\langle \cdots \rangle$ denote the average over $\rho(\boldsymbol{p})$. The physical significance of each order parameter is clear: Q measures the averaged square distance between $\boldsymbol{p}$ and $\hat{\boldsymbol{p}}$. R also quantifies the distance between the target distribution and the averaged $\boldsymbol{p}$ and is similar to Q, but R's meaning is clearer since it does not include the fluctuation of $\boldsymbol{p}$. On the other hand, D purely measures the fluctuation of $\boldsymbol{p}$ around the averaged value. We depict the phase space of good probability vectors with the order parameters in the right panel of Fig. 1.

The control and order parameters scale exponentially with N in the large N limit

$$M = e^{N\alpha},\ \Gamma = e^{N\gamma},\ Q = e^{Nq},\ R = e^{Nr},\ D = e^{Nd}. \tag{18}$$

Here we plot q, r and d of the ISM against α and γ in Fig. 2. Clearly $q = r$ holds, which is

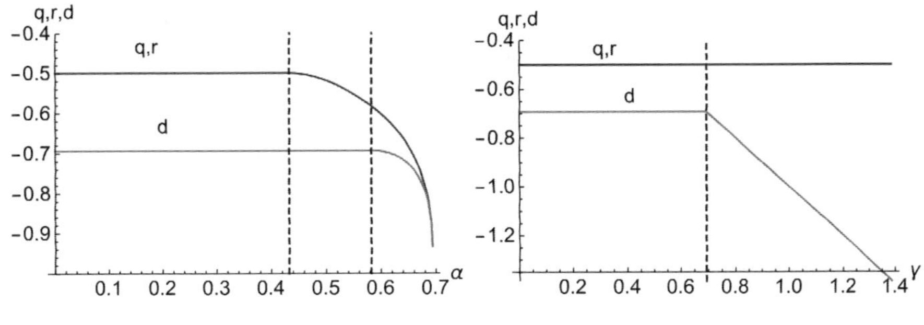

Figure 2. Plots of q, r and d against α at $\Gamma = 0$ (left) and against γ at a small value of α (right). Dotted vertical lines denote the transition points.

always true in the random case. Our crucial observation about the ME principle is that q (or r) does not depend on γ, that is, on the strength of the entropy bias at all as seen in Fig. 2, which is the case for all values of α. This means that the distances from the target distribution to the ME one and to a typical (randomly chosen) 'good' distribution are equal. Thus the ME does not help to infer the target distribution. This is the consequence of the random nature of the observables.

An interesting finding in the analysis of the random case was the existence of the so-called 'learning edge'. The learning edge ω_E is the scale separating accurately inferred values of $\omega_{\boldsymbol{s}}$ (see eq. (12)) from the ones which cannot be inferred: for target probability values with $\omega_{\boldsymbol{s}} \leq \omega_E$ we can infer those values, and for $\omega_{\boldsymbol{s}} > \omega_E$ we cannot. The learning edge is a strictly monotonically increasing function of α (see Fig. 3 as an example for the ISM), whose behaviour reveals the existence of phase transitions. When the learning edge takes values of ω_k with $k = 0, 1, 2$, the

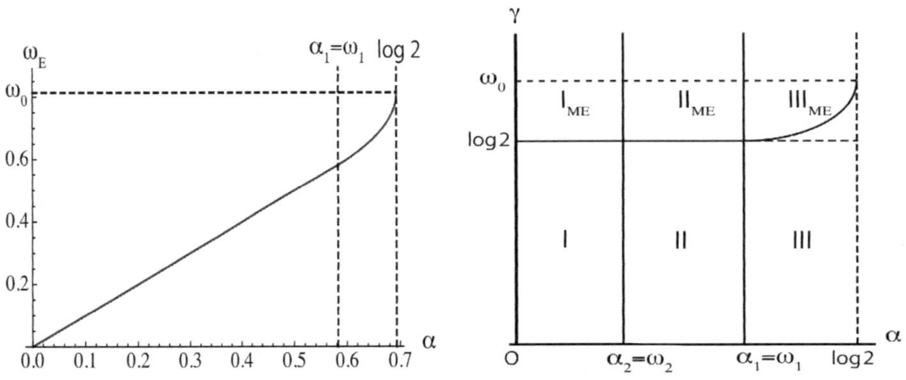

Figure 3. (Left) Plot of the learning edge ω_E against α at $\Gamma = 0$ for the ISM with $H = 0.5$. The dotted lines represent the transition points of the learning edge. At $\alpha = \log 2$, all the values of target probabilities are learned, implying the learning edge is not well defined for $\alpha > \log 2$. (Right) A shematic phase diagram of the random case. Each phase is named after the location of the learning edge.

phase transitions occur. We call each phase by the location of the learning edge: if $\omega_E < \omega_2$ then the phase is I, and phases II and III are defend as well (see the left panel of Fig. 1). The resultant phase diagram is given in the right panel of Fig. 3. In the phases subscripted by ME, the order parameter d is dominated by the ME bias Γ in contrast to the phases without the subscript.

An important finding of the calculation in the random case is that the marginal density of a small number (compared to 2^N) of probabilities of configurations factorizes a product of single-configuration probability densities $\rho(\boldsymbol{p}) \approx \prod_{\boldsymbol{s}} \rho_{\boldsymbol{s}}(p_{\boldsymbol{s}})$. Unfortunately, this nice property, at the origin of the existence of the learning edge, does not hold for the case of smooth observable, which makes the analysis much more complicated.

3. Numerical studies

3.1. Algorithm for sampling the probability distribution space

We briefly summarize our Monte Carlo (MC) algorithm. We randomly change the probability vector from $\boldsymbol{p}$ to $\boldsymbol{p}' = \boldsymbol{p} + \Delta\boldsymbol{p}$ in the phase space to perform sampling of the space of all distributions. The change of the vector, $\Delta\boldsymbol{p}$, must satisfy the following conditions

Orthogonality ($E = 0$)**:** $\Delta\boldsymbol{p} \cdot \boldsymbol{v}^\mu = 0, \forall\mu$.

Normalization: $\sum_{\boldsymbol{s}} \Delta p_{\boldsymbol{s}} = \Delta\boldsymbol{p} \cdot \mathbf{1} = 0$ where $\mathbf{1} = (1, 1, \cdots, 1)^t$ and the symbol ^{t} denotes the transpose.

Journal of Physics: Conference Series **638** (2015) 012018 doi:10.1088/1742-6596/638/1/012018

Positivity: $p_{\boldsymbol{s}} + \Delta p_{\boldsymbol{s}} \geq 0, \forall \boldsymbol{s}$.

Combining the orthogonality with the initial condition $\boldsymbol{p}(0) = \hat{\boldsymbol{p}}$, we see that the vector chosen in this algorithm always satisfies the constraints $\boldsymbol{v}^\mu \cdot \boldsymbol{p} = \boldsymbol{v}^\mu \cdot \hat{\boldsymbol{p}}, \forall \mu$, which is needed since we are basically interested in $E = 0$. These orthogonality and normalization conditions restrict the possible move directions. On the other hand, the positivity condition is maintained by appropriately choosing the move width x. With a chosen move direction $\boldsymbol{w}$, the minimum and maximum values of x are written as

$$x_{\max} = \max_{\boldsymbol{s}} \min \left(\frac{-p_{\boldsymbol{s}}}{\omega_{\boldsymbol{s}}}, \frac{1 - p_{\boldsymbol{s}}}{w_{\boldsymbol{s}}} \right), \; x_{\min} = \min_{\boldsymbol{s}} \max \left(\frac{-p_{\boldsymbol{s}}}{\omega_{\boldsymbol{s}}}, \frac{1 - p_{\boldsymbol{s}}}{w_{\boldsymbol{s}}} \right). \tag{19}$$

The intermediate values between these two bounds are randomly chosen with equal weight, and thus the trial move becomes $\Delta \boldsymbol{p} = x\boldsymbol{w}$. Finally, we calculate the entropy difference $\Delta S = S(\boldsymbol{p}') - S(\boldsymbol{p})$ and determines whether the move $\boldsymbol{p} \to \boldsymbol{p}'$ is accepted or not by the probability

$$p_{\text{accept}} = \min \left(1, e^{\Gamma \Delta S}\right). \tag{20}$$

3.1.1. Observables drawn from the Fourier basis In the analytical treatment, we put $\boldsymbol{v}$ as a random vector drawn from eq. (8), each component of which can take a continuous value. However for numerical implementation, it is more convenient to choose observables from a set of vectors constituting a complete orthogonal set, because the above orthogonal and normalization conditions can be easily satisfied in that way.

We choose the Fourier basis as such a set. The random case corresponds to the situation where all the modes (except for $\boldsymbol{w}^{\boldsymbol{0}}$) are drawn uniformly at random. In the studies reported below, we bias the weights of the Fourier modes based on the values of corresponding wavenumber $k = |\boldsymbol{k}|$, see eq. (9).

3.1.2. Lessons from the random case: equilibration and finite-size effects We define one MC step by one trial of changing the probability vector $\boldsymbol{p}$. As reported in [11], the total MC steps for sufficient equilibration and sampling grow fast as the system size increases. For example for $N = 10$, we need $O(10^8)$ MC steps typically, which requires a couple of days. Due to this expensiveness, we mainly run our simulations for sizes less than or equal to $N = 10$.

This system size might sound small, but our numerical simulation on the random case showed that even for $N = 10$ numerical result could be quantitatively compared to the analytical one. As a demonstration, we display the order parameters of the random case calculated by the MC simulation for finite N in Fig. 4, which exhibits a clear indication of the phase transitions even for size $N = 10$. This result in the random case provides a guide for simulating more general cases. Our simulations given below are based on the equilibration criterion invented in the random case, and we expect that the treated sizes $N = 8$ or 10 capture the behavior in the large-size-system limit.

3.2. Exponentially-decaying weight and the smoothness parameter
Here we assume that each mode of the Fourier basis is chosen with a probability weight $e^{-b|\boldsymbol{k}|}$, and we call b smoothness parameter. The pseudo code to construct the set of observables is as follows:

(i) Prepare a set of Fourier modes consisting of all the modes except for $|\boldsymbol{k}| = 0$. We call this set "pool". Similarly one prepares an empty set, called "observables".

(ii) Calculate the normalization constant as $Z = \sum_{\boldsymbol{k} \in \text{pool}} e^{-b|\boldsymbol{k}|}$.

(iii) Choose one mode of the wavenumber vector $\boldsymbol{k}$ from the pool with probability $p_{\boldsymbol{k}} = e^{-b|\boldsymbol{k}|}/Z$ and put the chosen mode into the observables set. Remove the mode from the pool.

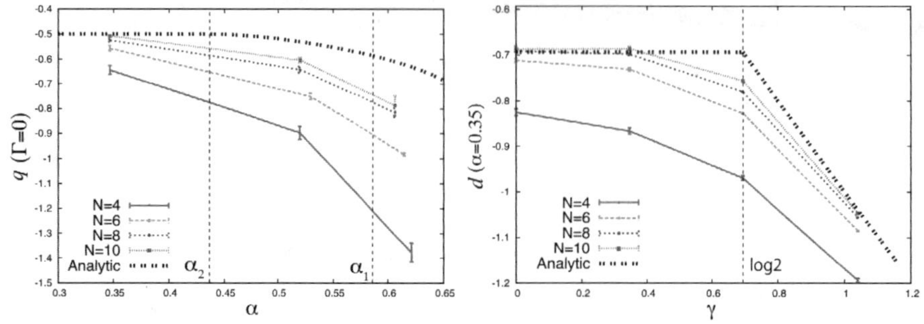

Figure 4. Numerically-evaluated order parameters q (left) and d (right) of the ISM with $H = 0.5$ plotted against α (left) and γ (right). Analytical predictions are given by black curves, which are well consistent with the numerical data. Clear indication of the phase transition from small to large Γ is seen in the right panel, and thus finite-size effects seem to be not so large.

(iv) Continue **(ii)**-**(iii)** until the size of the observables set becomes equal to M. At the end, what is left in the pool is our set of of possible MC-move directions.

We have to remove $\boldsymbol{w}^{\mathbf{0}} = \mathbf{1}/\sqrt{2^N}$ from the beginning since it is connected to the normalization condition and can neither be a nontrivial observable nor a MC-move direction. The exponentially-decaying weight in **(iii)** corresponds to putting $V_{\boldsymbol{k}}^{-1} = 2b|\boldsymbol{k}|$ in eq. (9).

As for the target model, we adopt the Boltzmann distribution $\hat{p}_{\boldsymbol{s}} = e^{-\beta\mathcal{H}}/Z$ generated from the p-spin Hamiltonian

$$\mathcal{H} = -\sum_{i_1<i_2<\cdots<i_p} J_{i_1 i_2\cdots i_p} s_{i_1} s_{i_2} \cdots s_{i_p}, \tag{21}$$

where we assume $J_{i_1 i_2\cdots i_p}$ is an i.i.d. random variable drawn from $\mathcal{N}\left(0, \frac{p!}{N^{p-1}}\right)$. Hereafter we call this model p-spin model (pSM). Parameter p measures the "smoothness" of the target distribution. In the large-p limit (after the large-N limit), this model converges to the random energy model (REM)[1]. The REM is a completely non-smooth model and we expect the smoothness of observables will not help inference of this model at all. On the other hand, if p is enough small, we expect that the ME distribution at large values of b preferentially has lower-spin multiplets in the effective Hamiltonian and matches to the target one, leading to an excellent performance in inference. Hence it is meaningful to compare different-p results for examining the effect of smoothness in the observables.

We below see the MC results with changing the parameters N, M, b, p and Γ. The quality of inference is quantified by the order parameter q as introduced in the random case. Error bar is estimated from the standard deviation σ_{sample} of the objective quantity among different samples as

$$\text{Error bar} = \frac{\sigma_{\text{sample}}}{\sqrt{N_{\text{sample}} - 1}}. \tag{22}$$

where N_{sample} is the number of simulated samples and are chosen to be 10 in most of the below results.

[1] The REM is defined by the Boltzmann distribution whose energies $\{E_{\boldsymbol{s}}\}_{\boldsymbol{s}}$ are drawn from an i.i.d. normal distribution $\mathcal{N}(0, 2/N)$ (see reprints in [12] for details).

3.2.1. Result in the smooth case We plot q of the 2SM in Fig. 5. We see that the larger b tends to give smaller q meaning better inference. Especially for $M = 64$ and 128 with enough large b, we find that q becomes smaller and smaller as Γ grows, implying that the ME distribution given in $\Gamma \to \infty$ realizes the perfect learning $q \to \infty$. Similar observation is found for the 3SM in Fig.

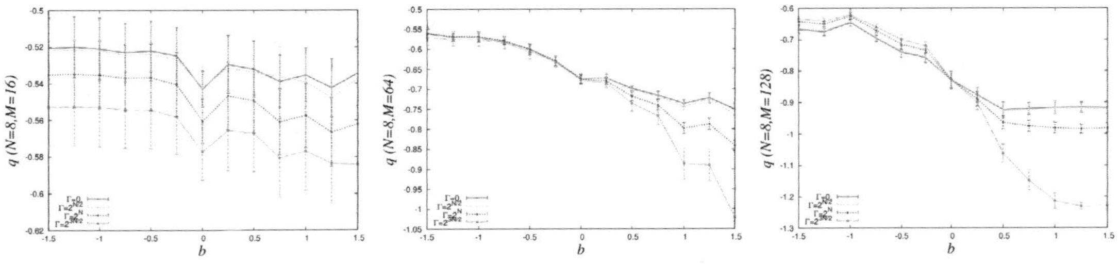

Figure 5. Plots of q for the $N = 8$ 2SM against b, for $M = 16$, 64, and 128 from left to right. Different colors correspond to different values of Γ.

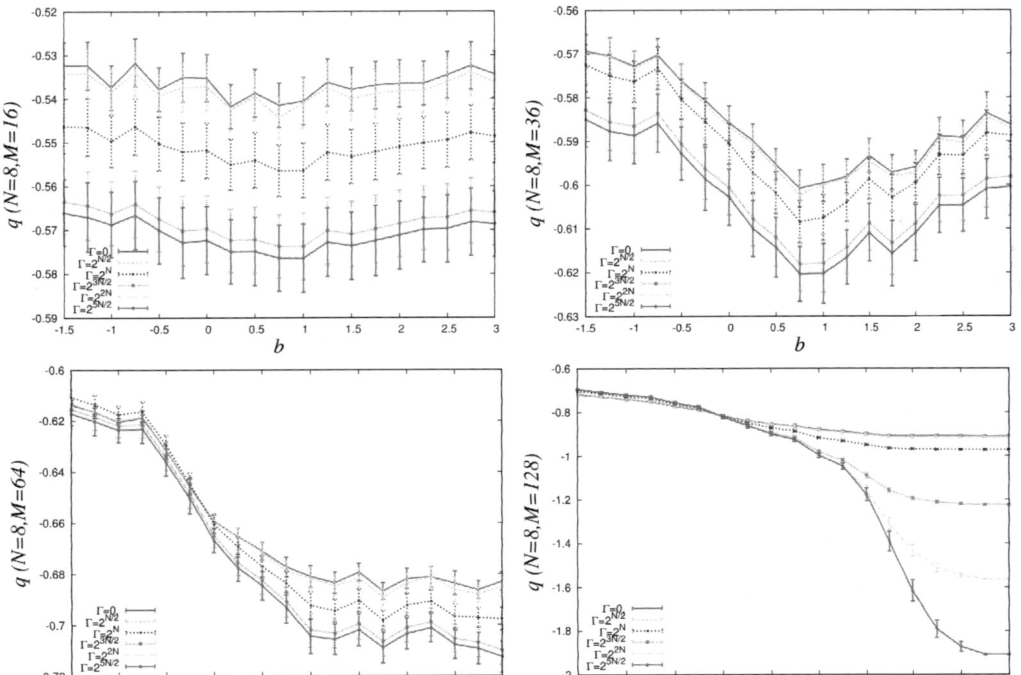

Figure 6. Plots of q for the $N = 8$ 3SM vs. b, for $M = 16, 36, 64$, and 128 from left to right and from upper to lower. The curve for $M = 36$ (right upper) shows a dip around $b = 0.8$, implying the existence of an optimal value of b.

6. In this case, we calculate a wider range values of b and Γ. The results of $M = 16, 64$, and 128 are very similar to the $2SM$ case. A characteristic dip is observed in $M = 36$, implying the existence of an optimal value of b, though the difference from other values of b is not so strong.

Journal of Physics: Conference Series **638** (2015) 012018 doi:10.1088/1742-6596/638/1/012018

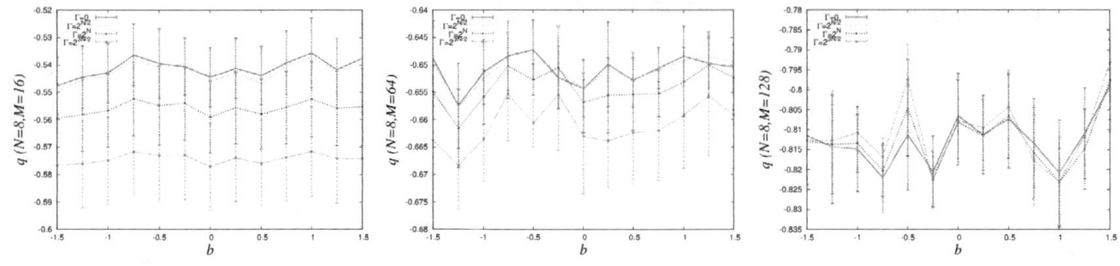

Figure 7. Plots of q for the $N = 8$ REM vs. b, for $M = 16$, 64, and 128 from left to right.

Fig. 7 shows the result for the REM. We see no clear dependence of q on b for any value of M, meaning that the smoothness does not help the inference at all, as expected.

The above results for the 2SM, 3SM, and REM clearly exhibit a certain compatibility between the observables and the target distribution. For further examining this, the 3SM result, especially the existence of an optimal value of b, can be a good starting point. To interpret this result, we introduce $N_k(\{\boldsymbol{v}_\mu\}_\mu)$ representing the number of modes with the wavenumber k in a given set of observables $\{\boldsymbol{v}_\mu\}_\mu$, and define the average occupation ratio of the k-modes by

$$u_k = [N_k(\{\boldsymbol{v}_\mu\}_\mu)]_b \binom{N}{k}^{-1}, \tag{23}$$

where the brackets $[\cdots]_b$ denote the average over the distribution with the smoothness b stated above. This quantity is easy to calculate because this is independent of the target distribution and moreover the MC simulation is not needed. The occupation ratios for $M = 16, 36, 64$, and 128 are shown in Fig. 8. A remarkable quantity for examining the 3SM result is u_3 which shows the following behavior

- For $M = 36$, it exhibits a clear peak around $b = 0.8$.
- For $M = 64$, it increases monotonically as b grows up to $b = 4$.
- For $M = 128$, it saturates to unity for $b \gtrsim 2$.

These observations are probably connected to the characteristic behavior of q: the dip for $M = 36$, the monotonic decrease of q with b for $M = 64$, and the significant decrease of q as Γ grows at large values of b for $M = 128$. It is also interesting to see how q behaves as Γ changes when $u_3 = 1$ is satisfied, which is demonstrated in the right panel of Fig. 9. For $b = 3$, where $u_3 = 1$, q seems to decrease as a linear function of γ in the large γ region, while at small values of b the dependence on γ is quite weak and can become positive (decreasing q) and negative (increasing q). For comparison, we also give the same plot of $M = 64$ in the left panel, in which no significant decreasing of q occur.

3.2.2. Consequences of the smooth-case results. The above results certainly demonstrate some situations where the ME distribution, and more generally, the introduction of an entropic bias helps to infer the target distribution. However, the interpretation based on u_k implies the close connection between target distribution and observables. If we know the "smoothness" of a target distribution, we can determine the appropriate observables for applying to the target system, but in practical situations the nature of the target distribution is unclear.

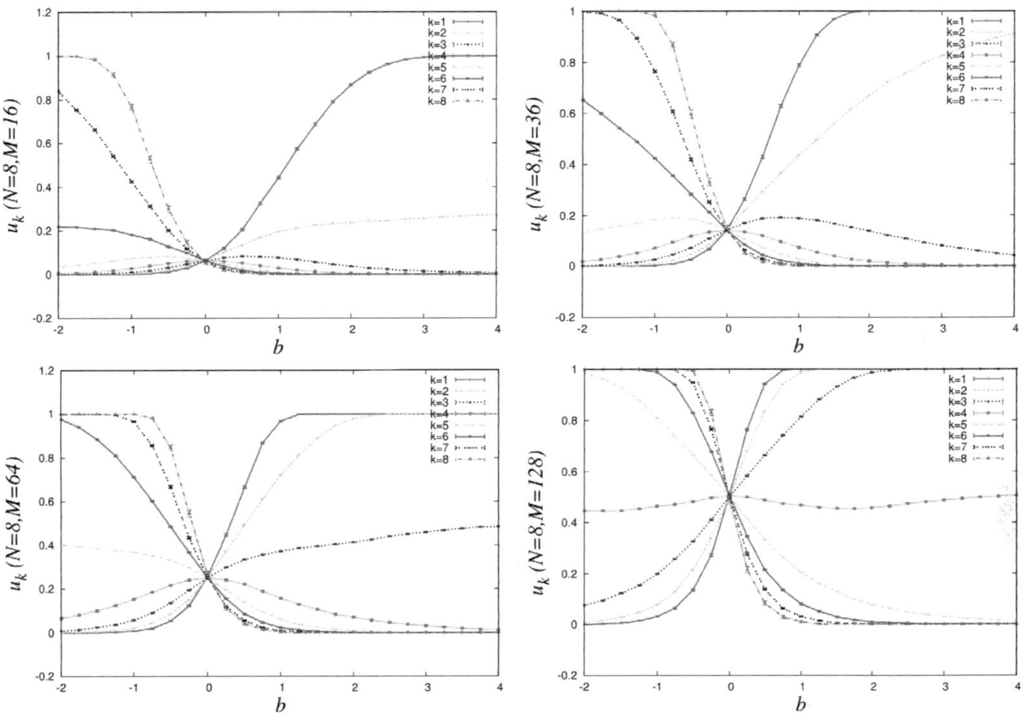

Figure 8. The occupation ratio of Fourier modes with wavenumber k for $M = 16, 36, 64$, and 128 in $N = 8$. Different colors correspond to different k.

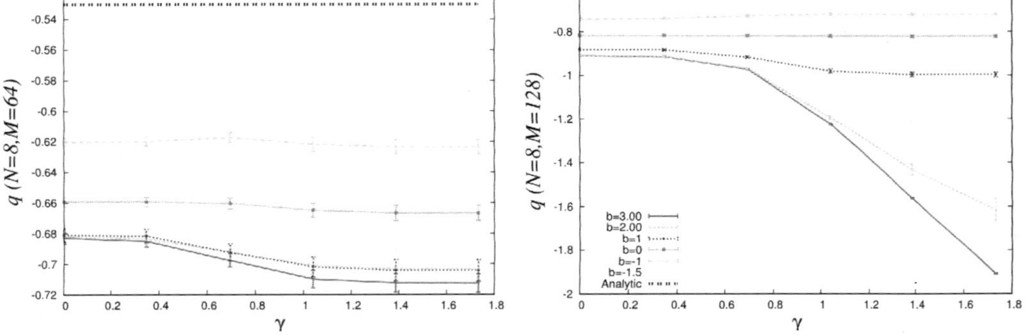

Figure 9. Plots of q for the $N = 8$ 3SM against γ, for $M = 64$ (left), and 128 (right). Black dotted lines denote the analytic solution derived in the random case. For $M = 128$ at $b = 3$, q seems to decrease linearly as γ grows at large values of γ.

3.3. On control of the ratio of "correct" observables.

Here, we examine a quantitative relation between q and the ratio of "correct" observables. A "correct" observable here is such that the corresponding interaction in the target energy, see eq. (21), is nonzero. For example in the 2SM, only observables with the wavenumber $k = 2$ are correct. Let us consider the 2SM and $M = \binom{N}{2}$ observables. We assume that the observables are a mixture only of $k = 2$ and 3 modes. We treat u_2 as a control parameter on this setup, in contrast to the smooth case where u_k is the function of b and M.

In Fig. 10, we give the plot of q against u_2 of the $N = 10$ 2SM. To examine the size dependence of q, we also show the u_2-q plots for $N = 6, 8$ and 10 at $\Gamma = 0$ and $2^{5N/2}$ in Fig. 11. In Fig.

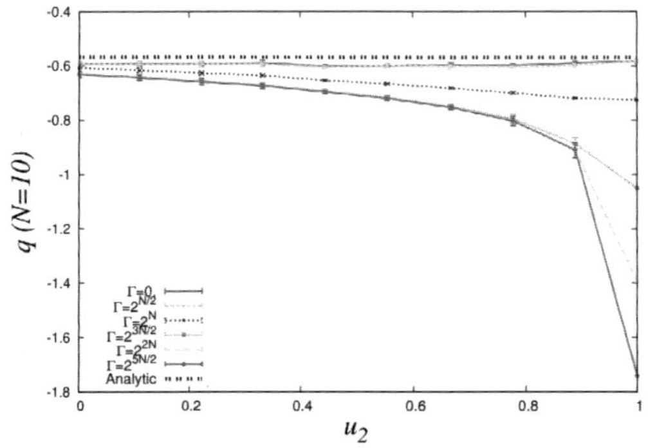

Figure 10. The order parameter q is plotted against u_2 in the $N = 10$ 2SM. Different colors correspond to different Γ values. At u_2=1, q will go to $-\infty$ in the limit $\Gamma \to \infty$.

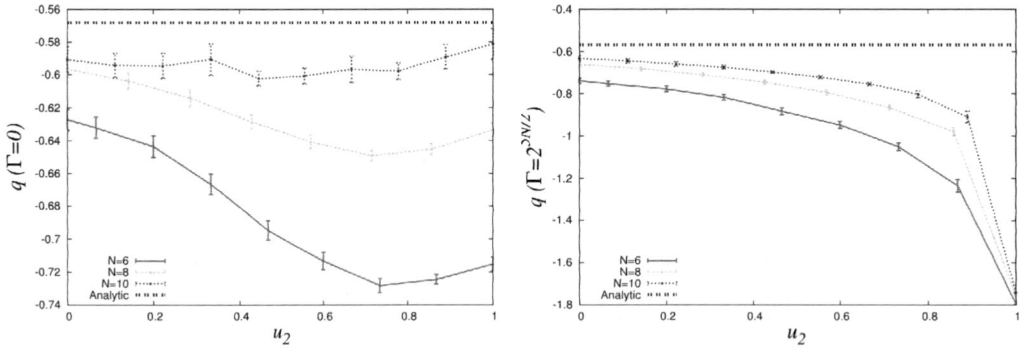

Figure 11. The u_2-q plots at $\Gamma = 0$ (left) and $2^{5N/2}$ (right). Different colors represent different system sizes. Black dotted lines denote the analytic solution in the random case.

10, we see that q decreases as u_2 increases at large values of Γ. Especially at $u_2 = 1$, the ME distribution appearing in the limit $\Gamma \to \infty$ recovers the target distribution and thus q should go to $-\infty$ in that limit. This tendency is actually observed in Fig. 10. On the other hand from Fig. 11, at $\Gamma = 0$ the u_2-q curves seem to converge to the analytical solution in the random case for any u_2 as the system size grows. A similar behaviour is expected even for large Γ if u_2 is not too large. These observations suggest to focus on u_2 close to, but slightly smaller than 1.

Comparing the size dependence of q at $u_2 = 1$ and around $u_2 = 0.9$ in Fig. 11, we find that the size dependence at $u_2 = 1$ seems to converge already for $N = 10$ and q in the limit $N \to \infty$ is expected to be different from the random-case one, while the one around $u_2 = 0.9$ is still significant and q seems to converge to the random-case value as N grows. This observation yields a pessimistic description: in the $N \to \infty$ limit, only the $u_2 = 1$ point is special and for all other values of $u_2 < 1$ the order parameter q converges to the random-case value irrespectively of the ME bias Γ.

These results imply that the ME principle does not work even as an approximation, at least in the present scale of the analysis, unless we exactly know what kind of interactions are needed to describe the target distribution. This consequence is only based on a limited result, and hence in the next subsection we reexamine it from another viewpoint.

3.4. Comparison of low- and high-$|\boldsymbol{k}|$ modes

As stated above, in realistic situations low-k modes are thought to be more important than high-k modes, and $k = 1$ and 2 modes are used in most applications. We now examine the possible difference in approximation performance between low-k and high-k modes. We choose 3SM as the target system. Due to computational limitations, we fix the system size to $N = 8$. The number of observables is fixed as $M = 36$ which is equal to $\binom{N}{1} + \binom{N}{2}$ with $N = 8$. Hence, we can include all $k = 1$ and 2 modes, the common choice in the applications, in a set of observables. Let us label this set by $a = 1$. We also define sets of observables labelled by $a = 2, 3, 4$, and 5. The sets $a = 2, 3$, and 4 consists only of $k = 3, 4$, and 5 modes, respectively, and the $a = 5$ one is a mixture of all the $k = 6$ and 7 modes. We compare the values of q among those different sets. Since the target system is the 3SM, the set $a = 2$ ($k = 3$) gives a better performance than the others, which is trivial. An interesting possibility is whether the results for $a = 1$ ($k = 1, 2$) are better than the ones for $a = 3, 4$, and 5. If so, it could give support to why the ME distribution constructed with $k =$ 1- and 2-observables works well in many applications. Unfortunately, we do not see any obvious difference in performance between $a = 1$ and $a \geq 3$ in Fig. 12.

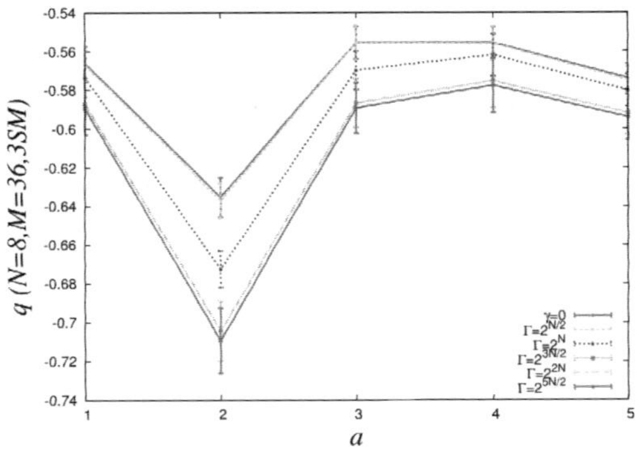

Figure 12. The order parameter q is plotted against u_2 in the $N = 10$ 2SM. Different colors correspond to different Γ values. At u_2=1, q will go to $-\infty$ in the limit $\Gamma \to \infty$.

4. Analytical treatment of the smooth-observables case: some results

In this section, we provide an analytical solution in some limited cases. We start from stating some assumptions and approximations for making the analysis amenable.

Journal of Physics: Conference Series **638** (2015) 012018 doi:10.1088/1742-6596/638/1/012018

4.1. Assumptions and approximations

4.1.1. Prior over the target distribution and the annealed approximation. We have introduced the prior distribution of observables (8) with the matrix M_{st} which controls the smoothness of observables. Similarly, we here introduce the prior over the target distribution. This was irrelevant in the random case due. In the presence of nontrivial M_{st}, however, we have to treat correlations among different spin configurations, which requires to specify the properties of the target distribution. The prior distribution of the target distribution is chosen as the following quadratic form

$$\hat{P}(\hat{\boldsymbol{p}}) = \hat{C}\prod_{\boldsymbol{s}}\theta(\hat{p}_{\boldsymbol{s}})\int_{-i\infty}^{i\infty} d\hat{\Lambda}\; e^{\hat{\Lambda}\sum_{\boldsymbol{s}}(\hat{p}_{\boldsymbol{s}}-1)-\frac{1}{2}\sum_{\boldsymbol{s},\boldsymbol{t}}\hat{M}_{\boldsymbol{st}}\hat{p}_{\boldsymbol{s}}\hat{p}_{\boldsymbol{t}}}. \tag{24}$$

The matrix $\hat{M}_{\boldsymbol{st}}$ controls the smoothness of the target distribution as $M_{\boldsymbol{st}}$ in eq. (8). The integration variable Λ keeps the normalization condition $\sum_{\boldsymbol{s}}\hat{p}_{\boldsymbol{s}} = 1$. The factor $\hat{C}$ is just the normalization.

As eq. (10) in the random case, we calculate the averaged logarithm of the partition function, F, by using the replica method with the RS ansatz. After some calculations, we get

$$\log\,[V^n]_{\text{smooth}} = \operatorname*{Extr}_{\Omega}\Bigg\{-\frac{M}{2}\left\{\log\left(1+\frac{Q+(n-1)R)}{E}\right)+(n-1)\log\left(1+\frac{Q-R}{E}\right)\right\}$$
$$+\frac{1}{2}nQ(Q'-R')-\frac{1}{2}n(n-1)RR'+\log\int\frac{\sqrt{\det M_{\boldsymbol{st}}}}{\sqrt{2\pi}^{2N}}\left(\prod_{s}dz_s\right)e^{-\frac{1}{2}\sum_{\boldsymbol{s},\boldsymbol{t}}M_{\boldsymbol{st}}z_{\boldsymbol{s}}z_{\boldsymbol{t}}}X,\Bigg\} \tag{25}$$

where the brackets $[\cdots]_{\text{smooth}}$ denote the average over eqs. (8,24), Ω represents the set of order parameters $\Omega = \{Q, Q', R, R', \Lambda, \Lambda'\}$, and

$$X = \int_0^\infty \prod_{\boldsymbol{s}} d\hat{p}_{\boldsymbol{s}}\hat{P}(\hat{\boldsymbol{p}})Y^n, \tag{26}$$

$$Y = \int_0^\infty \prod_{\boldsymbol{s}} dp_{\boldsymbol{s}} \exp\Bigg\{\sqrt{R'}\sum_{\boldsymbol{s}} z_{\boldsymbol{s}}(p_{\boldsymbol{s}}-\hat{p}_{\boldsymbol{s}}) - \Lambda\sum_{\boldsymbol{t}}(p_{\boldsymbol{t}}-\hat{p}_{\boldsymbol{t}})$$
$$-\frac{1}{2}Q'\sum_{\boldsymbol{s},\boldsymbol{t}} M_{\boldsymbol{st}}^{-1}(p_{\boldsymbol{s}}-\hat{p}_{\boldsymbol{s}})(p_{\boldsymbol{t}}-\hat{p}_{\boldsymbol{t}}) - \Gamma\sum_{\boldsymbol{s}} p_{\boldsymbol{s}}\log p_{\boldsymbol{s}}\Bigg\}. \tag{27}$$

Now, we should in principle take the quenched limit $n \to 0$, but unfortunately, the computation in this limit is far from being an easy task. Instead, we work in the annealed case $n = 1$, and the resultant formula becomes much simpler

$$\log\,[V]_{\text{smooth}} = \operatorname*{Extr}_{\Omega'}\left\{-\frac{M}{2}\log\left(1+\frac{Q}{E}\right)+\frac{1}{2}QQ'+\log X_{\text{anneal}}\right\}, \tag{28}$$

where $\Omega' = \{Q, Q', \Lambda, \Lambda'\}$

$$X_{\text{anneal}} = \int_0^\infty \prod_{\boldsymbol{s}} d\hat{p}_{\boldsymbol{s}}\hat{P}(\hat{\boldsymbol{p}})\int_0^\infty \prod_{\boldsymbol{s}} dp_{\boldsymbol{s}}\exp\Bigg\{-\Lambda\sum_{\boldsymbol{t}}(p_{\boldsymbol{t}}-\hat{p}_{\boldsymbol{t}})$$
$$-\frac{1}{2}Q'\sum_{\boldsymbol{s},\boldsymbol{t}} M_{\boldsymbol{st}}^{-1}(p_{\boldsymbol{s}}-\hat{p}_{\boldsymbol{s}})(p_{\boldsymbol{t}}-\hat{p}_{\boldsymbol{t}}) - \Gamma\sum_{\boldsymbol{s}} p_{\boldsymbol{s}}\log p_{\boldsymbol{s}}\Bigg\}. \tag{29}$$

Journal of Physics: Conference Series **638** (2015) 012018 doi:10.1088/1742-6596/638/1/012018

4.1.2. A simple choice of $M_{\boldsymbol{st}}$ and $\hat{M}_{\boldsymbol{st}}$. As noted above, $M_{\boldsymbol{st}}$ should only depend on the overlap $\sum_i s_i t_i$. Here we consider one of the simplest choice obeying this constraint:

$$M_{\boldsymbol{st}}^{-1} = \frac{e^{B \sum_i s_i t_i}}{(2\cosh B)^N}. \tag{30}$$

The parameter $B \in \mathbb{R}$ controls the smoothness of observables and the random case is reproduced in the limit $B \to \infty$. Note that B is defined from the inverse of $M_{\boldsymbol{st}}$. The matrix $M_{\boldsymbol{st}}$ itself can be written in the same form, but the parameter corresponding B becomes complex in general.

How about the matrix $\hat{M}_{\boldsymbol{st}}$? We do not want to bias a specific spin and thus $\hat{M}_{\boldsymbol{st}}$ should have the permutation symmetry of the spin index, meaning that $\hat{M}_{\boldsymbol{st}}$ is a function of the overlap $\sum_i s_i t_i$ and the two magnetizations $\sum_i s_i$ and $\sum_i t_i$ only. The symmetry between the two magnetizations should be also maintained. Here we choose

$$\hat{M}_{\boldsymbol{st}} = \hat{C} e^{\hat{B} \sum_i s_i t_i + \hat{H}(s_i + t_i)}. \tag{31}$$

As seen below, this functional form leads to the ISM as the target distribution.

4.1.3. A saddle-point approximation of the integration over $\boldsymbol{p}$. A notable point of the present formulation is that all the variables obey the exponential scale. Thanks to this, we can approximate the integration over $\boldsymbol{p}$ by a saddle-point (SP) technique. Actually, this approximation gives the exact result, which was demonstrated in [11] in the random case. We expect the same is true in the present case. Hence, instead of integrating out over $\boldsymbol{p}$ and $\hat{\boldsymbol{p}}$, we take the SP condition on the exponent of X_{anneal}. The result is

$$-\Lambda - Q' \sum_{\boldsymbol{t}} M_{\boldsymbol{st}}^{-1} p_{\boldsymbol{t}} + Q' \sum_{\boldsymbol{t}} M_{\boldsymbol{st}}^{-1} \hat{p}_{\boldsymbol{t}} - \Gamma(\log p_{\boldsymbol{s}} + 1) = 0, \tag{32}$$

$$\hat{\Lambda} + Q' \sum_{\boldsymbol{t}} M_{\boldsymbol{st}}^{-1} (p_{\boldsymbol{t}} - \hat{p}_{\boldsymbol{t}}) - \sum_{\boldsymbol{t}} \hat{M}_{\boldsymbol{st}}^{-1} \hat{p}_{\boldsymbol{t}} = 0, \tag{33}$$

Taking the extremization condition with respect to Q in eq. (28), we see Q' is proportional to M. Hence, for small M, we find that the SP solution of $\hat{\boldsymbol{p}}$ gives back the ISM

$$\hat{p}_{\boldsymbol{s}}^* = \hat{\Lambda} \sum_{\boldsymbol{t}} \hat{M}_{\boldsymbol{st}} = \frac{e^{H^* \sum_i s_i}}{(2\cosh H^*)^N}, \tag{34}$$

where H^* is a function of $\hat{B}$ and $\hat{H}$ in eq. (31). We expect the smallness of the second term in eq. (33) to hold for general M. Therefore, we hereafter choose the ISM as the target distribution. This is natural since the target distribution should not be changed as M increases. Thanks to this assumption, we can concentrate on solving only eq. (32) for $\hat{\boldsymbol{p}}$ defined by the ISM, see eq. (13).

Note that variable $p_{\boldsymbol{s}}$ is a probability value and cannot be negative, a fact ignored in deriving eq. (32). We give the name "SAT" to the set of spin configurations $\mathbf{s}$ whose probability values derived from the SP equation becomes positive, and "UNS" to the other configurations. Thus the correct SP equation is

$$\begin{cases} -\Lambda - Q' \sum_{\boldsymbol{t} \in \text{SAT}} M_{\boldsymbol{st}}^{-1} p_{\boldsymbol{t}} + Q' \sum_{\boldsymbol{t}} M_{\boldsymbol{st}}^{-1} \hat{p}_{\boldsymbol{t}} - \Gamma(\log p_{\boldsymbol{s}} + 1) = 0, & (\boldsymbol{s} \in \text{SAT}) \\ p_{\boldsymbol{s}} = f_{\boldsymbol{s}}(\Gamma), & (\boldsymbol{s} \in \text{UNS}) \end{cases} \tag{35}$$

This is the extension of the modified SP approximation used in [11]. For UNS, the function $f_{\boldsymbol{s}}(\Gamma)$ is expected to be a simple function of Γ. For example in the random case, $f_{\boldsymbol{s}}(\Gamma) = 0$ for $0 \le \Gamma < 2^N$ and $f_{\boldsymbol{s}}(\Gamma) = 1/2^N$ for $\Gamma \ge 2^N$.

Journal of Physics: Conference Series **638** (2015) 012018 doi:10.1088/1742-6596/638/1/012018

The difficulty in solving this set of coupled equations is the classification of the SAT and UNS configurations. In the random case $B \to \infty$, target probabilities are learned in descending order of their values, and the structure in the spin-configuration space is completely irrelevant: the classification of the SAT and UNS configurations is reduced to the self-consistent determination of the boundary value, namely, the learning edge, between the learned, large probabilities and the unlearned, small ones. However in the present case, due to the presence of the nontrivial matrix M_{st}, we have to treat the spin-configuration space itself, and the classification of the SAT and UNS configurations is very complicated.

In the following subsection, we assume that the learning edge can be uniquely determined for finite but still enough large B, which enables us to conduct the analysis. The applicable range of this assumption is also discussed.

4.2. Perturbation from large B at $\Gamma = 0$

In the discussion below, we put $\Gamma = 0$ in eq. (76) and use the following truncation with a fixed $\Xi \in \mathbb{N}$

$$\sum_{t} M_{st}^{-1}(\cdots) \approx \sum_{\{t|Q_{st}=N,N-2,\cdots,N-2\Xi\}} M_{st}^{-1}(\cdots), \tag{36}$$

where $Q_{st} = \sum_i s_i t_i$ and Ξ is the truncation range. Let us write the solution of the SP equation as

$$p_{s}^{*} = \hat{p}_{s} - \frac{\Lambda}{Q'} + \Phi_{s}. \tag{37}$$

For $s \in \mathrm{SAT}$ the residual term Φ_s is zero at the random case and we expect Φ is small for enough large B, while for $s \in \mathrm{UNS}$ $\Phi_s = \Lambda/Q' - \hat{p}_s$ since $p_s^* = 0$.

4.2.1. The $\Xi = 1$ case. Let us put $T_B = e^{BN}/(2\cosh B)^N$. Remembering that our target distribution is the ISM and depends only on the magnetization $M_s = \sum_i s_i = Nm_s$, we get from the third term in the upper line of eq. (76)

$$\sum_{\{t|Q_{st}=N,N-2\}} M_{st}^{-1}\hat{p}_t = T_B\left\{\hat{p}_{M_s} + e^{-2B}\left(D_s\hat{p}_{M_s+2} + U_s\hat{p}_{M_s-2}\right)\right\}, \tag{38}$$

where $U_s(D_s)$ is the number of up (down) spins of the configuration s. The variables M_s, U_s, and D_s are extensive and proportional to N. Hence, to validate the above truncation, we need

$$\epsilon \equiv Ne^{-2B} \ll 1. \tag{39}$$

Inserting the solution form eq. (37) into eq. (76), we get

$$\epsilon\frac{(1-m_s)}{2}\Phi_{M_s+2} + \Phi_{M_s} + \epsilon\frac{(1+m_s)}{2}\Phi_{M_s-2} = \frac{\Lambda}{Q'}\frac{1-T_B(1+\epsilon)}{T_B}. \tag{40}$$

The term in the righthand side appears due to the modification from the normalization

$$0 = 1 - \sum_t M_{st}^{-1} \approx 1 - T_B(1+\epsilon). \tag{41}$$

So, we can neglect the righthand side of eq. (40).

M_s-dependent solution. As Φ = const is an inappropriate solution to eq. (40), we look for a M_s-dependent solution. To lighten notations, we change the index of $\hat{p}$ from the magnetization M_s into the number of up spins, U_s, and carry out the same change of notations on the other variables. The equation we want to solve is now

$$\epsilon(1-u_s)\Phi_{U_s+1}+\Phi_{U_s}+\epsilon u_s\Phi_{U_s-1}=0, \tag{42}$$

where $u_s = U_s/N$. The general solution of this equation is not simple due to the u_s-dependence of the coefficients. Thus we work on the large N limit and assume the scale $\Phi(U_s)=e^{N\phi(u_s)}$. Correspondingly,

$$\Phi_{U_s\pm 1}=e^{N\phi(u_s\pm 1/N)}\approx e^{N\phi(u_s)}e^{\pm\frac{\partial\phi}{\partial u}|_{u=u_s}}. \tag{43}$$

The SP equation becomes quadratic of $y=e^{\frac{\partial\phi}{\partial u}|_{u=u_s}}$ and the solution is

$$y=\frac{-1\pm\sqrt{1-4\epsilon^2u_s(1-u_s)}}{2\epsilon(1-u_s)}. \tag{44}$$

In the limit $\epsilon\to 0$, we expect $\Phi(U_s)=0$ if U_s, implying $y=0$. Thus we need to take the plus sign in eq. (44). Using this solution and integrating $\log y$ with respect to u, we get

$$\phi(u)=\phi_0+u\log\left\{\frac{-1+\sqrt{1-4\epsilon^2u(1-u)}}{2\epsilon(1-u)}\right\}-\frac{1}{2\epsilon}\log\left\{-\epsilon+2\epsilon u+\sqrt{1-4\epsilon^2u(1-u)}\right\}$$
$$+\frac{1}{2}\log\left\{1-2\epsilon^2+2u\epsilon^2+\sqrt{1-4\epsilon^2u(1-u)}\right\}\equiv\phi_0+\phi_c(u). \tag{45}$$

Since our target distribution is the ISM, there is the edge value of U, $U_E=Nu_E$, corresponding to the learning edge ω_E. The constant $\Phi_0=e^{N\phi_0}$ is determined from the boundary condition at $U_s=U_E$. We can put $\Phi_{U_E-1}=\hat{p}_{U_E-1}-\frac{\Lambda}{Q'}$, hence from the SP equation at $U_s=U_E$, we get

$$\Phi(u)=-y(u_E)\left(\hat{p}_{U_E-1}-\frac{\Lambda}{Q'}\right)e^{N(\phi_c(u)-\phi_c(u_E))}, \tag{46}$$

where we used a simple relation

$$\frac{2\epsilon u_E}{1+\sqrt{1-4\epsilon^2u_E(1-u_E)}}=-y(u_E). \tag{47}$$

The corresponding p^*_U thus becomes

$$p^*_U=\hat{p}_U-\frac{\Lambda}{Q'}-y(u_E)\left(\hat{p}_{U_E-1}-\frac{\Lambda}{Q'}\right)e^{N(\phi_c(u)-\phi_c(u_E))}. \tag{48}$$

It is easy to confirm that $\mathrm{sign}(y(u))=-1$ and $0<|y(u)|<1$ for any $0<\epsilon<1$ and $0<u<1$. Hence, eq. (48) has a good interpretation. The factor $y(u_E)e^{N(\phi_c(u)-\phi_c(u_E))}$ controls the changing speed when we change U around U_E. Namely,

$$p^*_{U_E+K}\approx\hat{p}_{U_E+K}-\frac{\Lambda}{Q'}+(\mathrm{sign}(y(u_E)))^{K+1}|y(u_E)|e^{K\log|y(u_E)|}\left(\frac{\Lambda}{Q'}-\hat{p}_{U_E-1}\right). \tag{49}$$

If the increasing speed $e^{K\log|y(u_E)|}$ is slower than the one of $\hat{p}_{U_E+K}=\hat{p}_{U_E}e^{-\frac{\partial\omega}{\partial u}K}$, we can see all $U\geq U_E$ are fine and the corresponding $\{p^*_U\}_{U\geq U_E}$ take positive values, then the learning edge is well-defined. This is the case for $\epsilon < 1$ since $\log|y|$ is always negative and smaller than $-\frac{\partial \omega}{\partial u} = \frac{H^*}{2} > 0$. Thus our analysis is consistent with the assumptions.

We should clarify the magnitude relation between Λ/Q' and $\hat{p}_{U_E}$. We expect the righthand side of eq. (49) to be positive if $U \geq U_E$, and zero or negative for $U \leq U_E - 1$. The definition of y leads to a relation $e^{N\{\phi_c(u_E-1/N)-\phi_c(u_E)\}} \approx y^{-1}(u_E)$, which simplifies the inequalities for $U = U_E$ and $U = U_E - 1$:

$$\hat{p}_{U_E} \geq \frac{\Lambda}{Q'}\{1 - y(u_E)\} + y(u_E)\hat{p}_{U_E-1}, \tag{50}$$

$$\hat{p}_{U_E-1} \leq \frac{\Lambda}{Q'}\{1 - y(u_E)y^{-1}(u_E)\} + y(u_E)y^{-1}(u_E)\hat{p}_{U_E-1} = \hat{p}_{U_E-1}. \tag{51}$$

Hence the SP equation at $U_{\boldsymbol{s}} = U_E - 1$ is marginally satisfied, implying $p^*_{U_E-2} < 0$. These inequalities imply

$$\hat{p}_{U_E-2} < \frac{\Lambda}{Q'} < \hat{p}_{U_E}. \tag{52}$$

Effective number of observables. In the usual situation, we calculate the learning edge ω_E for a given number of observables M, but here we perform the opposite analysis, namely we calculate the effective number of observables M_{eff} for a given ω_E. If $M_{\text{eff}}(\omega_E)$ is smaller than the one in the random case with the same value of ω_E, it means that the inference improves as a result of the introduction of the nontrivial matrix $M_{\boldsymbol{st}}$.

From the extremization conditions with respect to Q and Q' in eq. (28) with negligibly small E, we get

$$Q = \left\langle \sum_{\boldsymbol{s},\boldsymbol{t}} M^{-1}_{\boldsymbol{st}}(p_{\boldsymbol{s}} - \hat{p}_{\boldsymbol{s}})(p_{\boldsymbol{t}} - \hat{p}_{\boldsymbol{t}}) \right\rangle, \tag{53}$$

$$Q'Q = M. \tag{54}$$

The summation over $\sum_{\boldsymbol{s},\boldsymbol{t}}$ in the first equation can be divided into three categories

$$\sum_{\boldsymbol{s},\boldsymbol{t}}(\cdots) = \left(\sum_{\boldsymbol{s},\boldsymbol{t}\in\text{SAT}} + \sum_{\boldsymbol{s},\boldsymbol{t}\in\text{UNS}} + 2\sum_{\boldsymbol{s}\in\text{SAT},\boldsymbol{t}\in\text{UNS}} \right)(\cdots). \tag{55}$$

Let us assume the average over ρ, $\langle\cdots\rangle$, can be replaced by the SP values of $\boldsymbol{p}$. Then we get

$$\sum_{\boldsymbol{s},\boldsymbol{t}\in\text{SAT}} M^{-1}_{\boldsymbol{st}}(p^*_{\boldsymbol{s}} - \hat{p}_{\boldsymbol{s}})(p^*_{\boldsymbol{t}} - \hat{p}_{\boldsymbol{t}}) = \sum_{\boldsymbol{s},\boldsymbol{t}\in\text{SAT}} M^{-1}_{\boldsymbol{st}} \left(\Phi_{\boldsymbol{s}} - \frac{\Lambda}{Q'}\right)\left(\Phi_{\boldsymbol{t}} - \frac{\Lambda}{Q'}\right) \doteq e^{N\hat{\sigma}}\left(\frac{\Lambda}{Q'}\right)^2, \tag{56}$$

where the symbol $\doteq$ denotes the equality in the exponential scale. To derive this, we assume $\Lambda/Q' > |\Phi_{\boldsymbol{s}}|$ which is true in the random case and is strongly expected in the present case too. Similarly, we have

$$\sum_{\boldsymbol{s},\boldsymbol{t}\in\text{UNS}} M^{-1}_{\boldsymbol{st}}(p^*_{\boldsymbol{s}} - \hat{p}_{\boldsymbol{s}})(p^*_{\boldsymbol{t}} - \hat{p}_{\boldsymbol{t}}) \approx \sum_{\boldsymbol{s},\boldsymbol{t}\in\text{UNS}} M^{-1}_{\boldsymbol{st}}\hat{p}_{\boldsymbol{s}}\hat{p}_{\boldsymbol{t}} \doteq \sum_{U_{\boldsymbol{s}}<U_E} \hat{p}^2_{U_{\boldsymbol{s}}}, \tag{57}$$

$$\sum_{\boldsymbol{s}\in\text{SAT},\boldsymbol{t}\in\text{UNS}} M^{-1}_{\boldsymbol{st}}(p^*_{\boldsymbol{s}} - \hat{p}_{\boldsymbol{s}})(p^*_{\boldsymbol{t}} - \hat{p}_{\boldsymbol{t}}) \doteq e^{N\sigma_E}\left(\frac{\Lambda}{Q'}\right)^2, \tag{58}$$

where $\sigma_E = \sigma(\omega_E)$. Collecting and comparing all these terms, we see the dominant term is

$$Q \doteq \sum_{s \in \mathrm{UNS}} \hat{p}_s^2 \doteq \left\{ \begin{array}{ll} e^{N(\sigma(\omega_2)-2\omega_2)} & (\omega_E \leq \omega_2) \\ e^{N(\sigma_E - 2\omega_E)} & (\omega_E > \omega_2) \end{array} \right. = e^{Nq(\omega_E)}. \tag{59}$$

We expect $\Lambda = 2^N$ in an interesting situation. Hence, we can calculate M_{eff} from a given value of $\Lambda/Q' = \mathcal{L}$ and from eq. (54) as

$$M_{\text{eff}} = \frac{2^N}{\mathcal{L}} Q \doteq e^{N(\log 2 - \omega_E + q(\omega_E))}. \tag{60}$$

The last inequality comes from $\mathcal{L} \doteq e^{-N\omega_E}$ derived from eq. (52). This effective M_{eff} is identical to the one in the random case for a given ω_E, implying that the inference is not improved at least in the exponential scale, unfortunately.

4.2.2. General $\Xi = O(1)$. Next, we proceed to the case of larger values of ϵ. Now, we need higher orders in the truncation. We assume Ξ is a $O(1)$ constant and is not extensive. As in the $\Xi = 1$ case, the SP equation is expressed as

$$\sum_{i=0}^{\Xi} \epsilon^i \sum_{j=0}^{i} \frac{u_s^j}{j!} \frac{(1-u_s)^{i-j}}{(i-j)!} \Phi_{U_s+i-2j} = 0. \tag{61}$$

To derive this, we neglect sub-leading factors in the large-size limit as

$$\binom{U_s}{j} \approx N^j \frac{u_s^j}{j!}. \tag{62}$$

This truncation is justified if eq. (39) is satisfied. Fortunately, we can solve this equation for general $\Xi = O(1)$. We state the outline of deriving the solution.

First we need to solve the characteristic equation

$$\sum_{i=0}^{\Xi} \epsilon^i \sum_{j=0}^{i} \frac{u_s^j}{j!} \frac{(1-u_s)^{i-j}}{(i-j)!} y(u_s)^{i-2j} = 0. \tag{63}$$

This has 2Ξ different roots. The appropriate roots are the ones going to zero as $\epsilon \to 0$. As far as we searched, the appropriates roots consist of some real-valued ones and some complex-conjugate pairs. For example in $\Xi = 3$, we found that the appropriate roots are constituted by one real-valued root y_1 and one pair of complex roots y_2 and y_3 which are in the complex conjugate relation. Using these roots, we can write

$$\Phi(u) = \sum_{i=1}^{\Xi} A_i e^{N\phi_i(u)}, \tag{64}$$

where

$$\phi_i(u) = \int^u dx \log y_i(x). \tag{65}$$

Now, the coefficients $\{A_i\}$ are determined from the boundary conditions $U_s = U_E, \cdots, U_{E+\Xi-1}$, leading to

$$\frac{\Lambda}{Q'} - \hat{p}_{U_E-\xi} = \Phi(U_E - \xi) \approx \sum_{i=1}^{R} A_i y_i^{-\xi} e^{N\phi_i(u_E)} \quad (\xi = 1, \cdots \Xi). \tag{66}$$

Thus means $p^*_{U_E-\xi}=0$, $(\xi=1,\cdots,\Xi)$ is maintained by the boundary conditions. From the boundary conditions, we also find the magnitude relation between Λ/Q' and $\hat{p}_{U_E}$

$$\hat{p}_{U_E-\Xi-1}\le\frac{\Lambda}{Q'}\le\hat{p}_{U_E}. \tag{67}$$

These are the construction of the solution of eq. (61).

We confirmed up to a reasonable value of Ξ that the real part of the suitable $\{y_i(u)\}$ are decreasing function with respect to u for a certain moderate range of $\epsilon>0$, and thus $\phi_i(u)$ too. Hence, $\Phi(U_E+K)$ is expected to decrease as K grows by the factor $e^{K\log|y_i(u_E)|}$, and the learning edge is well-defined.

The effective number of observables can be also evaluated in this case. Unfortunately, as long as $\Xi=O(1)$, $M_{\text{eff}}(\omega_E)$ cannot be different in the exponential scale from $M(\omega_E)$ in the random case, since both Q and $\mathcal{L}=\Lambda/Q'\doteq e^{-N\omega_E}$ are essentially the same as the ones of the random case. This naturally motivates us to treat $\Xi=O(N)$, but that is not easy since it becomes even nontrivial to write down the SP equation. We leave this as a future work. Instead in the next subsection, we try to examine the effect of the ME bias Γ in the same setting.

4.3. Perturbation from large B at finite Γ

4.3.1. Review of the random case. We start by reviewing the random case with relatively small M. The SP equation is

$$p_{\boldsymbol{s}}=\hat{p}_{\boldsymbol{s}}-\frac{\Lambda}{Q'}-\frac{\Gamma}{Q'}(1+\log p_{\boldsymbol{s}})\,. \tag{68}$$

For large $\Gamma\ge 2^N$, the order parameter Λ behaves as

$$\Lambda=(N\log 2-1)\Gamma, \tag{69}$$

and the solution of the SP equation is

$$p^*_{\boldsymbol{s}}=\begin{cases}\hat{p}_{\boldsymbol{s}}-\frac{\Gamma\log 2^N\hat{p}_{\boldsymbol{s}}}{Q'}, & (\boldsymbol{s}\in\text{SAT})\\ p_0+\Delta_{\boldsymbol{s}}, & (\boldsymbol{s}\in\text{UNS})\end{cases}, \tag{70}$$

where

$$p_0=e^{-1-\frac{\Lambda}{\Gamma}}=\frac{1}{2^N},\ \Delta_{\boldsymbol{s}}=\frac{Q'p_0}{Q'p_0+\Gamma}(\hat{p}_{\boldsymbol{s}}-p_0)\,. \tag{71}$$

The expression of $\Delta_{\boldsymbol{s}}$ can be derived from the perturbation with respect to Δ in eq. (68). Inequalities

$$p_0>\Delta_{\boldsymbol{s}},\ \Gamma>Q'\hat{p}_{\boldsymbol{s}}\ ({}^\forall\boldsymbol{s}\in\text{UNS}) \tag{72}$$

are needed to justify this perturbation. To make the above expressions consistent, the learning edge should be

$$\omega_E=q'-\gamma=q'-\lambda. \tag{73}$$

For $\Gamma<2^N\doteq\Lambda(\Gamma=0)$, the situation is essentially identical to the case $\Gamma=0$. This can be understood by comparing each term of the righthand side in eq. (68). For $\boldsymbol{s}\in$ SAT the dominant term is the first term $\hat{p}_{\boldsymbol{s}}$ and for $\boldsymbol{s}\in$ UNS the second one is dominating. The last term proportional to Γ/Q' is always smaller than the other two in the exponential scale and hence essentially does not change the SP equation.

4.3.2. General $\Xi = O(1)$. A lesson from sec. 4.2.2 tells that the introduction of the nontrivial M_{st} does not change the scale of the problem if $\Xi = O(1)$, and another lesson from sec. 4.3.1 says that the effect of the ME bias Γ appears when Γ becomes comparable with Λ in the scale. These two lessons imply that the ME bias and M_{st} cannot provide a meaningful effect on the inference in the present setting.

As an example, let us see the SP equation for the truncation range $\Xi = 1$ with Γ which can be written as

$$\Phi(u_s)\left(\epsilon(1-u_s)+1+\epsilon u_s\right) = -\frac{\Gamma}{Q'}\left(1+\log p_s\right), \tag{74}$$

We found that $\Phi(u_s) = p^*_{u_s} - \hat{p}_{u_s} + \frac{\Lambda}{Q'}$ is the decreasing function of u, and the leading-order terms in Φ are $p^*_s \doteq \hat{p}_s$ for $s \in \mathrm{SAT}$ and Λ/Q' for $s \in \mathrm{UNS}$. The scale of Φ itself is smaller than $p^*_s \doteq \hat{p}_s$ for $s \in \mathrm{SAT}$ due to the cancellation in $p^*_s - \hat{p}_s$. If Γ/Q' is smaller than any of these two terms, the situation will be identical to $\Gamma = 0$ as the random case with $\Gamma < 2^N$. This is consistent to the leading order of eq. (74). As Γ grows and $\Gamma \doteq \Lambda(\Gamma = 0) \doteq 2^N$, the situation is changed: Λ scales as Γ for $\Gamma > 2^N$ and the learning edge determined by the balance between Λ/Q' and $\hat{p}_s$ starts to decrease since Q' is kept to be constant against Γ. Thus introducing the ME bias worsens the inference performances, as found with the annealed approximation in the random case.

This situation seems to remain the same as long as $\Xi = O(1)$ where the coefficient of Φ is $O(1)$. Hence, from the annealed calculations, we cannot find any good effect on inference produced by the ME bias.

4.3.3. Some comments on the quenched case. A crucial difference between the quenched and annealed cases is the behavior of Q'. The extremization condition over Q' in eq. (25) after taking the $n \to 0$ limit becomes

$$Q' = \frac{M}{D}, \tag{75}$$

and D is scaled as $D = \Gamma^{-1}$ for $\Gamma > 2^N$ in the random case. Hence, Q' increases with the growth of Γ as Λ. This is good for the ME principle since the learning edge $\omega_E = q' - \lambda$ does not change as Γ grows.

On the other hand, the SP condition with respect to $\boldsymbol{p}$ becomes

$$\begin{cases} -\Lambda - Q'\sum_{t\in\mathrm{SAT}} M^{-1}_{st} p_t + Q'\sum_t M^{-1}_{st}\hat{p}_t + \frac{\sqrt{R'}}{Q'} z_s - \Gamma(\log p_s + 1) = 0, & (s \in \mathrm{SAT}) \\ p_s = f_s(\Gamma), & (s \in \mathrm{UNS}) \end{cases}. \tag{76}$$

The only difference with the annealed case is the presence of the term $\frac{\sqrt{R'}}{Q'} z_s$ and we expect that the nature of the SP equation does not change so much.

4.4. A semi-analytical treatment: numerical evaluation of the SP equation

In the above subsections, we have concentrated on studying the truncated SP equation. For general B, $\epsilon = Ne^{-2B}$ can be large so the truncation cannot be justified. To examine this case, we numerically evaluate (76). Actually, directly evaluation of (76) is not easy, and instead of that, we minimize the following cost function

$$L = \frac{1}{2}\sum_{s,t} M^{-1}_{st}(p_s - \hat{p}_s)(p_t - \hat{p}_t) + \frac{\Lambda}{Q'}\sum_t (p_t - \hat{p}_t) + \frac{\Gamma}{Q'}\sum_s p_s \log p_s, \tag{77}$$

for given Λ/Q' and Γ/Q' under the condition $p_{\boldsymbol{s}} \geq 0$. This is a convex optimization problem with constraints and can be solved by certain known algorithms such as the interior-point method. By this minimization, we can obtain the solution of the SP equation, $\boldsymbol{p}^*$, and thus related quantities can be evaluated. In solving this, we assumed the symmetry that $p^*_{\boldsymbol{s}}$ having the same magnetization value share the same probability value as well as $\hat{p}_{\boldsymbol{s}}$, which reduces a lot of computational costs.

An interesting finding might be the absence of the unique learning edge, which is demonstrated in Fig. 13. In the figure, the absolute values of the lefthand side of eq. (32)

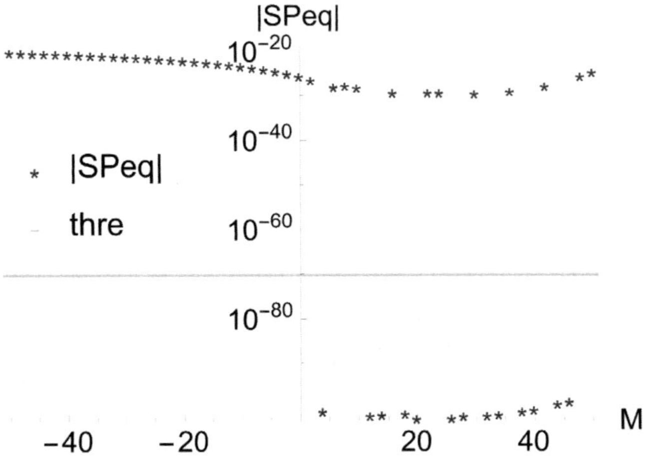

Figure 13. The absolute values of the lefthand side of eq. (32) of the upper line are plotted against the magnetization $M_{\boldsymbol{s}} = \sum_i s_i$. The target model is the ISM with $H = 0.5$, and other parameters are $N = 50, B = 0.2, H = 0.5, \Gamma = 0$ and $\Lambda/Q' = 5.0 \times 10^{-20}$. If $\boldsymbol{s} \in$ SAT (UNS), the corresponding value denoted by a blue point becomes smaller (larger) than the chosen threshold value 10^{-70} denoted by the green line. We see that the blue points locating below the threshold value are distributed in a patchy fashion, implying the absence of the learning edge.

of the upper line are plotted against the magnetization with the parameters: $N = 50, B = 0.2, H = 0.5, \Gamma = 0$ and $\Lambda/Q' = 5.0 \times 10^{-20}$. Namely, if $\boldsymbol{s} \in$ SAT, the corresponding value should be enough small (here we set the threshold value as 10^{-70}). In the random case $B \to \infty$, there exists an edge value of M, M_E corresponding to the learning edge ω_E, and all the SP-equation values for $M \geq M_E$ are smaller than the threshold, while the ones for $M < M_E$ are larger. In the present case $B = 0.2$, however, we see that the blue points locating below the threshold value are distributed in a patchy fashion and we cannot find such a clear edge value for M. This results implies that no learning edge can be defined, which makes the resolution of the SP equation directly (76) very difficult, as pointed out in sec. 4.1.3. Note that this calculation requires a very multiple precision because our variables obey the exponential scale and can be very large and small. For $N = 50$, we used 170 digits to express a real number.

5. Discussion and summary

In the present work, we have investigated the space of probability distributions constrained by smooth observables. We have first introduced a Monte Carlo procedure to sample the space of probability distribution over the set of configurations made of a small (in practice ≤ 10) number of spins. In this numerical study, we have used the order parameter q measuring the distance between the target distribution and the averaged one to quantify the quality of inference. Our numerical investigations have revealed that the ME principle can work better than other typical distributions, in particular when the the target distribution and the observables have comparable smoothnesses. As a consequence, once we know the nature of the interactions present in the target distribution, we can construct the ME distribution precisely describing the target system. However, our numerical simulations also show that a few errors in specifying the type of interactions leads to a significant worsening of the ME distribution. This suggests that the ME principle is not good in approximating a target distribution. In practical situations, it does not seem easy to precisely know the type of the interactions of the target system, and compatibility is a serious issue.

To get deeper insight, we have also carried out some analytical investigations. We have assigned a specific functional form to the matrix controlling the smoothness of the observables and introduced a prior distribution over the target distribution. In this setting, we have analyzed the perturbative regime from the random case, combining this approach with the annealed approximation and the saddle-point approximation. We were able to construct a solution in this case, but the solution was not informative. The annealed approximation, indeed, corresponds to 'bad' choices of the constraints, making the volume V increase compared to its typical value for quenched, random observables.

One of the main outcomes of the present work is that, in the presence of smooth observables, no learning edge, separating small from large target probabilities cannot be defined, see Fig. 13, as used to be the case in the non-smooth case [11]. This result highlight the complexity of learning in the smooth case, and its collective nature. While the probabilities of the different spin configuration were essentially learned independently of each other in the non-smooth case, this is clearly not the case anymore. How to describe in a mathematically controlled and precise way this collective phenomenon remains a challenge. We hope that the remarks presented here will be useful for further progresses in this direction.

Acknowledgments

T. O. is grateful to E. Aurell for a fruitful discussion at KITPC in China. He also acknowledges the support by Grant-in-Aid for JSPS Fellows, as well as the JSPS Core-to-Core Program "Non-equilibrium dynamics of soft matter and information". A part of numerical calculations were carried out on the TSUBAME2.5 supercomputer in the Tokyo Institute of Technology. R.M. was partly funded by the Agence Nationale de la Recherche Coevstat project (ANR-13-BS04-0012-01) and the [EU-]FP7 FET OPEN project Enlightenment 284801.

References

[1] Jaynes E T 1957 *Phys. Rev.* **106** 620
[2] Shneidman E, Berry M J, Segev R, Bialek W 2006 *Nature* **440** 1007
[3] Tkačik G, Marre O, Mora T, Amodei D, Berry II M J, Bialek W 2013 *J. Stat. Mech.* P03011
[4] Cocco S, Leibler S, Monasson R 2009 *Proc. Nat. Acad. Sci. USA* **106** 14058
[5] Seno F, Trovato A, Banavar J R, Maritan A 2008 *Phys. Rev. Lett.* **100** 078102
[6] Weigt M, White R A, Szurmant H, Hoch J A, Hwa T 2009 *Proc. Nat. Acad. Sci. USA* **106** 67
[7] Mora T, Walczak A M, Bialek W, Callan C G 2010 *Proc. Nat. Acad. Sci. USA* **107** 5405
[8] Cocco S, Monasson R, Weigt M 2013 *PLOS Comput. Biol.* **9** e1003176; *J. Phys. Conference Series* **473** 012010
[9] Bialek W, Cavagna A, Giardina I, Mora T, Silvestri, E, Viale M, Walczak A 2012 *Proc. Nat. Acad. Sci. USA* **109** 4786
[10] Kataoka S, Yasuda M, Furtlehner C, Tanaka K 2014 *Inverse Problems* **30** 025003
[11] Obuchi T, Monasson R, and Cocco S, 2015 Learning probabilities from random observables in high dimensions: the maximum entropy distribution and others, submitted to *J. Stat. Phys.*
[12] Mézard M, Parisi G, Virasoro M A 1987 *Spin Glass Theory and Beyond* (Singapore: World Scientific)
[13] Bishop C M 2006 *Pattern Recognition and Machine Learning* (New York: Springer)

Institute of Physics
Dirac House, Temple Back
Bristol BS1 6BE UK

ISSN: 1742-6588
ISBN 978-1-5108-1196-6

CURRAN ASSOCIATES INC.
proceedings.com